QUICK INDEX

ALPHABETICAL QUICK INDEX

Apalis	152	Hamerkop	110	Quelea	174
Avocet	74	Harrier	88	Raven	128
Babbler	146	Harrier-Hawk	94	Reed-Warbler	142
Barbet	26	Hawk	82	Robin-Chat	162
Bateleur	88	Hawk-Eagle	92	Roller	32
Batis	126	Helmet-Shrike	124	Ruff	72
Bee-eater	38–40	Heron	104–108	Rush-Warbler	142
Bishop	176	Hobby	98	Sanderling	72
Bittern	108	Honey-Buzzard	82	Sandgrouse	64
Boubou	122	Honeyguide	22	Sandpiper	70–72
Brownbul	138	Hoopoe	30	Saw-wing	136
Brubru	120	Hornbill	28	Scimitarbill	30
Buffalo-Weaver	174	House-Martin	134	Scops-Owl	52
Bulbul	138	Hyliota	146	Scrub-Robin	162
Bunting	194	Ibis	110	Secretarybird	96
Bush-Shrike	122–124	Indigobird	184	Seedeater	192
Bustard	64	Jacana	66	Shikra	90
Buttonquail	16	Kestrel	98–100	Shrike	122, 126–128
Buzzard	82, 90	Kingfisher	34–36	Snake-Eagle	88
Camaroptera	152	Kite	82–84	Snipe	66
Canary	192	Korhaan	64	Sparrow	186
Chat	164	Lapwing	78	Sparrowhawk	90–92
Cisticola	148–150	Lark	154–156	Sparrowlark	156
Cliff-Chat	164	Longclaw	188	Spinetail	48
Coot	68	Mannikin	182	Spoonbill	112
Cormorant	102	Martin	132	Spurfowl	14
Coucal	46	Masked-Weaver	172	Starling	166–168
Courser	80	Moorhen	68	Stilt	74
Crake	66–68	Mousebird	40	Stint	72
Crested-Flycatcher	118	Neddicky	150	Stonechat	164
Crombec	140	Nicator	138	Stork	114–116
Crow	128	Night-Heron	108	Sunbird	170
Cuckoo	42–44	Nightingale	164	Swallow	132–136
Cuckooshrike	130	Nightjar	56	Swamp-Warbler	142
Darter	102	Openbill	114	Swift	48–50
Dove	60–62	Oriole	118	Tchagra	120
Drongo	118	Osprey	84	Teal	18
Duck	18–20	Ostrich	14	Tern	80
Eagle	94–96	Owl	52–54	Thick-knee	74
Eagle-Owl	54	Owlet	52	Thrush	158
Egret	104	Oxpecker	168	Tinkerbird	26
Eremomela	140	Painted-snipe	66	Tit	130
Falcon	98–100	Palm-Swift	50	Tit-Babbler	146
Finch	178, 186	Palm-Thrush	164	Tit-Flycatcher	160
Finfoot	68	Paradise-Flycatcher	118	Trogon	30
Firefinch	182	Paradise-Whydah	184	Turaco	62
Fish-Eagle	84	Parrot	46	Turtle-Dove	60
Fishing-Owl	54	Pelican	112	Twinspot	178–180
Flamingo	112	Penduline-Tit	130	Vulture	84–86
Flycatcher	158–160	Petronia	186	Wagtail	188
Francolin	14–16	Pigeon	62	Warbler	142–144
Gallinule	68	Pipit	190	Wattle-eye	126
Go-away-bird	62	Plover	76	Waxbill	178–180
Goose	20	Pochard	20	Weaver	172–174
Goshawk	90–92	Pratincole	80	White-eye	146
Grass-Owl	52	Prinia	152	Whydah	184
Grebe	102	Puffback	120	Widowbird	176
Greenbul	138	Pygmy-Goose	18	Wood-Dove	60
Green-Pigeon	62	Pygmy-Kingfisher	34	Wood-Hoopoe	30
Greenshank	70	Pytilia	180	Wood-Owl	54
Ground-Hornbill	28	Quail	16	Woodpecker	24
Guineafowl	16	Quailfinch	178	Wren-Warbler	152

ROBERTS BIRD GUIDE

Kruger National Park and Adjacent Lowveld

A guide to more than 420 birds in the region

Original frontispiece from the first edition of *The Birds of South Africa* by Austin Roberts.
Illustrated by Norman C.K. Lighton.

THE JOHN VOELCKER BIRD BOOK FUND

In December 1935 the South African Bird Book Fund was formed to fund a complete and up-to-date new bird book for southern Africa under the authorship of Austin Roberts, and illustrated by Norman C.K. Lighton. The successful and popular first edition, titled *The Birds of South Africa*, was published on 8 June 1940. The affairs of the fund were handled by the secretary, John Voelcker, who became chairman after the untimely death of Austin Roberts in 1948. He, together with the other Trustees of the Fund, became the driving force that ensured that the original edition was revised and updated. In 1957, seventeen years after the release of the first edition, the Trustees published the second edition under the authorship of G.R. McLachan and R. Liversidge, who also revised the third (1970) and fourth (1978) editions, titled *Roberts Birds of South Africa*. The Trust had by now become known as the John Voelcker Bird Book Fund, a non-profit organisation with limited funding derived from subscriptions and minimal profit from the sale of books. The fifth and sixth editions (1985 and 1993) were revised by Gordon L. Maclean. The seventh edition (2005), entrusted to the Percy FitzPatrick Institute of African Ornithology, was completely rewritten under the editorship of P.A.R. Hockey, W.R.J. Dean, and P.G. Ryan. The dual publication of the large format, 1 300 page *Roberts VII* and the subsequent publication of the *Roberts Bird Guide* (2007) would not have been possible without the generous financial support of Impala Platinum Holdings Limited.

With well over 300 000 copies of *Roberts* books sold to date, the John Voelcker Bird Book Fund is committed to remaining the forerunner in the publication of up-to-date bird information in the region, and to making publications as affordable as possible to the southern African birding community.

ROBERTS BIRD GUIDE
Kruger National Park and Adjacent Lowveld

A guide to more than 420 birds in the region

Hugh Chittenden and Ian Whyte

Distribution maps and plate layout **Guy Upfold**

Much of the information supplied in this guide is derived from *Roberts Birds of Southern Africa VII*, compiled under the auspices of the Percy FitzPatrick Institute of African Ornithology (a DST/NRF Centre of Excellence).

PUBLISHED BY THE
TRUSTEES OF THE
**JOHN VOELCKER
BIRD BOOK FUND**

Published by the Trustees of the
John Voelcker Bird Book Fund
9 Church Square, Cape Town, 8001

Distribution maps Guy Upfold (www.birdinfo.co.za)
Plate 23 information supplied by Des Jackson
Plate design Guy Upfold (*www.birdinfo.co.za*)

Artists I.B. Weiersbye, G. Arnott, P.R.M. Meakin, R.J. Cook, C.S. van Rooyen, A. Barlow and A.C.V. Clarkson
Cover illustration Ingrid Weiersbye, from a photograph by Hugh Chittenden

© Text and illustrations, the John Voelcker Bird Book Fund

All rights reserved. No part of this publication may be reproduced, stored in a retrieval system
or transmitted, in any form or by any means, without the prior written permission of the copyright
holder. However, text extracts of fewer than 100 words will be deemed as fair use and may be
reproduced without prior permission as long as such extracts are duly acknowledged.

ISBN 978-1-77009-638-7

Printed and bound by Tien Wah Press, Malaysia

CONTENTS

FOREWORD 2

CONTRIBUTING ARTISTS 3

ACKNOWLEDGEMENTS 4

INTRODUCTION 5
   KNP ROAD MAP 6
   ENDEMISM 7
   ECOLOGY OF BIRDS IN THE KNP 8
   MAP OF MAJOR HABITAT TYPES IN THE KNP 9
   ADDITIONAL READING 12

SPECIES ACCOUNTS 14

GLOSSARY 196

ILLUSTRATED GLOSSARY 202

VAGRANTS AND HISTORICAL LISTS 204

ETYMOLOGY AND FOREIGN BIRD NAMES 206

INDEXES 228
   ALPHABETICAL QUICK ii
   PORTUGUESE 228
   GERMAN 232
   FRENCH 236
   ZULU 240
   XHOSA 242
   TSONGA 243
   SOUTH SOTHO 246
   NORTH SOTHO 247
   AFRIKAANS 248
   SCIENTIFIC 251
   COMMON NAMES 253

HOW TO USE THIS FIELD GUIDE 259

FOREWORD

Kruger National Park has almost 500 birds on its checklist. Some of these are accidental vagrants, but most are either resident or regular annual visitors to the Park. Over the years, there have been several efforts to document the birds of Kruger or of the Lowveld, but as human understanding and awareness of birds increases, the need to breathe new life into the available birding literature on Kruger and its surrounds is an ongoing pursuit.

The involvement of my ex-colleague, Dr Ian Whyte, who spent a lifetime in the Park as a research scientist, can only add immense value to this publication. Ian is a birder of great renown, and I doubt anyone has a more extensive and intimate knowledge of the avian profile of Kruger birds than him.

Birds are enigmatic and unpredictable and prone to shifts in distribution and density. Compiling a field guide that provides users with current and accurate information on facets, such as distribution, status, breeding cycles and feeding habits, is a very useful endeavour for experienced and novice birders like me.

And this is a book for people. Birds attract birders, both expert and beginner, and because Kruger is home to such a wide diversity of birds and represents the most accessible bastion for many localised or formerly more widespread species, the Park is a magnet for many *avitourists* – people travelling for the purpose of observing and often photographing birds.

Kruger is of course one of Africa's great national parks. Birds are obviously important to SANParks' management from the tourist component described above and from a conservation perspective as both individual species and important indicators of health of habitat. But birds are also an essential component in African folklore and thus have a focal place in the make-up of all of us who call ourselves African. This guide is a journey into the future and I implore all of us to begin the long walk.

Dr David Mabunda
CEO South African National Parks
and former Director of Kruger National Park

CONTRIBUTING ARTISTS

Contributing artists to this guide, and to the seventh edition of *Roberts Birds of Southern Africa*:

Ingrid Weiersbye was born in London, but schooled and lived in Rhodesia/Zimbabwe for 24 years, a country which shaped her deep interest in all aspects of the natural environment, its colours and diversity. Ingrid specialises in bird painting and strives to capture life, movement and light. She has had 15 solo exhibitions, and exhibited in the UK for seven years, including the annual Society of Wildlife Artists in London and the internationally renowned British Bird Watching Festival where she received best Art Exhibit award. Her work is included in several major publications, most recent being the 20 plates for the seventh edition of *Roberts Birds of Southern Africa* (Hockey et al.). From her studio in KwaZulu-Natal she travels extensively for her field references. Ingrid is married to Roger Porter, a specialist scientist employed in nature conservation, and has two sons.

Graeme Arnott spent his boyhood on a farm near Hwange National Park, Zimbabwe. He graduated from Rhodes University and taught at a school in Zimbabwe for 10 years. His passion for painting wildlife, especially birds, was fulfilled in 1975 when he left teaching to pursue painting on a full-time basis. Graeme lives at Kenton-on-Sea in the Eastern Cape from where he travels to do fieldwork on his subjects. He has illustrated the *Birds of Prey of Southern Africa* (Peter Steyn, 1982), *Shrikes of Southern Africa* (Tony Harris, 1988), *Robins of Africa* (Terry Oatley, 1998), and contributed 20 plates for the seventh edition of *Roberts Birds of Southern Africa* (Hockey et al.). In addition to book illustrating he paints for private and corporate collectors.

Penny Meakin was born in England and her early years were spent growing up in East Africa. She lives in Pietermaritzburg and is a professional artist with a particular interest in and a passion for game birds; however, her artistic repertoire encompasses all aspects of African fauna. In addition to many commissions by Rowland Ward and Struik Publishers, previous publications include *Wild Ways – the behaviour of African mammals* (Peter Apps, 2000), *The World's Greatest Wingshooting Destinations* (Chris Dorsey, 2002) and *Under a Hunter's Moon* (Nino Burelli, 2002). Penny was appointed sole artist for the acclaimed *Agred's Gamebirds of South Africa: Field Identification and Management* (Dr P.J. Viljoen, 2005) and is one of the seven contributing artists for the seventh edition of *Roberts Birds of Southern Africa* (Hockey et al.).

Chris van Rooyen was born in Johannesburg on 3 April 1964 and has lived there ever since. He received very little formal art training, but has always been passionate about birds and drawing and specialised in bird art from a very early age. He uses a combination of gouache and water colour to paint birds and has a special liking for birds of prey. He painted nine plates for the seventh edition of *Roberts Birds of Southern Africa* (Hockey et al.), covering all the raptors, as well as the front cover flight image. Chris currently works for the Endangered Wildlife Trust as a specialist advisor on the impact of industry on birds, and uses his spare time to paint and photograph birds. His wife, Izelle, is equally passionate about birds and the couple spend as much time as possible in the bush with their two sons.

Ronald Cook was born in Auckland, New Zealand, in 1940. His interest in birds began as a youth while out deep-sea fishing when he became enthralled by the close proximity of many seabird species. He studied portraiture, still life and landscape painting, has a diploma in Fine Art and in the early 1980s was elected a Fellow of the South African Art Aviation Guild. He has been painting professionally since 1984, works both in water colours and oils, exhibits regularly, specialising in wildlife, especially birds in their environment. Ron painted nine plates for the seventh edition of *Roberts Birds of Southern Africa* (Hockey et al.) and currently lives in Sandton, Gauteng.

Andrew Barlow was born in Harare, Zimbabwe, in 1970, educated at St Andrew's College in Grahamstown and studied Fine Art at the University of Stellenbosch. Since graduating in 1992, he has specialised in animal portraiture and illustration. He has contributed illustrations towards *Waders of Southern Africa* (Hockey & Douie, 1995) and the seventh edition of *Roberts Birds of Southern Africa* (Hockey et al.). Andrew currently lives in Somerset West, specialises in equine portraiture and in his spare time makes red wine.

Tony Clarkson was born in Durban in 1943 and as a scholar was influenced to study taxidermy and bird painting by Dr Phillip Clancey. While working in Ladysmith he sustained a sporting injury to his left hand and took up painting, concentrating on birds, which have been of lifelong interest to him. He illustrated the *Natal Bird Atlas* (Cyrus & Robson, 1980) and *Top Birding Spots* (Hugh Chittenden, 1992). Tony and his wife, Lorraine, live in Ladysmith, KwaZulu-Natal, where he practises as an architect and is an enthusiastic birder and bird painter.

ACKNOWLEDGEMENTS

We are indebted to a number of people who helped improve the original drafts of this guide, which is a regional companion book to the handbook *Roberts – Birds of Southern Africa*, VII ed., Hockey PAR, Dean WRJ, Ryan PG (eds) 2005. We are greatly indebted to Guy Upfold, who not only scanned all the original artwork and laid out the plates, but also solved seemingly insurmountable problems by producing all the distribution maps, re-drawing the habitat and road maps, and seeing to it that many other tasks, such as the design of the breeding bars, ran smoothly. To him and his kind and helpful wife Lee-Anne, we are extremely grateful.

We are grateful to Dr David Mabunda, CEO of SANParks for writing the foreword. We sincerely appreciate the time he sacrificed for this in his very busy schedule.

We also relied on a number of unselfish friends who helped improve the original drafts. Peter Lawson, Duncan McKenzie and Nick Squires are thanked for the time and effort they put into reading and improving the original draft, and sharing their considerable knowledge on the birdlife of Kruger and surrounding region. Bruce Lesley and Warren McCleland both made significant contributions and we are grateful for their efforts. Don English assisted in more subtle ways – the sharing of his knowledge through discussion and shared experiences over many years has contributed significantly to the completeness of this book.

A major contributor to the southern African Roberts Bird Guide was Des Jackson whose work on nightjars is included in this work. His knowledge in this field is acknowledged, and birders in this region will benefit enormously from his input. David Allen is thanked for his ongoing support and help with making skins available from the Durban Museum. We thank Ingrid Weiersbye for her input, and for illustrating the book front cover. Morné de la Rey supplied the Afrikaans names list used in this guide. Sandra MacFadyen of SANParks Geographic Imformation Systems is thanked for willingly supplying the original road and habitat maps. The staff of Jacana Media are thanked for their efforts in producing a quality final product.

The artwork for this guide was produced by Ingrid Weiersbye, Graeme Arnott, Penny Meakin, Chris van Rooyen, Ronald Cook, Tony Clarkson and Andrew Barlow. We thank those who supplied additional images for this guide.

Finally, we would like to thank our wives, Loueen (Chittenden) and Merle (Whyte) for their support and help throughout the duration of this project.

INTRODUCTION

This book is intended to serve as a condensed version of the southern African Roberts Bird Guide, targeting species that visitors to the Kruger National Park are likely to encounter. It is hoped that the colour maps give a better understanding of the distribution of the more than 420 species that regularly occur within the Park.

Breeding seasons
Data for breeding seasons (breeding bars on the right of the maps) was extracted mainly from *Roberts Birds of southern Africa* Vol VII, Hockey et al, 2005 and also from records published in *Nests and Eggs of Southern African Birds*, Tarboton W, 2001. Where necessary, breeding seasons from eastern Zimbabwe and north-eastern southern Africa (related habitats) were also taken into account, and should give a fairly accurate reflection of breeding seasons within the Park.

This breeding data should serve as in indication as to when birds could breed in the Kruger National Park and adjacent lowveld w, and NOT as actual recorded breeding data for the Park. Breeding bars are also given for species that do not breed in the Park. This data is included to reflect breeding seasons for non-breeding visitors and will help give a better understanding of the timing and movements of nomadic species. Data from adjacent regions is used to reflect the possible full and peak breeding seasons.

Maps

Dark green – Common or fairly common resident breeding species

Light green – Uncommon to fairly uncommon resident breeding species

Dark orange – Common or fairly common breeding intra-African migrant

Light orange – Uncommon to fairly uncommon breeding intra-African migrant

Dark blue – Common or fairly common non-breeding Palaearctic migrant

Light blue – Uncommon or fairly uncommon non-breeding Palaearctic migrant

Green/yellow-striped – Species with partial resident populations, or populations that are seasonally supplemented from regions away from the Park

Orange/blue-striped – Summer visitors that breed in North Africa and the Palaearctic region

Green/blue-checked – Southern African resident species that are seasonal visitors, usually uncommon and in general do not breed in the Park

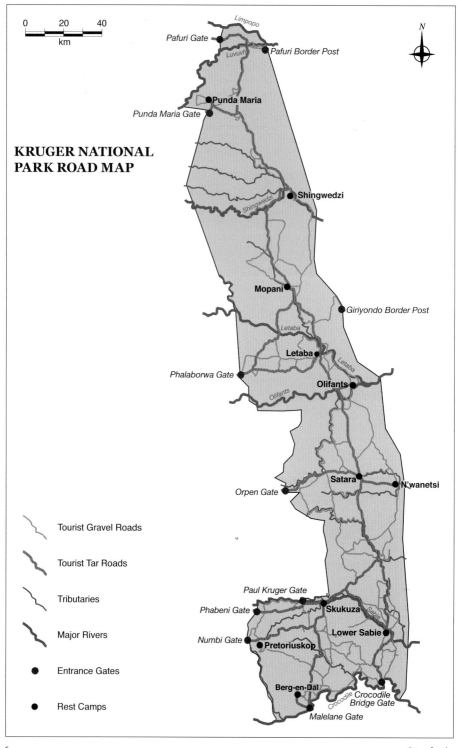

ENDEMISM

Endemics: The common names of all endemics in southern Africa are highlighted in green. Species considered as endemics are those whose breeding and non-breeding ranges are entirely within the southern African sub-region (with occasional records out the region).

Near-endemics: A species is regarded as a near-endemic if its range is largely restricted to southern Africa, but extends no more than 10-15% outside the region's borders.

Breeding endemics: Breeding endemics are those species that only breed within the region's borders but move or migrate away from the region in the non-breeding season.

In all, only 27 of the more than 420 bird species that regularly occur in the KNP are endemics or near-endemics, making up about 6% of the total.

RED DATA SPECIES

The Red Data status for rare or threatened species in this guide is derived mainly from published data in *Roberts VII* and *The Eskom Red Data Book of Birds of South Africa, Lesotho and Swaziland*. Barnes, K. N. (ed) 2000. The designation of Red Data status reflects the viewpoint mainly from a South African perspective and this data should be viewed with caution because national and international lists vary considerably and are also reviewed on a regular basis.

BIRD SIZE AND WEIGHTS

The source of information for the length and weight of birds is derived mainly from *Roberts VII*, but in some instances measurements from museum specimens and other sources have been used. Weight comparisons (e.g. ♂>♀) between male and female are given as a guide where there is sufficient data to show that one sex weighs on average 5% or more than the opposite sex. Males are generally heavier and larger than females, but in some family groups (e.g. raptors) females may weigh up to 30% more than their breeding partners.

BEHAVIOUR

Where space permits, bird behaviour and other characteristics, especially those that help aid identification, are placed under the heading 'STATUS'.

USEFUL ADDRESSES

BirdLife Botswana
Private Bag 003
Suite 348
MOGODITSHANE
BOTSWANA
Tel/Fax + 267 319 0540
blb@birdlifebotswana.org.bw

BirdLife Zimbabwe
PO Box RVL 100
Runiville
HARARE
ZIMBABWE
Tel/Fax + 263 4 481 496/ 490 208
birds@zol.co.zw

Zambian Ornithological Society
PO Box 33944
LUSAKA 10101
ZAMBIA
No tel available.
www.wattlecrane.com
zos@zamnet.zm

BirdLife South Africa
PO Box 515
Randburg
2125
South Africa
Tel + 27 (0)11 789 1122
Fax + 27 (0)11 789 5188
info@birdlife.org.za
www.birdlife.org.za

Introduction

ECOLOGY OF BIRDS IN THE KRUGER NATIONAL PARK

About half (49%) of the bird species utilising the Kruger National Park (KNP) can be classified as resident. The rest are either breeding (23%) or non-breeding (28%) visitors. Many visiting species are regular summer migrants, either intra-African (39% of breeding visitors and 7.5% of non-breeding visitors) or Palaearctic migrants (40.1% of non-breeding visitors). Other species visit the KNP occasionally and their visits depend on less predictable (stochastic) events such as droughts or very wet years.

There are many environmental factors which influence the distribution and numbers of these birds in the KNP, both spatially and temporally, and in either positive or negative ways. These factors include geographic position of KNP, its basic geology and soils, vegetation and habitats, rainfall in both its seasonality and abundance, the availability and distribution of surface water, other climatic forces, fire, and the impacts of large herbivores.

Geographic location

Joubert (2007) pointed out that the Lowveld region of South Africa, of which the (KNP) is a part, lies at the convergence of three of southern Africa's climatic influences. These are a mesic temperate influence from the south, a more humid and tropical influence from the north, and an arid influence from the west.

The majority of resident species are those favouring the southerly temperate influences, and most of these can be encountered throughout the year, but many are annual migrants and spend Austral summers in the Park. However, some of the South African special species occur in northern KNP because of the tropical influences prevailing there, and these species reach the southern limit of their distributional range here. These include Grey-headed Parrot, Tropical Boubou, Meves's Starling, African Yellow White-eye, Collared Palm-Thrush, Arnot's Chat, Mosque Swallow, Southern Hyliota, Racket-tailed Roller, Three-banded Courser, and Mottled and Böhm's spinetails. Senegal Coucal (not included in the species accounts) only marginally extends into the northern extremities of the KNP between the Luvuvhu and Limpopo river systems. Some species from the dry western parts of South Africa reach the eastern limits of their distribution in KNP. These include Meyer's Parrot and Grey-backed Camaroptera. Others species from the drier west enter the KNP area during more extreme droughts, some even achieving irruptive proportions (see "Rainfall" on page 11). River Warbler (also not included in the species accounts) is probably more common than records suggest, but because of its skulking behaviour in rank river and stream fringing thickets, it is unlikely to be seen by most visitors to the Park that are confined to vehicles. This Palearctic breeding migrant gives a sustained cricket-like call prior to departure in March/April.

Geology and soils

The geological history of KNP has in general terms resulted in the western half of the Park being constituted mainly by sandy granitic soil, while the eastern half consists of more clayey basaltic soils. In the southern half of the Park, the granites and basalts are separated by a narrow strip of sedimentary soils (the Karoo sediments). The granitic areas are largely undulating in nature, while the basalts are flatter and plains-like. Later volcanic activity formed the granitic koppies on the granitic soils, and fissures resulting from the separation of the continents as the Gondwanaland super continent broke up, produced rhyolitic dykes which formed the Lebombo Mountains on the eastern boundary of the Park. The western granites tend to support broad-leaved woodland and sour grasses on the crests of hills due to the very sandy nature of the soils there, and *Acacia* in the valleys where the clays and leached nutrients have accumulated over the years. The eastern basaltic soils tend to support a good grass cover (particularly in wetter years) and it is here that the grassland-favouring species are mainly to be found.

The Lebombo Mountains on the eastern boundary of the Park support some unique habitats such as the Lebombo Ironwood *(Androstachys johnsonii)* forests and the gorges where the larger rivers cut through the mountains.

Vegetation and habitats

A total of 35 different Landscape types have been identified in the Park, but this classification is too fine for highly mobile animals like birds. From an avian perspective therefore, the Park can be divided more simply into 13 habitats (see Habitat map on page 10). The most fundamental of these are the **Aquatic Habitats**. These include the major rivers, seasonal pools and pans, and the man-made dams. The more permanent rivers tend to support a resident suite of birds such as Pel's Fishing-Owl, Goliath and Grey herons, Little Egret, White-crowned Plover and Black-winged Stilt, while the seasonal pans are important for many summer migrants such as Lesser Moorhen, Allen's Gallinule and Little Grebe. As a result of the permanent open water, the man-made dams have had a major impact on the waterfowl such as ducks and geese and many of the Palaearctic waders.

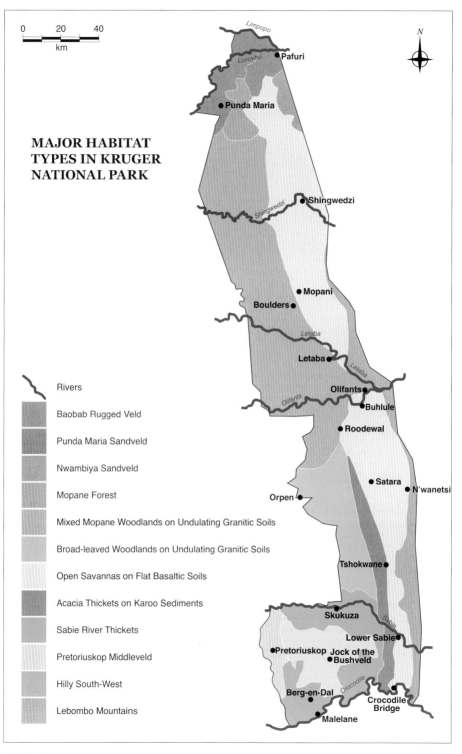

The granitic western habitats generally tend to be poorer from a birding perspective, but the northern **Undulating Mixed Mopani Woodland** habitats have been shown to be the habitat of choice for the Yellow-billed Oxpecker. Further south in the **Broad-leaved Woodlands on Undulating Granitic Soils**, riparian woodlands are important nesting sites for raptors such as Wahlberg's Eagle and the Bateleur.

The **Open Savannas on Flat Basaltic Soils** support rich grasslands where the grassland specialist bird species are to be found. These include Common Ostrich, Harlequin Quail, Kurrichane Buttonquail, Kori and Black-bellied bustards, Montagu's and Pallid harriers, African Quailfinch, Chestnut-backed Sparrowlark and Temminck's Courser. The basaltic soils described above can further be split into northern and southern halves as the Mopane tree *(Colophospermum mopane)* tends to dominate in the north, while it is absent in the south.

The northern parts of the Park have a further four habitats. Two of these are sandveld communities. The **Punda Maria Sandveld** is the only habitat where Southern Hyliota has been recorded, while the **Nwambiya Sandveld** is home to Pink-throated Twinspot, Rudd's Apalis and Eastern Nicator, and Crested Guineafowl are also common. Both of these habitats are favoured by Fawn-coloured Larks.

The tall **Mopane Forests** near Punda Maria constitute a specialised habitat for Arnot's Chat, while the northern **Baobab Rugged Veld** is an important habitat for Grey-headed Parrot, Mosque Swallow, Racket-tailed Roller, Three-banded Courser, and Mottled and Böhm's spinetails.

The **Thickets on Karoo Sediments**, which separate the granites from the basalts, support unique plant communities, mainly *Acacia* species and Many-stemmed Albizia *(Albizia petersiana)* which host many of the small-leaf gleaners such as apalises and eremomelas.

The **Lebombo Mountains** support major Lebombo Ironwood communities. These have been shown to be important for Crested Guineafowl and Yellow-bellied Greenbul, while further south they are the prime habitat for Shelley's Francolin.

The **Sabie River Thickets** are dense thorny thickets where many records of Violet-eared Waxbill have been made. They are also important small-leaf gleaners and bird parties.

The **Pretoriuskop Middleveld**, being at a higher altitude with a higher rainfall, supports some species which rarely occur elsewhere in KNP. These include Red-throated Wryneck, Amethyst Sunbird, Broad-tailed Warbler, Lazy Cisticola and Yellow-fronted Tinkerbird.

The **Hilly South-West** consists of many granite domes and steep wooded gulleys. Narina Trogon and Gorgeous Bush-Shrike occur here as well as Red-winged Starling, Mocking Cliff-Chat, Amethyst Sunbird, Broad-tailed Warbler and Lazy Cisticola.

Rainfall

Rainfall is of course the main driver of the ecology of birds as it influences both habitat quality and food availability. In KNP, the rainfall shows two distinct patterns or cycles. The first is the usual annual cycle in which the large majority of rain (about 85%) falls in the summer months between October and March. This annual pattern is the stimulus for the annual migrants (both intra-African and Palaearctic) to visit the area as well as the majority of breeding species.

The second cycle is one that is driven by the effects of the El Niño Southern Oscillation (ENSO – commonly referred to simply as El Niño). El Niño is a global phenomenon which is driven by ocean-atmosphere interactions. El Niño changes the distribution of rainfall, causing floods in some areas and drought in others. In this part of the world this manifests typically as a cycle of approximately 20 years in duration – 10 years of above average rain followed by 10 years of below, though the durations are variable. Dry years may sometimes also be experienced in a wet cycle, and wet years in a dry one.

Typically, the wet years (wet cycles) are characterised by flowing rivers, full dams and pans, and an abundant grass biomass. The full dams and pans attract birds like Lesser Gallinule, Lesser Moorhen, Lesser Jacana, White-backed Duck, Dwarf Bittern and Pygmy Goose, while the abundant grass provides ideal conditions for species such as Harlequin Quail, Kurrichane Button Quail, Black Coucal, Corncrake, Monotonous Lark, and the Pallid and Montague's harriers.

Dry cycles have a depleted grass biomass through the actions of herbivores, fires and termites. Many rivers cease to flow, and pools, pans and dams dry up. During droughts, the typical wet cycle birds are absent, but other species such as Lark-like Bunting, Grey-backed Sparrowlark, White-browed Sparrow-Weaver, Red-headed Finch, Marico Flycatcher and Chestnut-vented Tit-Babbler may expand their ranges into KNP from the drier west. The first two of those mentioned even reach irruptive proportions in some years. Lark-like Buntings were so common in KNP one year that they were reported to have vacated their usual habitats to the west. The longer-term effects of such droughts are a greater concern. During the droughts of the early 1990s, the Luvuvhu River ceased to flow, and to the east of the Lanner Gorge there was no water at all in the river bed. The water levels were so depleted that the vast majority of Sycomore Figs (*Ficus sycomorus*) died from lack of water. These figs are the major food resource for many frugivorous birds such as Grey-headed Parrots and Trumpeter Hornbills, and the Pafuri populations of these species have declined dramatically.

Global warming is of course a much talked about phenomenon, and the KNP will not escape its effects. These will most likely be expressed mainly through the rainfall. The predictions for the eastern Lowveld (which includes KNP) are that while the 10-year wet and dry cycles are likely to continue, the wetter years will be wetter, and the drier years drier. The long-term impacts that this may have on the area's avifauna can only be speculated upon. The dry years may see more of the western species penetrating KNP and in even larger numbers, while the wetter years will tend to favour waterbirds and the grassland specialists. But the more important questions are how the drier periods will affect the wetland species, and what the impacts may be of wetter years to those species favouring drier conditions. It may be that future, more extreme conditions (wetter or drier) will result in losses of some species from the system.

Tropical cyclones
Tropical cyclones generate huge winds and copious rainfall. When such cyclones make a landfall, the winds may cause terrific damage. KNP is situated too far from the coast for such winds to be a real threat, but the rainfall they generate can be responsible for considerable flood damage. To a natural system such flooding is not usually destructive, and in fact such floods are important as they stimulate fish migrations, scour out developing reedbeds, fill dams and pans, and flush the alien invasives such as Water Hyacinth *(Eichhornia crassipes)* and Water Lettuce *(Pistia stratiotes)* out of the system. However, when extreme rainfall events occur (such as in 1925 and 2000) the impacts are significant. Many tall trees in the riparian zones are swept away by the flood leaving only isolated trees in the riparian zones.

Other climatic factors
During extremely cold winters, many species which usually frequent the higher altitudes of the escarpment area to the west move down to the lower altitudes in the KNP. These are mainly forest species such as Cape Batis and Grey Cuckooshrike, but Fiscal Flycatcher, African Stonechat, White-starred Robin and even Gurney's Sugarbird have also been recorded.

Fire
Fire is a natural part of the KNP's ecosystems. Lightning from early summer thunderstorms sets the veld on fire almost every year somewhere in KNP, particularly after wet years which have resulted in a large standing crop of dry grass. These fires occur at the beginning of the wet season and the impacts of the removal of the grass do not persist for long and are therefore minimised. Out-of-season fires, however, can occur early in the dry season. These fires are caused mainly by illegal immigrants from Mozambique who walk through the Park and must light fires for protection when over-nighting on passage. The impacts of these fires often persist for months, the veld lying bare while awaiting the arrival of the first rains. The effects of such fires on the grassland birds is then very significant, and can result in the almost complete absence of some species as they leave the Park in search of more suitable habitats.

Large herbivores
The impact of larger herbivores (particularly elephants) on habitats is a concern for managers. Over the longer term, elephants at higher densities tend to change woodlands to grasslands as taller trees are removed either by being pushed over or through ring-barking which kills the tree. Many birds depend on tall trees either in the open savannas or in the riparian zone for nesting and roosting, and the loss of these trees may yet pose a threat to the continued presence of these birds in the Park.

ADDITIONAL READING

BARNES, K.M. (ed.). 2000. *The Eskom Red Data Book of Birds of South Africa*. Birdlife South Africa, Johannesburg.

CHITTENDEN, H. 2007. *Roberts Bird Guide*. The Trustees of the John Voelcker Bird Book Fund, Cape Town.

HARRISON, J.A., ALLAN, D.G., UNDERHILL, L.G., HERREMANS, M., TREE, A.J., PARKER, V. & BROWN, C.J. (Eds.). 1997. *The Atlas of Southern African Birds. (Volumes 1 & 2)*. Birdlife South Africa, Johannesburg.

HOCKEY, P.A.R., DEAN, W.R.J. & RYAN, P.G. (eds.). 2005. *Roberts Birds of Southern Africa*. VIIth Edition. The Trustees of the John Voelcker Bird Book Fund, Cape Town.

JOUBERT, S.C.J. 2007. *The Kruger National Park – a History (Volumes 1, 2 & 3)*. High Branching, Johannesburg.

KEMP, A.C. 1974. *The Distribution and Status of the Birds of the Kruger National Park*. Koedoe Monograph No. 2. National Parks Board of Trustees, Pretoria.

KEMP, A.C., DEAN, W.J.R., WHYTE, I.J., MILTON, S.J. &. BENSON, P.C. 2003. Birds of the Kruger National Park: possible responses and contributions to ecological heterogeneity. In: *The Kruger Experience – Ecology and Management of Savanna Heterogeneity*. Eds. du Toit, J. Rogers, K.H. & Biggs, H.C. Washington, Island Press. Pp 276-291.0000

KEMP, A.C., HERHOLDT, J.J., WHYTE, I.J. & HARRISON, J. 2001. Birds of the two largest National Parks in South Africa: a method to generate estimates of population size for all species and assess their conservation ecology. *South African Journal of Science* 97 (9/10): 393-403.

NEWMAN, K. 1980. *Birds of Southern Africa. 1: Kruger National Park*. MacMillan South Africa, Johannesburg.

SCHMIDT, E., LÖTTER, M. & McCLELAND, W. 2002. *Trees and Shrubs of Mpumalanga and Kruger National Park*. Jacana Media, Johannesburg.

SINCLAIR, I. & WHYTE, I.J. 1991. Field Guide to the Birds of the Kruger National Park. Struik, Cape Town.

TARBOTON, W.R., KEMP, M.I. & KEMP, A.C. 1987. *Birds of the Transvaal*. Transvaal Museum, Pretoria.

A male Black-bellied Bustard giving unique territorial frog-like 'quark' call. Immediately after call, pulls head down into shoulders, then extends neck and gives an explosive kww-ick *that sounds like a cork being pulled from a bottle.*

PLATE 1 Ostrich, Francolin, Spurfowl p206

COMMON OSTRICH *Struthio camelus* (1) ♂ **up to 2 m; 70 kg**
SEXES ♂>♀. *Largest and heaviest living bird; flightless.* **M** *Mostly black with white wings and white or buff tail. Neck and legs almost naked.* **F** *Plumage greyish-brown.* **JUV** *Resembles ♀, but smaller. Chicks unmistakable with 'hedgehog' black and white down.* **STATUS** From the original wild population. Confined mainly to the eastern basaltic plains, especially north of the Letaba River, but recorded widely elsewhere. Usually in pairs, but form small flocks in the non-breeding season. Closure of waterholes appears to have benefitted this species in the north through diminished predator pressure. Waterhole closure has resulted in withdrawal of other common prey species from the grasslands, and therefore also of predators. **HABITAT** Prefers short grassland and open woodland. Avoids dense woodland. **FOOD** Almost entirely plant material, pulled out of ground or stripped off plants. **CALL** Deep booming *boo boo boooooh hoo* given by ♂ mainly in br season (not unlike a lion's roar!). **BR** Polygamous. Nest a scrape in ground; eggs laid by one or more females. Males incubate at night and females mostly by day. (**Volstruis**) ALT NAME *Ostrich*

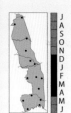

COQUI FRANCOLIN *Peliperdix coqui* (188) 24 cm; 250 g
SEXES ♂>♀. *Smallest southern African francolin.* **M** *Head and neck pattern distinctive, making identification easy.* **F** *Superficially resembles Shelley's Francolin but is overall paler, has broader white eye-stripe and is smaller.* **JUV** *Resembles ♀ but paler above with rufous-buff mottling.* **STATUS** An uncommon breeding resident with a patchy distribution, forming coveys of 6–8 birds in non-br season. Found throughout, but usually only seen crossing roads, and also gives its presence away when calling. **HABITAT** Tall grassland and well-grassed woodland. **FOOD** More seeds, shoots and small fruits than underground corms and bulbs; increases insect intake in summer. **CALL** High-pitched 2-syllable *co-qui, co-qui ...* Dominant call by ♂ is *ter, ink, ink, terra, terra, terra, terra.* **BR** Monogamous. Shallow grass-lined scrape well concealed among grass tufts. (**Swempie**)

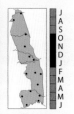

SHELLEY'S FRANCOLIN *Scleroptila shelleyi* (191) 33 cm; 475 g
SEXES *Alike,* ♂>♀. *Boldly marked black border to white throat, and black speckling to lower breast and belly are most distinctive features.* **JUV** *Resembles adult black and white breast barring irregular.* **STATUS** An uncommon resident, sedentary and forming coveys of 4–8 birds during non-br season. Not easy to see owing to its secretive habits, but often heard calling, especially in the early morning and late afternoon. **HABITAT** Favours tall moist sweetveld grassland and open wooded savanna, often on stony terrain or among rocky outcrops, especially the rhyolitic Lebombo Mountains. **FOOD** Corms, bulbs and seeds in winter, with increased insect intake during br season. **CALL** Very distinctive 4-note call *I'll drink YOUR beer,* repeated 3–4x. **BR** Monogamous. Usually lays 4–5 eggs in shallow scrape well concealed in grassland or under bush. (**Laeveldpatrys**)

NATAL SPURFOWL *Pternistis natalensis* (196) 34 cm; 410 g
SEXES *Alike,* ♂>♀. *With lack of bare red or yellow skin around eyes, it is unlikely to be confused with Swainson's Francolin.* **JUV** *Bill dull greenish and generally paler than adult.* **STATUS** The most common spurfowl in the park; often enters the camps and gardens. A near-endemic usually found in pairs or small coveys. **HABITAT** Variety of wooded habitats, favouring riparian margins. **FOOD** Winter diet dominated by plant matter such as bulbs, roots, seeds and some fruits, supplemented by insects in summer. Scratches through herbivore dung in search of undigested seeds. **CALL** Loud and raucous *kak-keek, kak-keekekeek* and similar variations thereof. **BR** Monogamous. Usual clutch of 5–6 eggs placed in grass-lined scrape well concealed beneath dense plant cover or in rank grass. Only ♀ incubates. (**Natalse Fisant**) ALT NAME *Natal Francolin*

SWAINSON'S SPURFOWL *Pternistis swainsonii* (199) 36 cm; 605 g
SEXES *Alike,* ♂ *considerably larger than* ♀. *The only spurfowl in the Park with a distinctive red face and neck. Legs black.* **JUV** *Duller overall, with less bare red skin on face; throat with white feathers and bill brown with yellow base.* **STATUS** A common resident found throughout, particularly on the eastern grasslands; usually in pairs or small parties. **HABITAT** Variety of woodland, savanna and grassland. Partial to tall grassland. **FOOD** Bulk of winter diet bulbs, tubers, roots, seeds and berries, supplemented by insects in summer. **CALL** Repetitive harsh *krrraa, krrraa, krrraa, krrraa ...* **BR** Monogamous. Usual clutch 5–7 eggs; nest a scrape well concealed in dense cover. (**Bosveldfisant**) ALT NAME *Swainson's Francolin*

PLATE 2 Quail, Francolin, Guineafowl p206

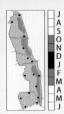

HARLEQUIN QUAIL *Coturnix delegorguei* (201) 18 cm; 75 g
SEXES ♀>♂. **M** *Easily distinguished from other quail by white eyebrow and throat, black belly and conspicuous chestnut flanks.* **F** *Darker buff underparts than Common Quail (unlikely to be seen in the Park), and black necklace visible on lower throat. Upperparts show conspicuous black spots.* **JUV** *Similar to adult ♀ but paler and lacks black spots on upperparts.* **STATUS** Abundant in wetter years, but can be absent in droughts. Br intra-African migrant (many overwinter, particularly after summers of good rainfall). Historical records suggest that Common Quail *Coturnix coturnix* occurred in the Park, but in the past two decades there has been no definite evidence of this, and it's now suspected that this species has been misidentified for ♀ Harlequin Quail. Common Quail is a common inhabitant of the eastern grasslands of s Africa, has been recorded in the adjacent Nelspruit region, and is likely to occur within the KNP only as a very rare passage migrant. **HABITAT** Mainly eastern basaltic grasslands, but also floodplains, and well grassed open woodland savanna. **FOOD** Predominantly insects; also seeds, green shoots and leaves. **CALL** 2–3 note *whit whit whit*, similar to Common Quail but more metallic. **BR** Probably monogamous. Usual clutch of 4–8 eggs laid in shallow grass-lined scrape and incubated by ♀ alone. (**Bontkwartel**)

KURRICHANE BUTTONQUAIL *Turnix sylvaticus* (205) 15 cm; 43 g
SEXES *Similar,* ♀ *larger and more brightly coloured than* ♂. *In flight contrasting dark flight feathers and light-coloured wing coverts distinguish it from quails.* **M** *Black spots on sides of orange-rufous breast, and pale eye diagnostic. Distinctive dark crown stripe.* **F** *Similar to* ♂ *but more richly coloured above, buffier on throat and darker black spotting.* **JUV** *Similar to adult but greyer; brown eyes.* **STATUS** Locally common to abundant br visitor in wetter years, some birds may overwinter after summers of good rainfall; generally nomadic. **HABITAT** Grassland and open savanna woodland. **FOOD** Insects and seeds. **CALL** Deep, haunting, flufftail-like *hoom hoom hoom ...* given as advertising call by ♀. **BR** Polyandrous. 3–4 eggs laid in grass-lined scrape. (**Bosveldkwarteltjie**)

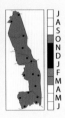

CRESTED FRANCOLIN *Dendroperdix sephaena* (189) 33 cm; 350 g
SEXES *Similar,* ♂>♀. *Unmistakable bantam-like build and broad white eye-stripe make it easy to identify. Has habit of holding its tail cocked. Plumage on back of ♀ lightly barred.* **JUV** *Resembles adult but paler above.* **STATUS** Common resident, found in coveys of 2–5 birds in non-br season. **HABITAT** Essentially a bird of the bushveld preferring dense stands of scrub and thicket, often with sparse ground cover. **FOOD** Primarily corms, bulbs, green shoots, fruit and berries in non-br season but changing to an insect-dominated diet in summer. **CALL** Well-synchronised duet between ♂ and ♀: *ki kerrik, ki kerrik ...* or *pump handle, pump handle ...* or *eat a caterpillar, eat a caterpillar ... ,* which sounds as though only one bird is calling. **BR** Monogamous. Clutch usually 5–6 eggs laid in grass-lined scrape, incubated by ♀ only. (**Bospatrys**)

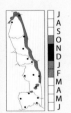

CRESTED GUINEAFOWL *Guttera edouardi* (204) 50 cm; 1,3 kg
SEXES *Alike,* ♂>♀. *Unmistakable with thick crest of curly black feathers on crown. Pale outer secondaries show as a narrow white stripe in flight. Crimson eye and black neck distinctive.* **JUV** *Duller than adult and has fine black and white barring (above and below).* **STATUS** Br resident, locally fairly common. **HABITAT** Mainly Nwambia sandveld, *Androstachys* forests on Lebombo range and riparian thickets at Pafuri. Recent records suggest its range in KNP may be expanding. **FOOD** Omnivorous. Wide range of plant and animal food includes seeds, fruit, berries, shoots, bulbs, roots and insects. **CALL** Noisy and loud clucking call *chuk-chuk-chukchuker-chuk-chuk ...* **BR** Monogamous. Shallow scrape concealed in thick grass on forest verges. Incubation by ♀ only. (**Kuifkoptarentaal**)

HELMETED GUINEAFOWL *Numida meleagris* (203) 56 cm; 1,35 kg
SEXES *Alike,* ♂>♀. *Bony casque on head, red facial skin and blue upper neck easily distinguish it from Crested Guineafowl.* **JUV** *Browner body with reduced helmet; upper neck and throat feathered.* **STATUS** Locally common resident. Numbers fluctuate, more common in wetter years. **HABITAT** Widespread, but less common in the grasslands. **FOOD** Omnivorous: insects, seeds, bulbs, tubers and berries. **CALL** Staccato cackling *kek, kek, kek, kek, kaaaaaa ka ka ka ...* alarm; also 2-note *buck-wheat, buck-wheat* contact call. **BR** Monogamous. Nest a shallow scrape, well concealed under grass or in thicket. Incubation by ♀ only. (**Gewone Tarentaal**)

PLATE 3 Whistling Duck, Ducks, Pygmy Goose p206

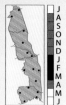

FULVOUS DUCK *Dendrocygna bicolor* (100) 46 cm; 733 g
SEXES Similar. Golden-brown head and underparts distinctive; unlikely to be confused with any other duck in region. **JUV** Paler and without rufous scaling on upperparts. **STATUS** Rare vagrant; many birds are summer br migrants to the region; highly nomadic and often in association with White-faced Duck. May breed in the Park in wetter years. **HABITAT** Inland water bodies, especially with abundant aquatic vegetation. **FOOD** Mainly aquatic plant matter; also aquatic insects. **CALL** 2-syllabled whistle *tooo-ee* or *tou-ee*, usually in flight. **BR** Monogamous. Nest well concealed in long grass near water. (**Fluiteend**)

WHITE-FACED DUCK *Dendrocygna viduata* (99) 47 cm; 702 g
SEXES Alike. Very distinctive with white face, overall dark colour, barred flanks and erect stance. **JUV** Duller than adult; 'dirty' brown face and throat. **STATUS** Common resident especially in wetter years, but highly nomadic. Gregarious in non-br season and favours more permanent water bodies during the flightless winter moult period. Usually in flocks or small family groups. **HABITAT** Wide range of aquatic habitats preferring those with extensive shallows and emergent vegetation. **FOOD** Aquatic vegetation, seeds, tubers, insects and molluscs; forages in both day and night. **CALL** Very vocal, especially in flight; characteristic 3-note whistle *swee-swee-sweeu*. **BR** Monogamous. Nest well hidden in waterside grasses or sedges. Clutch usually 9–11 eggs. (**Nonnetjie-eend**)

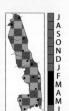

WHITE-BACKED DUCK *Thalassornis leuconotus* (101) 43 cm; 660 g
SEXES Similar. Small, large-headed duck that sits low in the water with humpback appearance. Conspicuous white patch at base of bill; white on back usually only visible when flying. **JUV** Duller than adult, white loral patch smaller and duller. **STATUS** Very rare vagrant in KNP, mostly recorded in wetter years. **HABITAT** Favours open well-vegetated water bodies. **FOOD** Mainly seeds, especially water-lily seeds. Feeds mainly early morning and in the evenings, sleeping or roosting in the heat of the day. **CALL** Squeaky 2-note whistle *tit-weet*. **BR** Monogamous. Nests over water in emergent vegetation, but no records for KNP. (**Witrugeend**)

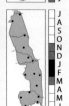

AFRICAN PYGMY-GOOSE *Nettapus auritus* (114) 33 cm; 262 g
SEXES ♂>♀. Small; orange body and dark green back diagnostic. **M** Well-defined head markings with vertical black ear-stripes and bright yellow bill. **F** Duller and lacks bold head markings; bill buff coloured. **JUV** Less distinctly marked, resembles ♀. **STATUS** Mostly an uncommon vagrant; usually present in small numbers in wetter years. Listed as Near-threatened in s Africa. **HABITAT** Prefers permanent clear waters with floating and emergent vegetation, especially water lilies. **FOOD** Mainly ripe seeds and flower parts of water-lilies, also other aquatic plants. **CALL** Soft twittering 2-note calls: *cho-choo* or *pee-wee*. **BR** Monogamous. Recorded breeding at Leeupan; nests in tree cavities. (**Dwerggans**) **ALT NAME** *Pygmy Goose*

RED-BILLED TEAL *Anas erythrorhyncha* (108) 46 cm; 570 g
SEXES Similar, ♂>♀. Only pink-billed duck in region, with distinctive dark crown. ♀ has steeper forehead than ♂, and bill has more black on ridge than ♂. **JUV** Greyer, duller and with less distinct scaling. **STATUS** A common duck species in s Africa but an uncommon visitor to the Park, present mainly in wetter years. Usually found in pairs. **HABITAT** Most inland water bodies. **FOOD** Mainly grass and aquatic seeds and fruits, also aquatic insects. **CALL** ♂ gives soft swizzling *whizzt* or *zee*, ♀ a loud *quaaak*. **BR** Monogamous. Prefers to breed on the margins of shallow ephemeral pans and dams. Nest usually placed on dry ground close to water. (**Rooibekeend**)

HOTTENTOT TEAL *Anas hottentota* (107) 35 cm; 250 g
SEXES Similar. Smallest duck in region. Blue-grey bill, dark crown and small size diagnostic. ♀ has slightly browner (lighter) crown. In flight male has green speculum, and trailing edge to secondaries white. **JUV** Duller, plainer above and less spotted below. **STATUS** Rare vagrant in KNP, recorded mainly in the northern parts. **HABITAT** Prefers permanent or semi-permanent shallow freshwaters. **FOOD** Both plant and animal matter. **CALL** ♀ gives harsh *ke-ke*; ♂ in courtship gives soft ticking notes. **BR** No records for KNP, the attached records reflect breeding from elsewhere in ne southern Africa. (**Gevlekte Eend**)

PLATE 4 Geese, Ducks pp206–7

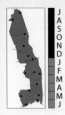

EGYPTIAN GOOSE *Alopochen aegyptiaca* (102) 68 cm; 2,1 kg
SEXES *Distinctive dark patches on eye and breast, as well as head and neck pattern.* **SEXES M** *Larger, with thicker neck and different call.* **F** *Smaller and with slimmer neck.* **JUV** *Lacks eye and breast patches, and forewing panel grey, not white.* **STATUS** Common to abundant resident with widespread movement. Usually found in pairs or small family groups but after breeding they sometimes congregate on sandbanks on the major river beds in flocks of up to a few hundred birds. The majority of these are probably unpaired immature birds. **HABITAT** Fairly common along the margins of both dams and rivers, visiting ephemeral water bodies during the wet seasons. **FOOD** Mainly vegetable matter, and primarily a grazer of grass shoots close to the water's edge; also rhizomes and tubers. **CALL** ♀ gives coarse strident *pur-pur-pur* honking; ♂ hisses loudly. **BR** Monogamous. Nests on ground or on old stick nests of other species, e.g. Hamerkop. (**Kolgans**)

SPUR-WINGED GOOSE *Plectropterus gambensis* (116) 98 cm; ♂ 5,1 kg, ♀ 3,5 kg
SEXES *Largest goose-like duck in region. Could be confused with much smaller Comb Duck in flight but white on forewing and underwing conspicuous.* **SEXES** ♂>♀. **M** *Red facial skin more prominent and extends beyond eye. Large carpal spur sometimes seen on perched bird or even in flight.* **F** *Similar to ♂ but duller and facial skin less extensive.* **JUV** *Browner than adult and without red facial skin.* **STATUS** Uncommon vagrant that occasionally breeds in the Park. Mainly recorded in wetter years. Undergoes flightless moult during the mid-winter period. **HABITAT** Large water bodies, dams and floodplains. **FOOD** Mainly plant material, especially rhizomes, stolons and leaves of aquatic plants. Also grass shoots, and seeds. Limited animal matter eaten includes termite alates and aquatic insect larvae. May feed on plant matter and invertebrates disturbed by Hippopotamus. **CALL** Weak, high-pitched *cherwit*, mostly in flight. **BR** Monogamous; usually nests in dense cover near water. (**Wildemakou**)

COMB DUCK *Sarkidiornis melanotos* (115) 67 cm; 1,60 kg
SEXES *Similar, ♂>♀. In flight lacks white on forewing and has more white on underparts than Spur-winged Goose.* **M** (**BR**) *Large laterally compressed knob (comb) on forehead. Orange-buff on head, neck and undertail variable.* **M** (**NON-BR**) *Comb much reduced and loses orange-buff coloration.* **F** *Duller than ♂, bill grey and lacks comb.* **JUV** *Dark brown upperparts, light brown underparts, head becoming mottled brown with white eye-stripe.* **STATUS** Intra-African partial migrant, fairly common in wetter years. Nomadic, often undertaking extensive movements. Usually in pairs or small family groups during the br season. Often perches in trees. **HABITAT** Marshes and well-vegetated temporary pans in savannas and floodplains. **FOOD** Grass seeds, water-lily fruits, leaves and aquatic insects. **CALL** Various harsh frog-like *guk-guk* calls by ♀, and hisses by ♂. **BR** Polygynous. Nests in tree cavities, also on top of old Hamerkop nests. (**Knobbeleend**) ALT NAME *Knob-billed Duck*

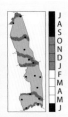

AFRICAN BLACK DUCK *Anas sparsa* (105) 55 cm; 1,0 kg
SEXES *Alike, ♂>♀. Dark-plumaged duck with grey/black bill and white speckles on back.* **JUV** *Browner than adult with whitish underparts.* **STATUS** Fairly common resident, usually in pairs along KNP's major permanent rivers; generally sedentary, territorial and often seen flying fast and low over streams and rivers. **HABITAT** Prefers fast-flowing perennial streams; rarely seen on still waters. **FOOD** Predominantly aquatic invertebrates, plant material and seeds; also fruit and berries from plants overhanging water. Other animal food includes molluscs, crabs and fish fry. **CALL** Loud *quack* by ♀; ♂ gives wheezy *peep* ... **BR** Monogamous. Nests in dense grass or flood debris close to water's edge. (**Swarteend**)

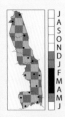

SOUTHERN POCHARD *Netta erythrophthalma* (113) 50 cm; 780 g
SEXES ♂>♀. *Both sexes have large white wingbar in flight.* **M** *Dark brown head and upperparts, red eye and pale blue-grey bill (with blackish nail) distinctive.* **F** *Overall paler plumage with whitish crescent extending behind eye (below eye in ♀ Maccoa Duck).* **JUV** *Resembles ♀ but pale with less distinct facial markings.* **STATUS** Rare non-br visitor to KNP, present mainly in wetter years. Highly aquatic, seldom found ashore. **HABITAT** Prefers fairly deep water bodies, but found in seasonal and permanent wetlands with emergent vegetation, and floodplains. **FOOD** Mainly aquatic plant matter but also insects to a lesser extent. **CALL** Not very vocal; ♂ gives soft *prerr prerr*, ♀ nasal *krrrow* in flight. **BR** Monogamous. but no records for KNP. (**Bruineend**)

PLATE 5 Honeyguides, Honeybirds p207

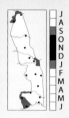

SCALY-THROATED HONEYGUIDE *Indicator variegatus* (475) 19 cm; 48 g
SEXES *Alike. Fine blackish streaks on throat and light scaling on breast diagnostic. White outer tail feathers conspicuous in flight.* **JUV** *Bolder black stripes on throat and upper breast.* **STATUS** Uncommon to rare but generally an overlooked resident usually located by call. Recorded at a few widely scattered localities. **HABITAT** Forest and thickets; favours mature well-developed riverine woodland where barbets and woodpeckers are common. **FOOD** When available, beeswax, honeybees and their larvae, also other insects (often taken on the wing) and occasionally seeds and fruit. **CALL** Fast churring trill, likened to a finger run up a hair comb, or a helicopter starting up. Call-sites maintained through successive generations from same tree(s) or regularly used branches. **BR** Polygynous brood parasite, hosted by woodpeckers and barbets. Main hosts in Park are Golden-tailed and Cardinal woodpeckers and Black-collared Barbet. (**Gevlekte Heuningwyser**)

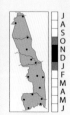

GREATER HONEYGUIDE *Indicator indicator* (474) 20 cm; 48 g
SEXES ♂>♀. **M** *Whitish cheeks, black throat and pink bill diagnostic. White outer feathers conspicuous.* **F** *Somewhat bulbul like; separated from Scaly-throated Honeyguide by plain breast, and lacks the moustachial stripe of Lesser Honeyguide.* **JUV** *Dark bill with pale yellowish underparts, and white outer tail feathers.* **STATUS** Generally uncommon but widely distributed resident; usually located by loud monotonous call. **HABITAT** Woodland savanna and riverine forest. **FOOD** When available, beeswax; and to a lesser extent bee larvae. Also bees and other insects such as ants and beetles. Sometimes hawks insects in flight. **CALL** Sustained and loud bisyllabic WHIT-purr, WHIT-purr ... (or VIC-tor, VIC-tor ...). Call sites used year-round, but mostly in br season. Well known for its habit of guiding people and mammals such as honey badgers to beehives. Initiates 'guiding' by attracting person or animal with continuous chattering call and conspicuous flight. Leading then follows with bird chattering between short flights from perch to perch in the direction of known hive. Male able to make a drumming sound like that of a snipe by vibrating fanned tail in display flights. **BR** Polygynous brood parasite, hosted by wide range of hole-nesting species, especially African Hoopoe, bee-eaters, barbets, kingfishers and starlings. (**Grootheuningwyser**)

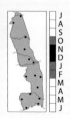

LESSER HONEYGUIDE *Indicator minor* (476) 15 cm; 28 g
SEXES *Alike. Only honeyguide in park with a dark malar stripe. Has a much more robust bill than Brown-backed Honeybird which has a slender bill. Characteristic white outer tail feathers distinctive in flight.* **JUV** *Lacks white loral stripe (between eye and beak), and black malar stripe.* **STATUS** Fairly uncommon resident that is locally nomadic in the non-br season. Sedentary, usually solitary and commonly located by sustained monotonous call. Flight fast, direct and undulating. Drinks at bird baths. **HABITAT** Wooded savanna and riverine evergreen forest where host species are common. **FOOD** When available, beeswax and bee larvae. Also wasps and their larvae, and insects such as termite alates, ants and caterpillars. **CALL** Toneless and monotonous klew, klew, klew ... , about 1 note every second and repeated 10–40x from traditional song posts. **BR** Polygynous brood parasite, hosted mainly by barbets (Acacia Pied, Black-collared and Crested), but also on other hole-nesting species such as woodpeckers, Green Wood-Hoopoe, Striped Kingfisher, Little Bee-eater, and starlings. (**Kleinheuningwyser**)

BROWN-BACKED HONEYBIRD *Prodotiscus regulus* (478) 13 cm; 14 g
SEXES *Alike. Very similar to African Dusky and Spotted flycatchers but extensive white in outer tail distinctive, and most noticeable in flight. Lacks pale streaking on breast, and darker brown than Spotted Flycatcher.* **JUV** *Gape orange, outer tail feathers white, and yellowish below.* **STATUS** Uncommon local resident. Also an altitudinal migrant so numbers probably increase in KNP during winter. Perhaps more common than suspected owing to its retiring nature and confusion with similar species. **HABITAT** Open woodland and riparian bush. **FOOD** Gleans scale insects and woolly aphids from foliage; other insects also recorded. May take aerial insects on the wing like a flycatcher. **CALL** Insect-like dzreeeee, rising in pitch and lasting 3–4 sec. Call not loud like honeyguides. **BR** Brood parasite, probably polygynous, parasitising various cisticolas (such as Rattling and Neddicky), camaropteras and prinias. (**Skerpbekheuningvoël**) **ALT NAME** *Sharp-billed Honeyguide.*

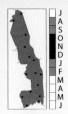

BENNETT'S WOODPECKER *Campethera bennettii* (481) 23 cm; 70 g
SEXES ♂>♀. *Spotted (not streaked) underparts distinguish it from similar-sized Golden-tailed Woodpecker and smaller Cardinal Woodpecker.* **M** *Forehead and crown crimson; malar stripe dark red, throat white.* **F** *Chestnut throat, chin and cheeks diagnostic. Forehead black speckled with white.* **JUV** ♀ *has chestnut throat and cheek, but black forehead speckled with buff; juv* ♂ *similar to adult* ♂. **STATUS** Locally common br resident, usually in pairs or small family groups. Probably permanently territorial, and pairs remain together year-round. **HABITAT** Tall deciduous broad-leaved woodland; less common in *Acacia* woodland. Usually easily seen in Shingwedzi restcamp. **FOOD** Mainly ants, their eggs and pupae (feeds mostly from ground). Termites and other insects occasionally taken. Ants and their larvae gathered with sticky tongue. **CALL** Characteristic loud ringing *wirrit, wirrit*, call repeated rapidly 6–8x. More vocal than other woodpeckers in the region, but seldom drums. **BR** Monogamous. Holes excavated in dead tree limbs, or uses other old woodpecker holes. (**Bennettse Speg**)

GOLDEN-TAILED WOODPECKER *Campethera abingoni* (483) 21 cm; 70 g
SEXES *Similar. Similar to smaller Cardinal Woodpecker with streaked, not blotched underparts, and greenish upperparts yellower. Cardinal Woodpecker has barred upperparts.* **M** *Malar stripe dark red with black speckling; forehead and crown grey-black, speckled with crimson; nape crimson.* **F** *Malar stripe absent in some subspecies; others black, speckled white. Forehead and crown black, speckled with white; nape crimson.* **JUV** *Similar to adult but duller, sexes resembling respective parents.* **STATUS** Fairly common resident, sedentary with pairs remaining together year-round. **HABITAT** Most woodland types and riparian forest. **FOOD** Mainly ants, termites and their eggs and pupae; also other insects. Entirely arboreal; prey gleaned from tree trunks and branches using sticky tongue (typical of *Campethera* woodpeckers). **CALL** Very distinctive *hair* or *weea* call uttered by both sexes. Drums weakly. **BR** Monogamous. Nests excavated in tree trunks with entrance holes 50–53 mm wide. Occasionally parasitised by Greater, Lesser and Scaly-throated honeyguides. (**Goudstertspeg**)

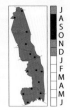

CARDINAL WOODPECKER *Dendropicos fuscescens* (486) 15 cm; 30 g
SEXES ♂>♀. *Conspicuously barred back and overall black and white appearance. The smallest woodpecker in the region and the only woodpecker in the Park with a brown forecrown (both male and female). Golden-tailed Woodpecker (much larger) only other woodpecker in KNP with streaked underparts.* **M** *Red crown and brownish forecrown. Heavily streaked cheeks and throat eliminate confusion with* ♂ *Bennett's Woodpecker.* **F** *Hind crown and nape black, not red.* **JUV** *Duller and greyer than adult; both sexes have red patch in centre of crown.* **STATUS** Common resident, sedentary and in pairs throughout year. Presence usually given away by tapping sound of foraging bird. **HABITAT** Most woodland types, especially *Acacia*. **FOOD** Barbed tongue (typical of *Dendropicos* woodpeckers) used to secure prey such as larvae and pupae of beetles and other insects. **CALL** Chittering rattle *dri-dri-dri-dri* ... and variations thereof. **BR** Monogamous. Nest holes excavated in dead tree trunks. Parasitised by Lesser and Scaly-throated honeyguides. (**Kardinaalspeg**)

BEARDED WOODPECKER *Dendropicos namaquus* (487) 24 cm; 80 g
SEXES ♂>♀. *Finely barred grey-white underparts, large size and white facial markings distinctive.* **M** *Crown red, nape and hindneck black.* **F** *Like* ♂ *but crown black, flecked white.* **JUV** *Both sexes have red on crown and forehead (more extensive than adult* ♂). **STATUS** Fairly common and widespread resident in suitable habitats; sedentary, pairs remain together year-round. The largest woodpecker in the region. **HABITAT** Open mixed deciduous woodland, especially with tall trees. **FOOD** Insect larvae and pupae extracted from crevices in wood and beneath bark using long barbed tongue. Occasionally takes lizards and spiders. **CALL** Series of variable *wik-wik-wik* notes, increasing in tempo. Loud drumming is characteristic and distinctive habit of this species; loud fast drumming anywhere in the Park likely to be this species marking its territory. **BR** Monogamous. Nest holes excavated in large dead tree stems. Not parasitised by honeyguides because of its strategy of breeding during the winter when honeyguides do not breed. (**Baardspeg**)

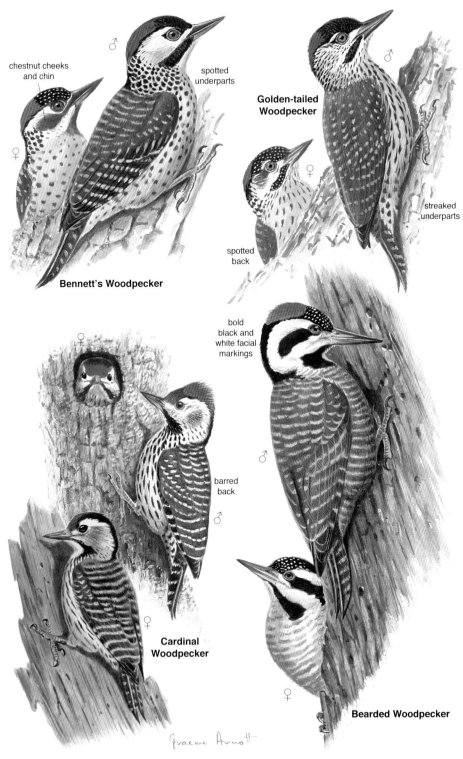

PLATE 7 Tinkerbirds, Barbets p207

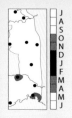

YELLOW-RUMPED TINKERBIRD *Pogoniulus bilineatus* **(471) 12 cm; 15 g**
SEXES *Alike. Two clear white lines through sides of black head diagnostic; yellow wing panels more evident than in Yellow-fronted Tinkerbird. Yellow rump patch not easy to see in field.* **JUV** *Lacks white stripes on side of face; bill initially horn-coloured.* **STATUS** Rare and occurs only at isolated localities (Crocodile River at Crocodile Bridge, and has colonised Skukuza in recent years). Sedentary and found singly or in pairs. **HABITAT** Riparian habitats and well-vegetated restcamps in the south. **FOOD** Mainly small fruit and berries, especially mistletoes. **CALL** Intermittent, monotonous, 4–6 notes *pot, pot, pot, pot*. Also gives series of trills. **BR** Monogamous. Excavates small hole in underside of slender dead branch or stump. (**Swartblestinker**) **ALT NAME** *Golden-rumped Tinker Barbet*

YELLOW-FRONTED TINKERBIRD *Pogoniulus chrysoconus* **(470) 12 cm; 14 g**
SEXES *Alike. Conspicuous yellow forehead is diagnostic feature, but this does sometimes vary in colour from pale yellow to deep orange (uncommon).* **JUV** *Forecrown initially black.* **STATUS** Fairly common resident, sedentary and found singly or in pairs. In KNP mainly confined to higher altitudes in the south-western regions, but also at Pafuri. **HABITAT** Mixed broad-leaved woodland, dry sparse savanna and riparian fringing forest, especially where mistletoes are common. **FOOD** Favours sticky fruit of mistletoes, also other small fruit, including figs, as well as insects. Regurgitated mistletoe seeds stick to branches, especially in process of bill-wiping, thus dispersing these parasitic plants. **CALL** Continuous and monotonous tinkering *pot, pot, pot* ... lasting for minutes. **BR** Monogamous. Excavates small hole about 20 mm in diameter in underside of dead branch or stump. (**Geelblestinker**) **ALT NAME** *Yellow-fronted Tinker Barbet*

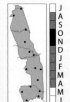

ACACIA PIED BARBET *Tricholaema leucomelas* **(465) 18 cm; 31 g**
SEXES *Alike. Larger than tinkerbirds, and as Red-fronted Tinkerbirds do not occur in KNP, there should be no confusion.* **JUV** *Lacks red patch on forecrown.* **STATUS** Fairly common near-endemic, sedentary and found singly or in pairs. Occurs throughout park, most common in drier regions. **HABITAT** Favours *Acacia*-dominated mixed woodland. **FOOD** Mainly small fruit, including mistletoes; also insects and nectar from aloes. **CALL** Distinctive nasal toy tin trumpet *pehp, pehp, pehp*, and a soft *hoop ... hoop*. **BR** Monogamous. Excavates hole in dead or growing tree stump, often in fairly hard wood, and occasionally utilises other birds' nests, e.g. woodpecker holes and swallow nests. Parasitised by Lesser Honeyguide, and very occasionally by Greater Honeyguide. (**Bonthoutkapper**) **ALT NAME** *Pied Barbet*

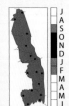

BLACK-COLLARED BARBET *Lybius torquatus* **(464) 20 cm; 54 g**
SEXES *Alike. Bright red face, throat and upper breast, bordered by broad black collar. Shade of red varies from bright scarlet to salmon pink. Rare* xanthochroic *(yellow replaces red) form not recorded in KNP. Large black bill.* **JUV** *Lacks red face (which is initially brownish at fledging).* **STATUS** Common and conspicuous resident occurring throughout Park. Usually in pairs. **HABITAT** Mostly well-wooded habitats, especially riparian, where fruit-bearing trees are abundant. **FOOD** Mainly frugivorous, and partial to figs. Also takes insects and nectar. **CALL** Synchronised duet, one bird giving *two*, the other *pouddle*, creating *two-pouddle, two-pouddle* ... repeated 10–20x with the birds usually facing each other and bobbing their heads. **BR** Probably monogamous; sometimes with nest helpers. Excavates hole (diameter usually 45–55 mm) in dead tree stump. Parasitised by Lesser Honeyguide, also to some extent by Greater Honeyguide. (**Rooikophoutkapper**)

CRESTED BARBET *Trachyphonus vaillantii* **(473) 24 cm; 74 g**
SEXES *Fairly similar.* **M** *Yellow face speckled with red, black crest and chest band make it easily identifiable. Largest barbet in the area and only one that is sexually dimorphic.* **F** *Similar to ♂ but less brightly coloured.* **JUV** *Resembles adult but browner on wings and has less red streaking below.* **STATUS** Common and widespread resident; usually in pairs that remain together throughout the year. **HABITAT** Favours drier woodland, especially *Acacia*. **FOOD** Omnivorous. Often takes insects on the ground; also feeds on wide range of fruit, and recorded taking nectar. **CALL** Loud and sustained unmusical trill *tr-r-r-r-r r-r-r-r-r-r* ... , likened to an alarm clock with the bell removed. **BR** Monogamous. Br in hole excavated in underside of a dead stump; occasionally parasitised by Greater and Lesser honeyguides. (**Kuifkophoutkapper**)

PLATE 8 Hornbills, Ground-Hornbill pp207–8

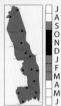

RED-BILLED HORNBILL *Tockus erythrorhynchus* **(458) 40 cm; 140 g**
SEXES *Alike, ♂>♀ and longer billed with black patch at base of lower mandible.* **JUV** *Bill smaller and more orange; eyes grey, turning brown then yellow.* **STATUS** Common resident, normally in pairs or small flocks. More common in restcamps during winter (non-br) months. **HABITAT** Open woodland, particularly well-grazed areas favoured by concentrations of larger herbivores. **FOOD** Mainly insects, also small reptiles and mammals, rarely seeds and fruit. **CALL** Series of clucking notes *kok kok kok kokok kokok ...* **BR** Monogamous. ♀ seals herself in natural tree cavity during incubation and early nestling period. (**Rooibekneushoringvoël**)

SOUTHERN YELLOW-BILLED HORNBILL *Tockus leucomelas* **(459) 50 cm; 190 g**
SEXES *Alike, ♂>♀ and with larger bill. Yellow bill and pink facial skin diagnostic. Eyes yellow.*
JUV *Bill smaller and more dusky; eyes grey becoming brown.* **STATUS** Common and widespread resident and near-endemic. Usually in pairs or small family groups in non-br season. **HABITAT** Wide range of savanna and more closed woodland. **FOOD** Insects, small animals and reptiles; also fruit and seeds. **CALL** Long series of clucking notes *kok kok kok korkorkorkorkor ...* **BR** Monogamous. ♀ seals herself in natural tree cavity, leaving narrow vertical feeding slit. Half-grown chicks reseal nest after she leaves nest to help with feeding. (**Geelbekneushoringvoël**)

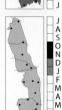

CROWNED HORNBILL *Tockus alboterminatus* **(460) 52 cm; 225 g**
SEXES *Alike, ♂>♀ and with slightly longer casque. Darker brown plumage and redder or deeper orange bill than Bradfield's Hornbill. Yellow stripe at base of bill diagnostic.* **JUV** *Bill pale yellow-orange.* **STATUS** Wide-ranging and uncommon resident, usually in pairs or small family groups. **HABITAT** Tall closed-canopy woodland and forest mainly along the Luvuvhu River and the Sabie River west of the Mkuhlu picnic site. **FOOD** Wide range of insects and small animal prey, also fruit in winter. **CALL** Loud high-pitched whistling notes. **BR** Monogamous. ♀ seals herself in natural tree hole, leaving vertical feeding slit. (**Gekroonde Neushoringvoël**)

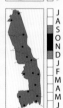

AFRICAN GREY HORNBILL *Tockus nasutus* **(457) 46 cm; 155 g**
SEXES *Similar, ♂>♀. With drab grey plumage and broad white eye-stripe, unlikely to be confused with any other species.* **M** *Bill blackish with white stripe below casque.* **F** *Bill tip red; rest of upper mandible pale yellow.* **JUV** *Bill smaller, lacking casque.* **STATUS** Locally common resident. **HABITAT** Throughout the KNP in woodland, common in Mopane. **FOOD** Mainly insects; also small animals and seeds. **CALL** Series of high-pitched notes *pi pi pi pi ... pipipiew*. **BR** Monogamous. ♀ seals herself in natural tree cavity or crevice in rocks. (**Grysneushoringvoël**) **ALT NAME** *Grey Hornbill*

TRUMPETER HORNBILL *Bycanistes bucinator* **(455) 58 cm; 644 g**
SEXES *Similar, ♂>♀ and with larger casque than ♀. ♀ can be distinguished from ♂ even in flight by much less prominent casque.* **JUV** *Bill and casque smaller, feathers around base of bill and face brown.* **STATUS** Fairly common resident especially in the riparian areas of the Sabie, Luvuvhu and Limpopo rivers. Found in family groups and flocks during non-br season. Numbers have perhaps declined in KNP owing to the decline in fig trees along some major rivers caused by both droughts and floods. **HABITAT** Favours riparian woodland. **FOOD** Mainly fruit but readily takes insects, birds' eggs, nestlings, wasp nests and even small crabs. **CALL** Loud nasal wailing *nhaa, nhaa ha ha ha ... like a crying baby*. **BR** Usually monogamous. Nests in natural cavities with entrance hole sealed by ♀ using mud brought by ♂. Narrow vertical slit left open for food deliveries. (**Gewone Boskraai**)

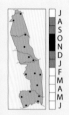

SOUTHERN GROUND-HORNBILL *Bucorvus leadbeateri* **(463) 110 cm; 3,77 kg**
SEXES *♂>♀. Differ slightly; ♀ has blue patch (variable in size) on bare skin under throat. Some males also have small blue spots on throat, so body and bill size (♂ larger) also important when distinguishing sexes. Bare red skin on upper neck inflatable, and all-white primary feathers normally only visible only in flight.* **JUV** *Browner than adult Facial skin initially pale grey-brown, turning yellow within a year, flecked red by 2 years, orange by 3 years and fully red only by 4–6 years.* **STATUS** In KNP fairly common; listed as Vulnerable in s Africa. Concern for decline in numbers, but in KNP the population stable at about 250 birds. Resident and territorial; usually found in groups of 3–5 birds. **HABITAT** Favours open woodland. **FOOD** Small animals up to squirrel size, reptiles such as lizards, snakes and tortoises, also insects, frogs and snails. **CALL** Deep booming 4-note duet (usually in early morning), ♂ at lower pitch than ♀. **BR** Monogamous, and br co-operatively. Nests usually in natural tree cavity, sometimes in crevice on rock face or on old stick nest. (**Bromvoël**) **ALT NAME** *Ground Hornbill*

PLATE 9 Hoopoe, Wood-Hoopoe, Scimitarbill, Trogon p208

AFRICAN HOOPOE *Upupa africana* (451) 26 cm; 53 g
SEXES Similar, ♂>♀. Conspicuous and unlikely to be confused with any other species. Crest usually held flattened, forming a long point behind the crown, and only raised when bird lands, in display or when alarmed. ♀ is duller rufous than ♂ and slightly greyer on face and breast. **JUV** Like ♀ but duller and with shorter crest. Bill initially shorter and less decurved. **STATUS** Fairly common resident. Usually in pairs. **HABITAT** Most open woodland, especially that with a short grass understorey; also common in restcamps. **FOOD** Insects and their larvae, earthworms, small frogs, snakes, lizards, and termites, which may be taken in flight. **CALL** Soft, low and sustained *hoo-poo* or *hoo-poo-poo*. **BR** Monogamous and occasionally br co-operatively. Uses variety of nest sites, including natural tree cavities, old woodpecker or barbet holes, underground termite mound chambers, stone walls and ground holes. Incubation period of 15–16 days is by ♀ only, who is fed by ♂ at nest during this period. Recorded nestling periods are 26–32 days. Parasitised by Greater Honeyguide. (**Hoephoep**) ALT NAME *Hoopoe*

GREEN WOOD-HOOPOE *Phoeniculus purpureus* (452) 33 cm; 80 g
SEXES Alike, ♂>♀. Female has slightly shorter and straighter bill. **JUV** Bill black, and plumage without iridescence. White spots in primaries and outer tail feathers present. **STATUS** Common resident, sedentary and in groups of 2–14 birds (average 4–5). Groups normally roost communally in natural tree holes. If threatened by predators, roosting birds position themselves with rear ends facing intruder and secrete foul-smelling brown fluid from preen gland. **HABITAT** Woodland and riparian forest. **FOOD** Insects (especially larvae), centipedes, geckos, frogs, also some fruit and nectar. **CALL** Distinctive cackling chorus, likened to laughter of women, performed by all members of group. ♂'s voice lower pitched than ♀'s. **BR** Monogamous and co-operative. Br pair assisted by rest of group members. Usually nests in natural tree cavities, but old woodpecker and barbet holes occasionally used. Incubation period 17–18 days by ♀ only. ♀ and chicks fed by all group members during nestling and post-nestling periods. Parasitised by Greater and Lesser honeyguides. Young honeyguides fed by group up to 3 months after fledging. (**Rooibekkakelaar**) ALT NAME *Red-billed Wood-Hoopoe*

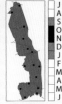

COMMON SCIMITARBILL *Rhinopomastus cyanomelas* (454) 26 cm; 32 g
SEXES Similar, ♂>♀. Could be confused with juv Green Wood-hoopoe that also has a black bill, and similar white wing and tail patches in flight. Bill, however, strongly decurved and more slender (at all ages). ♀ can be distinguished from ♂ by brown-tinged ear coverts, throat and breast. **JUV** Like ♀ but plumage duller and bill slightly less decurved. **STATUS** Fairly common resident, sedentary and usually in pairs, sometimes in post-br groups of up to 6. **HABITAT** Various woodland types, favouring riverine fringing bush; frequently seen in restcamp gardens. **FOOD** Mainly small insects and their larvae, occasionally nectar. **CALL** Plaintive, ventriloquial *poui-poui-poui ...* , repeated in phrases of 3–5 syllables; harsh chattering *ker-ker-ker* used in excitement or alarm. **BR** Monogamous. Nests in natural cavities in trees, less often in old woodpecker or barbet holes. Only ♀ incubates, fed by ♂. Incubation period 13–14 days, nestling period 21–24 days. Occasionally parasitised by Greater Honeyguide. (**Swartbekkakelaar**) ALT NAME *Scimitarbilled Wood-Hoopoe*

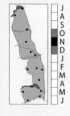

NARINA TROGON *Apaloderma narina* (427) 32 cm; 67 g
SEXES Because of their habit of sitting motionless for long periods, these strikingly beautiful birds are difficult to detect in their preferred forest habitat. **M** Crimson underparts and bright green throat and upperparts very distinctive. Bare skin on upper eyelids, ears and throat turns pale turquoise during br season (both sexes); ♂ inflates turquoise throat in display. **F** Forehead to lower breast cinnamon; breast pink, belly and undertail crimson. **JUV** Duller than ♀ with buff-spotted wings. **STATUS** Uncommon and localised resident mainly along the Sabie, Luvuvhu and Limpopo rivers. Individuals have been recorded at widespread unlikely localities, probably resulting from dispersal of juveniles or altitudinal movements from cooler forests to warmer low-lying areas in search of better food supplies in winter. Usually solitary or in pairs. **HABITAT** Mainly well-developed riparian woodland in the KNP. **FOOD** A perch-and-wait feeder, taking wide range of insect prey from the tree foliage, mainly smooth-skinned caterpillars, moths, spiders and even small chameleons. **CALL** Male gives series of growling hoots *hroo-hoo, hroo-hoo, hroo-hoo ...* , rising to a crescendo; blue throat skin may be inflated with each hoot. **BR** Monogamous. Nests in natural tree cavities (with wide entrance hole), or in hollow stump. Incubation by ♀ in early morning, evening and night, and by ♂ during day. (**Bosloerie**)

PLATE 10 Rollers p208

EUROPEAN ROLLER *Coracias garrulus* **(446) 31 cm; 122 g**
SEXES *Alike. Blue head, square tail with no tail streamers and light brown back distinguish it from Lilac-breasted and Racket-tailed rollers. Distinct demarcation between blue nape and brown mantle; build stockier than Racket-tailed Roller. In flight, bright azure blue underwing contrasts with blackish primaries.* **JUV** *Duller than adult, more olivaceous green than blue, and throat brownish.* **STATUS** Common non-br migrant, present in KNP from Nov. When present is the most common roller in the Park. Departs Mar/Apr. Frequently perches strategically on dead outer branches. **HABITAT** Dry open woodlands, especially those with grassy clearings. **FOOD** Insects, predominantly beetles, grasshoppers, crickets, ants and termites, occasionally small vertebrates; uses 'sit-and-wait' hunting technique. Most prey taken off ground, some in flight. **CALL** Generally silent in region, with occasional *raak* call. **BR** Br in Europe, Middle East and Asia. (**Europese Troupant**)
ALT NAME *Eurasian Roller*

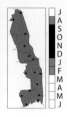

LILAC-BREASTED ROLLER *Coracias caudatus* **(447) 29 cm (36 cm incl streamers); 108 g**
SEXES *Alike. Only roller with lilac throat and breast and blue belly. Long, straight outer tail streamers absent during moult (post-br, usually Dec–May).* **JUV** *Like adult but duller with brownish wash on crown and nape; tail streamers absent.* **STATUS** Common, conspicuous and widespread resident. Generally sedentary; usually found singly or in pairs. **HABITAT** Wide variety of dry woodland habitats, generally avoiding dense riparian woodland. **FOOD** Insectivorous, taking wide variety of locusts, crickets, beetles, butterflies and lizards; also small snakes, frogs, birds and rodents. Most prey taken from the ground. Attracted to grass fires. **CALL** Not very vocal; loud harsh *ghak, ghak, gharrak* notes, mostly given in flight display during br season. **BR** Monogamous and fiercely territorial. Nests in natural tree cavities or sometimes in large woodpecker holes. (**Gewone Troupant**)

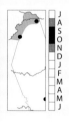

RACKET-TAILED ROLLER *Coracias spatulatus* **(448) 31 cm (37 cm incl streamers); 97 g**
SEXES *Alike. Overall bluer appearance than similar-sized Lilac-breasted Roller. Tail streamers have spatulate tips that may be absent during moult (post-br).* **JUV** *Lacks tail streamers. Throat duller and breast pinkish, streaked with white (not lilac as in Lilac-breasted Roller).* **STATUS** Uncommon br resident and subject to local movements in the non-br season. Pairs do not always return annually to previously used nest sites. Regularly recorded only in the Pafuri region with some records in the Punda Maria area. **HABITAT** Favours well-developed deciduous woodland with little or no understorey. **FOOD** Insectivorous, taking grasshoppers and beetles using 'sit-and-wait' strategy; also insects in flight. **CALL** Usually silent with loud harsh *cha* or *tchek* calls given at onset of br season. **BR** Monogamous, sometimes with nest helpers. Uses natural tree cavities and sometimes large woodpecker holes. (**Knopsterttroupant**)

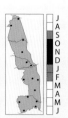

PURPLE ROLLER *Coracias naevius* **(449) 33 cm; 168 g**
SEXES *Alike. Broad cream eyebrow, purple-brown underparts heavily streaked with white, and stocky thickset build make it unmistakable. Black bill and light eye-stripe distinguish it from smaller Broad-billed Roller.* **JUV** *Duller than adult.* **STATUS** Fairly common and widespread in KNP. Resident but disperses after breeding; little evidence of local movements in KNP during the non-br season. **HABITAT** Dry woodland and savanna, occupying both small-leaved and broad-leaved habitats. **FOOD** Mainly insectivorous, but occasionally takes small reptiles, mice and young birds. **CALL** Calls are harsh but less strident than other rollers. **BR** Monogamous. Favours shallow natural cavities in tall trees in which to breed. (**Groottroupant**)

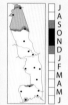

BROAD-BILLED ROLLER *Eurystomus glaucurus* **(450) 29 cm; 105 g**
SEXES *Alike. Smallest roller in region and only one with a yellow bill. Overall dark appearance, with cinnamon-brown upperparts and lilac-purple below. In flight shows brilliant ultramarine primaries and outer secondaries.* **JUV** *Bill yellow. Breast and belly dull greyish-blue, mottled grey.* **STATUS** Fairly common br inter-African migrant arriving late Sept and early Oct; last birds leave by April. Mainly along the Limpopo, Luvuvhu and Shingwedzi rivers, but there are recent records along the Sabie River and further south with breeding recorded in the Berg-en-dal area. **HABITAT** Mostly well-developed riparian woodland. **FOOD** Insectivorous, specialising in ant alates and other aerial insects such as beetles, bees, wasps, flies and grasshoppers. **CALL** Noisy during br season, with range of harsh, guttural and nasal notes. **BR** Monogamous; nest in natural cavity in large tree. (**Geelbektroupant**)

PLATE 11 Kingfishers p208

MALACHITE KINGFISHER *Alcedo cristata* (431) 14 cm; 17 g
SEXES *Alike. Common water's edge species. Key distinguishing feature barred greenish-blue and black crown plumage that extends down to eyes (African Pygmy-Kingfisher has orange eyebrow).* **JUV** *Bill initially black, becoming red. Distinguished from juv African Pygmy-Kingfisher by blue crown that extends to eyes (as with adult).* **STATUS** Most common small kingfisher in region, most likely to be seen on reeds or branches overhanging water. Resident, with regular local movements to exploit water and food availability in the non-br season. **HABITAT** Aquatic margins; seldom found away from water. **FOOD** Mostly small fish, also tadpoles, frogs, insects, small crabs and lizards. **CALL** Short, sharp *kweek* or *seek*, usually given in flight. **BR** Monogamous. Burrow excavated by both sexes in vertical bank. (**Kuifkopvisvanger**)

HALF-COLLARED KINGFISHER *Alcedo semitorquata* (430) 18 cm; 38 g
SEXES *Alike. Black bill and blue face diagnostic. Could be confused with smaller juv Malachite Kingfisher, which has black bill, but orange cheeks. ♀ has reddish base to lower mandible.* **JUV** *Orange underparts duller than adult.* **STATUS** Uncommon resident, with local seasonal movements and dependent on water availability. Regarded as Near-threatened in s Africa. Most likely to be seen along streams with well-wooded margins **HABITAT** Mostly along clear, well-vegetated, fast-flowing rivers and streams. **FOOD** Mainly fish, also crabs and aquatic insects. **CALL** High-pitched sibilant squeak *tseep*. **BR** Monogamous. Nest burrow excavated into vertical stream bank. (**Blouvisvanger**)

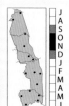

GREY-HEADED KINGFISHER *Halcyon leucocephala* (436) 21 cm; 45 g
SEXES *Alike. From front or side, chestnut belly diagnostic. From rear, similar to Woodland Kingfisher but has black (not blue) back and all-red bill. ♂ has duller chestnut underparts.* **JUV** *Bill blackish, duller than adult, and with dark grey mottling or barring.* **STATUS** Uncommon in KNP but may be recorded throughout the area, especially during spring and autumn migration. Br intra-African migrant present Sept–May, usually singly or in pairs. **HABITAT** Mature woodland and mostly near water, though not dependent on aquatic habitats. **FOOD** Insects, particularly grasshoppers, also lizards and mice; rarely frogs and fish. **CALL** Weak sibilant trill *ji-ji-ji-ji-chi*. **BR** Monogamous. Nest excavated in vertical bank. (**Gryskopvisvanger**) **ALT NAME** *Grey-hooded Kingfisher*

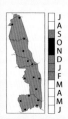

AFRICAN PYGMY-KINGFISHER *Ispidina picta* (432) 13 cm; 15 g
SEXES *Alike. Smallest southern African kingfisher. Broad orange eyebrow distinguishes it from Malachite Kingfisher, and violet patch on cheeks distinctive.* **JUV** *Bill initially black; orange eyebrow and violet ear-patch.* **STATUS** Fairly common br intra-African migrant, present Sept/Oct–Mar (some departing as late as May or early Jun). **HABITAT** Favours moist woodland, forest and riverine margins; not necessarily near water. **FOOD** Mostly insects, also small crabs, frogs and lizards. **CALL** Short, sharp *kweek* or *seek*, usually in flight. **BR** Monogamous. Burrow excavated by both sexes into earth bank or in side of Aardvark burrow. (**Dwergvisvanger**) **ALT NAME** *Pygmy Kingfisher*

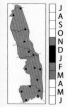

WOODLAND KINGFISHER *Halcyon senegalensis* (433) 23 cm; 63 g
SEXES *Alike. Black lower mandible diagnostic. Confusion in other regions has occurred with the similar-looking Mangrove Kingfisher (unlikely to be recorded in the Park) as rare individuals with all-red bills have been recorded. Can be separated by call and the black eye-patch that extends behind eye in Woodland Kingfisher but only as far as the eye in Mangrove Kingfisher, and their black (not brown) legs.* **JUV** *Bill black or mostly black, upper mandible becoming gradually redder from base; duller than adult.* **STATUS** Very common and conspicuous br intra-African migrant, present Nov–Apr; nocturnal migrant. **HABITAT** Open woodland savanna, particularly along river, swamp or wetland fringes. **FOOD** Mostly insects, especially grasshoppers. Wide range of other prey, including small birds, lizards, snakes and frogs. **CALL** Loud trilling song *kri-trrrrrrr*, descending and fading. A very noisy species that calls commonly in the restcamp gardens. **BR** Monogamous. Nests in old woodpecker, barbet or natural tree cavities. (**Bosveldvisvanger**)

PLATE 12 Kingfishers p209

BROWN-HOODED KINGFISHER *Halcyon albiventris* (435) 23 cm; 60 g
SEXES Similar, ♀>♂. Only kingfisher in region with brownish streaked head and upper breast. Confusion possible with smaller (and whiter) Striped Kingfisher, which has black/red bill. In flight, dark turquoise flight feathers conspicuous; underwing cinnamon. **M** Differs from ♀ by black (not brown) back and wing coverts. **F** Distinguishable from similar ♂ by brown (not black) wing coverts and back. **JUV** Resembles ♀, with brown wing coverts, but duller. Bill dull blackish becoming redder from base. **STATUS** Common resident, and mostly sedentary. Small proportion of nominate race from the Cape (more heavily streaked on flanks and breast, and partial migrants during non-br season) may be found in the Park during the winter months. **HABITAT** Most woodland types, favouring riparian margins; regular restcamp garden species. **FOOD** Mainly insects, but also geckos, small chameleons, skinks, young birds, snakes, fish, crabs and rodents. **CALL** Loud extended *chi-chi-chi-chi-chi* call, and descending trill, *Ki-ti-ti-ti*. **BR** Monogamous. Nest a horizontal tunnel excavated into vertical bank. Eggs laid on soil in chamber about 1 m from entrance hole. Incubation period about 14 days, by ♀ only. (**Bruinkopvisvanger**)

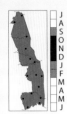

STRIPED KINGFISHER *Halcyon chelicuti* (437) 18 cm; 38 g
SEXES Similar, ♀>♂. Most likely to be confused with larger Brown-hooded Kingfisher, but smaller and has less brown and a streaked whitish breast, white neck collar and blue rump. In flight, conspicuous white underwing pattern distinguishes it from Brown-hooded Kingfisher, which has no white in wing. ♂ underwing pattern differs from ♀ by darker black carpal patch and subterminal wing band. **JUV** Duller and less streaked than adult. **STATUS** Locally common resident with some evidence of local altitudinal movements in winter. Usually in pairs; highly territorial. **HABITAT** Open woodland particularly on habitats with granitic substrates, Mopane, and less common in *Acacia*. **FOOD** Insects (mostly grasshoppers), small lizards, snakes and rodents. **CALL** Very vocal: loud ringing trill *keep-kirrrrr*, given by both sexes but mainly by ♂. Calls often elicit communal responses from others of the species in adjacent territories. **BR** Monogamous, and occasionally a co-operative breeder with additional helpers. Nests in old barbet and woodpecker holes, natural tree cavities and swallow nests. Parasitised by Lesser Honeyguide. (**Gestreepte Visvanger**)

GIANT KINGFISHER *Megaceryle maxima* (429) 44 cm; 364 g
SEXES ♂>♀. **M** Easily recognisable by its large size, rufous upper breast, shaggy crest and speckled slatey throat and belly. Underwing and lower belly white (not rufous as in ♀). **F** Underparts have reversed colour pattern with rufous on belly (not chest). Underwing and undertail coverts rufous. **JUV** Similar to adult but sides of chest mottled black and chestnut. Underwing coloration the same as respective adult. **STATUS** Common resident, sedentary but moves in response to water levels and food availability. Often seen on bridge railings over rivers. **HABITAT** Almost all permanent or semi-permanent water bodies that provide fish and have perches from which to hunt. **FOOD** Mainly fish, also crabs, frogs and aquatic invertebrates. Fish vigorously beaten against a solid object until dead, and carapaces of large crabs dismembered before being swallowed. Small crabs swallowed whole. **CALL** Loud, harsh and nasal *kek* or *kakh-kakh-kakh*. **BR** Monogamous. Nest tunnel dug by both sexes 1–3 m into vertical bank next to, or close to, the water's edge. Eggs incubated for 25–27 days by both adults. Nestling period about 37 days, nestlings fed mainly by ♂. (**Reusevisvanger**)

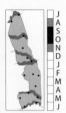

PIED KINGFISHER *Ceryle rudis* (428) 25 cm; 84 g
SEXES Similar. Only kingfisher in region with an all-black-and-white plumage. **M** Double breast band. **F** Easily distinguishable with single breast band. **JUV** Breast band greyish-black; feathers on face, throat and breast fringed buff. **STATUS** Common to very common and one of the 3 most numerous kingfishers in the world. Resident, but locally nomadic in response to changes in water level and food supply. Usually in pairs or small family groups. **HABITAT** All water bodies with fish; less common along wooded, fast-flowing streams. Unlikely to be found away from aquatic margins. **FOOD** Mainly fish; also invertebrates such as crabs, prawns and aquatic and terrestrial insects. Hover-hunts more than any other kingfisher, enabling it to hunt far from shore lines and perches. **CALL** Highly vocal with chattering high-pitched *kwik* or *kwik-kwik-kwik* calls. **BR** Monogamous; about 1 out of 3 pairs with nest helpers (always males). Br season varies regionally, peaking Aug–Nov, but as late as May/Jun on Botswana floodplains. Nest tunnel excavated by both adults close to water in suitable bank. Eggs incubated by both sexes for about 18 days; chicks fed by both adults and nest helpers if present. (**Bontvisvanger**)

PLATE 13 Bee-eaters p209

WHITE-FRONTED BEE-EATER *Merops bullockoides* (443) 23 cm; 35 g
SEXES *Alike. Only bee-eater in region with red/white throat combination and square-ended tail.* **JUV** *Red and blue colours paler, and crown green.* **STATUS** Locally common resident, with little evidence of local movements in KNP after the br season. Gregarious, almost always in small groups or large flocks. **HABITAT** Usually associated with river banks along watercourses in dry woodland. **FOOD** Exclusively flying insects, especially honeybees; also beetles, flies, dragonflies, moths and butterflies. **CALL** Nasal *gaaaa or gaaauuu*. **BR** Monogamous, colonial and usually a co-operative breeder. Pairs mate for life, and a new burrow is excavated each year. Tunnels are about 1 m long in vertical alluvial banks. Flocks sometimes remain in general vicinity of br territory year-round, roosting in nest holes at night. Parasitised by Greater Honeyguide. (**Rooikeelbyvreter**)

SWALLOW-TAILED BEE-EATER *Merops hirundineus* (445) 21 cm; 23 g
SEXES *Alike. Deeply forked blue tail diagnostic. Underparts unlike smaller Little Bee-eater with thin blue gorget that separates pale yellow throat from green (not buff) breast and pale blue lower belly.* **JUV** *Paler than adult and with underparts uniformly green.* **STATUS** Very uncommon with restricted distribution in KNP. Paler western population known to move eastward and suspected to be present in the Park during the winter months, but it is also possibly a resident species. Recent records in the Sabi-Sand Game Reserve to the west of KNP suggest it may be more widespread than previously believed. Usually in pairs or small flocks and a vagrant during non-br season. Roosts communally in tightly packed row on a perch when not br. **HABITAT** Semi-arid woodland to tall broad-leaved woodland. **FOOD** Flying insects, particularly bees. **CALL** High-pitched piping *kweep kweepy buzz, kweep*. **BR** Monogamous, solitary and occasionally a co-operative breeder. Nest tunnels excavated in low bank or in side/top of Aardvark burrow. Parasitised by Greater Honeyguide. (**Swaelstertbyvreter**)

LITTLE BEE-EATER *Merops pusillus* (444) 16 cm; 15 g
SEXES *World's smallest bee-eater. Confusion only likely with Swallow-tailed Bee-eater, but has black (not blue) gorget, buff (not green) belly and almost square orange/green tail.* **JUV** *Like adult but lacks gorget, and has faintly streaked breast.* **STATUS** Locally common resident, nomadic in non-br season; not gregarious and usually in pairs or small family groups. Pairs remain together year-round. **HABITAT** Woodland, often near water. **FOOD** Mostly flying insects taken on the wing, especially bees and wasps. **CALL** Disyllabic *s-lip* ... , and *sip-sip sip*. **BR** Monogamous and solitary. Nest tunnels excavated by both birds into low bank or road cutting often only 30 cm high. Also excavates nest hole in ceiling of Aardvark burrow. Nests not usually reused in successive years. 2–6 white eggs laid on sand in unlined nest cavity. Incubation period 18–20 days, by both sexes. Parasitised by, and one of the primary hosts of Greater Honeyguide. (**Kleinbyvreter**)

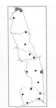

BLUE-CHEEKED BEE-EATER *Merops persicus* (440) 25 cm (30 cm incl streamers); 48 g
SEXES *Alike, ♂>♀, tail streamers project 55 mm (♂) and 40 mm (♀). Key identification feature for this species green crown and pale blue to whitish forehead. Birds arriving in spring are often in moult showing drab faded plumage, unlike birds on northward migration in late summer that are in fresh bright breeding plumage.* **JUV** *Tail streamers shorter; overall duller and with feathers tipped pale bluish, giving a scaly appearance.* **STATUS** Uncommon non-br Palearctic migrant with restricted distribution in KNP. Regularly occurs only at Pafuri and in the Nwanetsi area, particularly at the Sweni bird hide and in the vicinity of the Singita Lebombo lodges. Present from mid-Oct, with arrivals peaking early Nov, departing late Mar/Apr, some as late as early May. Highly gregarious, regularly in flocks of 20 or more. Roosts communally shoulder to shoulder, mostly in thorny or leafy trees. **HABITAT** Prefers moist woodland, avoiding very arid areas; also favours riparian margins. **FOOD** In Africa feeds mainly on dragonflies, damselflies and lacewings, also bees, wasps and other flying insects. **CALL** Most common call a pleasant polysyllabic *dirirup*, often given in flight. (**Blouwangbyvreter**)

PLATE 14 Bee-eaters, Mousebirds p209

EUROPEAN BEE-EATER *Merops apiaster* (438) 23 cm (25 cm incl streamers); 52 g
SEXES *Alike. Combination of yellow throat, black gorget, turquoise underparts and brown and yellow back unmistakable. Two separate populations occur in region, and are indistinguishable in the field: one from the Palearctic (non-br), and the other a br intra-African migrant. Birds arriving in spring are often in moult showing drab faded plumage, unlike birds on northward migration in late summer that are in fresh bright breeding plumage.* **JUV** *Lacks tail streamers, gorget less distinct and overall duller and greener.* **STATUS** Common and widespread. It is suspected that all the birds visiting the Park are non-br Palearctic migrants (not from the s African br population). Main arrivals in Oct, departure in late Mar or early Apr. Highly gregarious, often in flocks of 20–100. **HABITAT** Wide range of woodland and shrublands throughout KNP. Often forage at higher altitudes where they can be heard but not seen. **FOOD** Predominantly bees, wasps, flying ants and termites. **CALL** Various liquid and melodious trill notes *prruip, pruik* or *kruup*. **BR** Breeds extralimitally – no breeding records for KNP. (**Europese Byvreter**) ALT NAME Eurasian Bee-eater

SOUTHERN CARMINE BEE-EATER *Merops nubicoides* (441) 25 cm (35 cm incl streamers); 62 g
SEXES *Alike, ♂ streamer projection up to 120 mm, ♀ 105 mm. Largest African bee-eater; striking pinkish-red plumage distinctive.* **JUV** *Like adult but duller and with shorter tail streamers.* **STATUS** Locally common intra-African migrant. One of the later arrivals, reaching KNP only in Nov/Dec after leaving br grounds in Zimbabwe and further north; departs Mar/Apr. Normally highly gregarious, but sometimes solitary or in small groups in KNP. **HABITAT** Open woodland and savanna. **FOOD** Wide range of flying insects. **CALL** Loud *gro-gro-gro ...* or *rik-rik-rik ...* , notes and variations thereof. **BR** Breeds extralimitally to the north of the region – no breeding records for KNP. (**Rooiborsbyvreter**) ALT NAME Carmine Bee-eater

SPECKLED MOUSEBIRD *Colius striatus* (424) 33 cm; 55 g
SEXES *Alike. Overall brown appearance, black face, and black and white bill distinguish it easily from the Red-faced Mousebird. Flight slower and more laboured than Red-faced Mousebird. Seldom flies in flocks – individuals usually follow one another one by one between trees.* **JUV** *Similar to adult but lacks black face, tail is shorter and bill coloration reversed (pale greenish-white above and black below).* **STATUS** Very common resident. Locally nomadic in search of fruit. **HABITAT** Wide range of wooded habitats but mainly in the riparian zone of perennial and semi-permanent rivers. **FOOD** Wide range of fruit; also feeds on leaves, buds, flowers and nectar. Mistletoes an important source of fruit at certain times of the year. **CALL** Highly vocal especially when disturbed. Most frequent call sharp *chee chee chee*. Alarm call *zik zik*. **BR** Monogamous, polygamous and sometimes a communal and co-operative breeder. Untidy shallow bowl nest built 1–7 m high. Incubation period 12–15 days by both sexes. Nestling period 17–18 days. Occasionally parasitised by Jacobin Cuckoo. (**Gevlekte Muisvoël**)

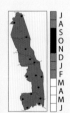

RED-FACED MOUSEBIRD *Urocolius indicus* (426) 32 cm; 56 g
SEXES *Alike, ♂>♀. Soft grey plumage has green sheen to upperparts if seen in good light. Bare red skin around eyes distinctive. Often flies fast and direct in small flocks of up to 15 birds; more musical call than Speckled Mousebird is often given in flight, an added useful aid to identification.* **JUV** *Ear coverts buff and bare facial skin greenish, bill greenish with dark tip and pink base.* **STATUS** Widespread and common resident. Locally nomadic with records of altitudinal movements in winter months. Sunbathes by perching in upright position, with legs apart and belly facing sun; roosts in clusters, dust-bathes frequently and drinks water regularly. **HABITAT** Acacia savanna, dry woodland, riparian bush and restcamp gardens. **FOOD** Mainly fruit; feeds on new leaves and shoot tips, also nectar from aloes. **CALL** Musical *ti-wi-wi, ti wi wi wi* or *tree-ree-ree* falling in pitch and often given in flight. **BR** Monogamous, sometimes br co-operatively. Untidy shallow cup nest built by both sexes, usually 1–7 m high in shrub or tree. Incubation 10½–15 days by both sexes. Nestling period 14–20 days. (**Rooiwangmuisvoël**)

PLATE 15 Cuckoos p209

JACOBIN CUCKOO *Clamator jacobinus* (382) 34 cm; 81 g
SEXES *Alike. Two colour morphs occur, pied morphs far more common in KNP. Black morph birds distinguished from Black Cuckoo by crest and white wing patch.* **JUV** *Dark brown above. Straw-yellow bill distinguishes it from juv Levaillant's Cuckoo (black bill).* **STATUS** Fairly common and conspicuous br intra-African migrant, present Oct–Apr. **HABITAT** Mostly dry woodland habitats but also along riverine margins. **FOOD** Spiny or hairy caterpillars. Also takes termites on the wing. **CALL** Loud *kleeu-wi-wip* as main call. Very vocal after arrival in spring and the early summer months, frequently calling from prominent perches at the tops of trees. **BR** Brood parasite hosting mainly on Dark-capped Bulbul, but also Sombre Bulbul and Terrestrial Brownbul. To a lesser extent also hosts on a wide range of other species including Speckled Mousebird, flycatchers and shrikes. (**Bontnuwejaarsvoël**)

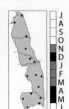

LEVAILLANT'S CUCKOO *Clamator levaillantii* (381) 39 cm; 122 g
SEXES *Alike. Larger than Jacobin Cuckoo; throat and breast light to heavily streaked. Two colour morphs occur. Uncommon black morph only distinguishable from smaller black Jacobin Cuckoo in the hand.* **JUV** *Upperparts and tail brown, underparts buff and variably streaked.* **STATUS** Far less common than Jacobin Cuckoo, but nevertheless a fairly common br intra-African migrant in KNP, arriving Oct, departing by May. Conspicuous when calling. **HABITAT** More heavily wooded areas than Jacobin Cuckoo, especially along watercourses. **FOOD** Mainly hairy caterpillars; also other insects. **CALL** Loud *kreeu, kreeu* followed by fast *tutututututu* as main calls. **BR** Brood parasite, hosting only on babblers, making Arrow-marked Babbler its target in the Park. (**Gestreepte Nuwejaarsvoël**) ALT NAME *Striped Cuckoo*

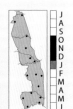

GREAT SPOTTED CUCKOO *Clamator glandarius* (380) 39 cm; 134 g
SEXES *Alike, ♂>♀. Overall pale grey plumage, long wings, spotted upperparts, grey crest and pale underparts distinctive. Gives itself away by loud call.* **JUV** *Crown blackish, spots on upperparts buffish.* **STATUS** Uncommon br intra-African migrant, arriving Sept, most departing by Mar–Apr, some as late as May. Recorded throughout KNP but more common in the northern parts, in Mopani and especially in the open *Acacia* habitats of the Pafuri region. **HABITAT** Open savanna woodland, especially *Acacia*. **FOOD** Mainly caterpillars, both hairy and hairless, also insects and lizards. In spring and summer also searches for caterpillars in the lush green grass on the ground. **CALL** Conspicuous loud and rasping *keeow*, repeated with acceleration to a raucous chatter. **BR** Brood parasite; in KNP primary hosts are the various starling species. (**Gevlekte Koekoek**)

RED-CHESTED CUCKOO *Cuculus solitarius* (377) 29 cm; 75 g
SEXES M *Only cuckoo in region with dull rufous breast. Upperparts darker than Common and African cuckoos.* **F** *Rufous breast slightly paler.* **JUV** *Upperparts and throat black, flecked white; belly barred black and white.* **STATUS** Widespread and common br intra-African migrant, present Sept–Mar. **HABITAT** Riverine forests and bush, closed woodland and well-wooded restcamps. **FOOD** Mainly hairy caterpillars, but also takes beetles, grasshoppers, termite alates and small vertebrates. **CALL** Main call persistent and monotonous *piet-my-vrou* ... **BR** In KNP, brood parasite primarily of robin-chats, especially White-throated Robin-Chat. Also hosts on scrub-robins, Cape Wagtail and to a lesser extent on African Stonechat. (**Piet-my-vrou**)

THICK-BILLED CUCKOO *Pachycoccyx audeberti* (383) 36 cm; 110 g
SEXES *Alike. Slate grey to blackish upperparts, pure white underparts, yellow base to lower mandible and lack of crest make it unmistakable. Legs and feet bright yellow.* **JUV** *Grey mottled underparts, white forecrown and yellow feet distinguish it from other juv cuckoos.* **STATUS** Uncommon to rare; seasonal movements unclear. Regular sighting in KNP only at Pafuri and in the area of the Sabie, Sand and Mbyamite rivers. Normally perches quietly under tree canopy where it may remain inconspicuous for long periods. **HABITAT** Tall, well-developed woodland and riverine forest. **FOOD** Caterpillars and other insects. **CALL** Loud, repetitive and high-pitched *were-wick*. **BR** Brood parasite; only known host in region is Retz's Helmet-Shrike. (**Dikbekkoekoek**)

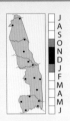

BLACK CUCKOO *Cuculus clamosus* **(378) 30 cm; 90 g**
SEXES *Differ slightly.* **M** *Only all-black cuckoo in region without a crest.* **F** *Similar to ♂ but frequently barred rufous below.* **JUV** *Duller than adult.* **STATUS** Fairly common br intra-African migrant, present late Sept–Apr. Gives its presence away by call, but difficult to locate in mid or upper canopy of tree or bush as it seldom perches on exposed outer tree branches. **HABITAT** Riparian forest, dense woodland and thornveld thickets. **FOOD** Mainly caterpillars, also other insects such as beetles and aerial insects taken on the wing. **CALL** Slow and mournful *I'm so sick, I'm so sick…*, also *whirly whirly whirly…* crescendo. Calls mainly through to end Dec or early Jan, then mostly silent till departure Mar–Apr. **BR** Brood parasite, mainly of Southern and Tropical boubous (i.e. *Laniarius* shrikes) in the Park. (**Swartkoekoek**)

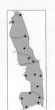

COMMON CUCKOO *Cuculus canorus* **(374) 33 cm; 105 g**
SEXES *Very similar.* **M** *Difficult to distinguish from African Cuckoo in field. Black bill with greenish-yellow base (African Cuckoo has more yellow at base of bill). Tail appears barred below, but not as broadly barred as in African Cuckoo.* **F** *Both grey and rufous (hepatic) morphs found. Rufous morphs (very rare) reddish-brown, barred blackish-brown above and underparts pale chestnut barred blackish.* **JUV** *Grey, rufous or intermediate forms occur, all with white spot on nape.* **STATUS** Uncommon non-br migrant, present Oct–Apr. **HABITAT** Found in various woodland types, but favours dry woodland; avoids forest thickets. **FOOD** Mainly caterpillars. **CALL** Silent in s Africa. On breeding grounds in the northern hemisphere gives loud repetitive *cuck-oo* call (of cuckoo-clock fame). (**Europese Koekoek**) ALT NAME *European Cuckoo*

KLAAS'S CUCKOO *Chrysococcyx klaas* **(385) 18 cm; 26 g**
SEXES M *Small white patch behind eye and green half-collar extending onto sides of breast diagnostic. Lacks white wing spots of Diderick Cuckoo.* **F** *Whitish fleck behind eye and without white wing spots.* **JUV** *Finely barred brown and white (African Emerald Cuckoo barred green and white); juv Diderick Cuckoo has red bill.* **STATUS** Fairly common br intra-African migrant (Sept–Apr), some birds overwintering. **HABITAT** Variety of woodland habitats, including drainage line woodland areas, and often found in the restcamp gardens. **FOOD** Mainly caterpillars and butterflies. **CALL** Distinctive 2-note *meit-jie* or *may-che*. **BR** Brood parasite of at least 18 species in s Africa. Hosts in the Park include sunbirds, Chinspot Batis, African Dusky Flycatcher, and Long-billed Crombec. (**Meitjie**)

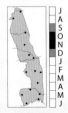

AFRICAN CUCKOO *Cuculus gularis* **(375) 32 cm; 105 g**
SEXES *Very similar. Breast of ♀ sometimes buff. Similar to Common Cuckoo but bill broader, deeper and with more extensive yellow at base. Broad whitish bars on undertail.* **JUV** *Broader white spots on outer tail webs.* **STATUS** Fairly common br intra-African migrant, present Sept–Apr. A shy species that is difficult to locate in the woodland. Seldom perches on exposed branches. **HABITAT** Open woodland and riparian zones. **FOOD** Mainly caterpillars, also other invertebrates and sometimes hawks termite alates. **CALL** *Coo-coo, Coo-coo…*, second syllable higher and louder. Calling ceases in December, thereafter difficult to separate from Common Cuckoo (that does not call in s Africa). Similar to call of African Hoopoe. **BR** Brood parasite of Fork-tailed Drongo. (**Afrikaanse Koekoek**)

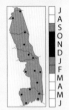

DIDERICK CUCKOO *Chrysococcyx caprius* **(386) 19 cm; 30 g**
SEXES M *Plumage variable from deep emerald green to bright coppery bronze above. Shades of bronze morph birds also vary considerably, from a light bronze wash to green upperparts, to deep iridescent bronze. Only about 15-20% of birds lack bronze sheen to upperparts. Forehead and supercilium white. Eyes red, white wing spots and green bars on flanks.* **F** *Barred flanks and white wing spots diagnostic.* **JUV** *Resembles ♀ but with coral red bill.* **STATUS** Common br intra-African migrant, present Sept/Oct–Feb/Mar. Occasionally overwinters. Fairly sedentary after arrival in breeding grounds. **HABITAT** Mostly adjacent water bodies and riparian woodland, especially near Lesser Masked-Weaver and Village Weaver breeding colonies. **FOOD** Mainly caterpillars, also termites, termite alates, grasshoppers and butterflies. **CALL** 5–7 plaintive high-pitched notes *dee-dee-deederik*. **BR** Brood parasite of wide range of species, especially bishops, weavers and sparrows. (**Diederikkie**) ALT NAME *Diederik Cuckoo*

PLATE 17 Coucals, Parrots p210

BLACK COUCAL *Centropus grillii* (388) 35 cm; 100 g
SEXES *Alike, ♀>♂. Only coucal with seasonal plumage variation.* (**BR**) *Black head and body with chestnut wings very distinctive. Most likely to be seen in this dress from Nov–Mar.* (**NON-BR**) *Adult moults into brown plumage (buff below) and retains black tail.* **JUV** *Similar to non-br adult: brown above, whitish-buff below; tail brown and barred.* **STATUS** Scarce br migrant, but more common in wetter years when grass cover is good. Regarded as Near-threatened in s Africa. **HABITAT** Favours dense short grassland and tall, rank grasses on vlei margins or ephemeral pans on eastern grasslands. **FOOD** Mainly insects; also small reptiles. **CALL** Fast double-note call *dod-der, dod-der* ... **BR** Polyandrous. In ideal conditions females mate with 2 or more males. Males build nest and incubate and care for young alone. Cup nest with grass pulled over the top, placed in dense grass. (**Swartvleiloerie**)

BURCHELL'S COUCAL *Centropus burchellii* (391) 41 cm; 170 g
SEXES *Alike, ♀>♂. The only common coucal in the Park. Similar to Senegal Coucal (isolated records between Luvuvhu and Limpopo rivers in far north of Park) but rump and tail coverts finely barred (not plain). Northern race C. b. fasciipygialis with more heavily streaked neck and flanks, likely only to be found in the northern tip of the park.* **JUV** *White eyebrow and streaking on brownish crown.* **STATUS** Near-endemic. Common resident and generally sedentary, usually in pairs. **HABITAT** Favours riparian margins and moist grassland on the edge of vleis or dams. **FOOD** Wide range of prey including small mammals, insects, reptiles, frogs, small birds and eggs. **CALL** Typical cascade of bubbling 'water bottle' notes. **BR** Monogamous. Bulky, loosely domed nest placed 0,5–10 m high in thicket or low vegetation. (**Gewone Vleiloerie**)

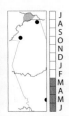

MEYER'S PARROT *Poicephalus meyeri* (364) 23 cm; 115 g
SEXES *Alike, ♂>♀. Combination of brown (not grey) head and neck, dark grey-brown back and wings, and turquoise underparts distinguishes it from similar-sized Brown-headed. Yellow on crown of eastern race (present in the north-western corner of the Park) not always present, and unlikely to be found on imm birds.* **JUV** *Lacks yellow markings and generally more greenish-brown than adult.* **STATUS** Only recorded in the Pafuri area where it reaches the eastern extremity of its range, and where it is thought to hybridise with the Brown-headed Parrot. Uncommon resident, usually in pairs or small groups. **HABITAT** Dry savanna woodland. **FOOD** Fruit, nuts and seeds, also flowers and buds. **CALL** Highly vocal with harsh loud screeching *klink-kleep, cheewe-cheewe* and similar notes. **BR** Monogamous. Natural tree cavities used as well as woodpecker and barbet holes that are enlarged. (**Bosveldpapegaai**)

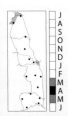

GREY-HEADED PARROT *Poicephalus fuscicollis* (-) 32 cm; 300 g
SEXES *Similar, ♀ sometimes with small patch of orange on forecrown. The only large parrot in the Park; grey head distinctive. Much smaller Meyer's and Brown-headed parrots have yellow (not red) shoulder or underwing patches.* **JUV** *Head paler grey, both sexes with orange-red on forehead, lost by ♂ at 6–8 months.* **STATUS** Uncommon resident in KNP occurring only in the baobab areas of Punda Maria and Pafuri, nomadic in non-br season in search of food. Earlier records for the Olifants Gorge area have not been repeated in recent years. **HABITAT** Riparian and lowland woodland, usually with Baobab trees present. **FOOD** Primarily kernels of unripe fruit such as Marula, Nyala-tree, corkwoods (*Commiphora*), and terminalias or cluster-leaf trees (*Terminalia* spp); also fruit flesh. **CALL** Loud and raucous *tzu-weee* call notes. **BR** Monogamous. Nests high in tree cavities, particularly Baobab trees. (**Savannepapegaai**)

BROWN-HEADED PARROT *Poicephalus cryptoxanthus* (363) 23 cm; 145 g
SEXES *Alike, ♂>♀. Predominantly green back, wings and breast diagnostic. Meyer's Parrot has brown or dark grey-brown back and wing coverts. Almost no yellow on shoulder patch visible, and underwing coverts almost entirely yellow (not partially yellow as in Meyer's Parrot). Confusion only likely at Pafuri where both species may occur.* **JUV** *Duller than adult; neck and upper breast yellowish-olive.* **STATUS** Fairly common resident, with numbers increasing in the latter half of winter. Nomadic in response to food availability; usually in pairs or small groups. **HABITAT** Open lowland woodland and riparian fringes. **FOOD** Fruit (especially figs), kernels, seeds, flowers, green shoots and also nectar. **CALL** Strident *kreeek*; flight call *chreeo-chreeo*. **BR** Monogamous. Nests in natural tree cavities, 4–10 m above ground. (**Bruinkoppapegaai**)

PLATE 18 Spine-Tails, Swifts p210

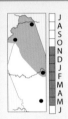

MOTTLED SPINETAIL *Telacanthura ussheri* (422) 14 cm; 34 g
SEXES *Alike. Mottled (not white) throat and white vent line key features that distinguish it from Little Swift. Short secondaries and longer primaries also give it a different flight appearance.* **JUV** *Upperparts have grey edging to feathers.* **STATUS** Very uncommon and localised resident, regarded as sedentary. Likely to occur only in the Pafuri region of the Park and during the summer months usually found not far from suitable Baobab tree nesting sites, one of which is located in the Nyalaland Wilderness Trails Camp. Found singly or in small groups. **HABITAT** Favours dry deciduous woodland with scattered Baobab trees. **FOOD** Aerial arthropods. **CALL** Various twittering calls; mostly silent. **BR** Monogamous; nests singly or in small colonies (2–5 pairs) in suitable Baobab tree hollows. (**Gevlekte Stekelstert**)

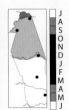

BÖHM'S SPINETAIL *Neafrapus boehmi* (423) 9 cm; 14 g
SEXES *Alike. Conspicuous short tail gives it a bat-like appearance. The only swift in the region with a dark throat and white belly and vent. Fluttering bat-like flight diagnostic.* **JUV** *Similar to adult primaries with white outer edges.* **STATUS** Localised and very uncommon. Most likely to be seen close to the Luvuvhu river, and surrounding riparian woodland. Resident and sedentary; may disperse after breeding, hence records further south to Shingwedzi River. **HABITAT** Deciduous and evergreen woodland; usually linked to Baobab trees. **FOOD** Aerial arthropods. **CALL** 4-syllabled twitter *tri-tri-tri peep*. **BR** Monogamous. Nests in Baobab tree cavities in S Africa; further north of the Park also uses deep pits, wells and mine shafts. (**Witpensstekelstert**)

LITTLE SWIFT *Apus affinis* (417) 13 cm; 25 g
SEXES *Alike. Square-ended tail (appearing rounded when spread) and broad white rump diagnostic. Lacks white vent bar and has different flight appearance to Mottled Spinetail.* **JUV** *Duller than adult feathers pale fringed.* **STATUS** Common resident, particularly around nesting sites where they also roost. In KNP associated with man-made structures such as larger bridges and buildings in restcamps. These structures have enabled the population to increase and led to a more widespread distribution in KNP. Atlas data suggests some populations to be partially migratory, but there is little evidence of this in KNP. Highly gregarious, often seen in mixed-species flocks. **HABITAT** Throughout Park where suitable nesting and roosting sites available. **FOOD** Aerial insects. **CALL** Noisy, especially around br sites; call a shrill scream. **BR** Monogamous and colonial; nest a closed bowl of feathers and grass glued together with saliva. Has extended breeding season from Sept–Mar/Apr. (**Kleinwindswael**)

HORUS SWIFT *Apus horus* (416) 15 cm; 26 g
SEXES *Alike. Distinguished from White-rumped Swift by stockier build, less deeply forked tail, and broader white rump patch that extends onto flanks.* **JUV** *Similar to adult; body feathers with pale tips.* **STATUS** Uncommon br resident and br visitor, usually in small groups of 2–6 birds. **HABITAT** Wide range of habitats, often close to br sites during the late summer months. Nests are excavated into banks mainly along the major rivers and drainage lines, so this species is more likely to be encountered away from man-made structures, unlike White-rumped and Little swifts. **FOOD** Aerial arthropods. **CALL** Not very vocal; gives buzzing *preeeooo-preeeooo*. **BR** Monogamous. Unlike other swifts in region, breeds in tunnels excavated by other bird species in vertical banks. (**Horuswindswael**)

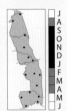

WHITE-RUMPED SWIFT *Apus caffer* (415) 16 cm; 24 g
SEXES *Alike. Long, slender, deeply forked tail and narrow white rump band (almost 'V' shaped) are key features.* **JUV** *Similar to adult.* **STATUS** Common br intra-African migrant, present Aug–May (many birds overwinter in KNP). Often in mixed-species foraging flocks. Not uncommon in and around restcamps where it breeds. **HABITAT** Wide range of habitats; generally near br sites. **FOOD** Aerial arthropods. **CALL** Screams and chattering; not as shrill as other swifts. Silent while foraging but vocal near br and roosting sites. **BR** Monogamous (often colonial); uses medium-sized road culverts rather than the larger bridges used by Little Swifts; also uses Little Swift or swallows' nests. (**Witkruiswindswael**)

PLATE 19 Palm-Swift, Swifts p210

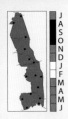

AFRICAN PALM-SWIFT *Cypsiurus parvus* (421) 15 cm; 14 g
SEXES *Alike. Small, slim, plain-plumaged swift with distinctive long, narrow, scythe-shaped wings and deeply forked tail.* **JUV** *Rufous-tipped feathers and less deeply forked tail.* **STATUS** Common resident; the apparent recent use of buildings and other structures for nesting has greatly aided population expansion in KNP. Pair members roost on nest pad throughout br cycle. Fairly common in most restcamps. **HABITAT** Most woodland regions, especially where tall palms or buildings are present. **FOOD** Aerial arthropods, feeding mainly at tree-top height. **CALL** Twittering given near nesting sites; also a high-pitched scream. **BR** Monogamous, solitary or colonial. Nest a small pad of feathers and dry plant material, which together with the eggs is glued to palm frond or building. Eggs cannot be turned during incubation period. (**Palmwindswael**) **ALT NAME** *Palm Swift*

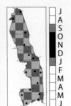

ALPINE SWIFT *Tachymarptis melba* (418) 20 cm; 77 g
SEXES *Alike. The largest swift in the region and the only one with white underparts and brown breast band.* **JUV** *Upperparts darker with white fringing to plumage.* **STATUS** Fairly common br migrant (in eastern regions, incl KNP, present Aug–Mar). Usually seen in non-br months when they are not bound to nesting sites on cliffs along the escarpment. Mostly in small groups, and readily joins mixed-species flocks. **HABITAT** Forages over wide range of habitats and often at high altitudes; may be found anywhere in the Park. **FOOD** Aerial insects. **CALL** Loud high-pitched trill, especially near br sites. **BR** Monogamous and colonial; nests glued with saliva to sides of vertical cracks in cliffs, but breeding not recorded in KNP. (**Witpenswindswael**)

COMMON SWIFT *Apus apus* (411) 17 cm; 37 g
SEXES *Alike. Difficult to distinguish from African Black Swift and can look different depending on different light conditions. Common Swift has little contrast in colour between upperparts, and secondary and primary flight feathers, which are uniformly dark. Tail more deeply forked than in African Black Swift. Narrower and more pointed wings give it a less stocky appearance than African Black Swift; throat patch too is less distinct.* **JUV** *Throat patch whiter and larger.* **STATUS** Common and widespread non-br Palearctic migrant, present late Oct–Mar; usually in flocks of tens or even hundreds. Permanently airborne in region, roosting on the wing at high altitude. Large flocks of all-black swifts seen in spring and mid-summer months most likely to be this species and not African Black Swifts that are likely to be breeding at that time of the year and foraging singly or in small groups. Active in the vicinity of thunderstorms. **HABITAT** Occurs over all KNP's habitats. **FOOD** Aerial arthropods. **CALL** High-pitched scream *shreee*, but mostly silent in s Africa. (**Europese Windswael**) **ALT NAME** *European Swift*

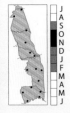

AFRICAN BLACK SWIFT *Apus barbatus* (412) 19 cm; 42 g
SEXES *Alike. From above, appears overall slightly browner with less black in appearance than Common Swift. Black back contrasts with paler (brownish) head and rump, and lighter coloured secondaries and greater wing coverts form pale triangle on inner wing. White throat patch distinct. Broader wings and shorter, less deeply forked tail and slightly stockier appearance than Common Swift.* **JUV** *Forehead and crown feathers fringed white.* **STATUS** Resident only in the Lanner Gorge area of Pafuri. Elsewhere in KNP usually only a winter visitor after breeding season on cliffs along the escarpment is complete. Usually in flocks; foraging birds may cover 1 000 km/day. **HABITAT** Mostly mountainous regions in br season, but wide-ranging during non-br season and may be found anywhere in the Park. **FOOD** Aerial arthropods, including termite alates, flies, beetles, ants and bees. **CALL** High-pitched screaming *shreee*. Very vocal during aerial displays. **BR** Monogamous; nest a pad of feathers and grass, glued with saliva in horizontal cracks in cliffs. (**Swartwindswael**) **ALT NAME** *Black Swift*

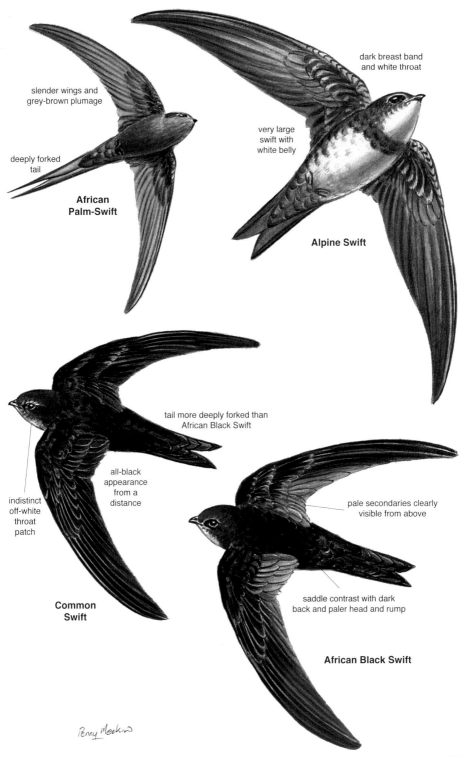

AFRICAN GRASS-OWL *Tyto capensis* (393) 36 cm; 420 g
SEXES Alike. *Most resembles Barn Owl but darker brown on upperparts. In flight, legs protrude past short tail.* **JUV** *Similar to adult buffier below.* **STATUS** Uncommon resident occurring throughout Park, but may be more common than expected especially on the eastern grasslands; listed as Vulnerable in s Africa. Probably reasonably common in wetter years when grass is abundant, but is almost absent during droughts. **HABITAT** Favours tall rank, or dense short, grassland. **FOOD** Predominantly large vlei rats (*Otomys* species), also other rodents, birds and insects. **CALL** Like shortened version of Barn Owl screech; also rapid frog-like clicks. **BR** Monogamous. Little or no nest: a flimsy pad at end of tunnel in dense grass. (**Grasuil**) **ALT NAME** Grass Owl

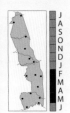

BARN OWL *Tyto alba* (392) 32 cm; 335 g
SEXES *Alike, but ♀ often slightly darker. Similar to African Grass-Owl that also has a heart-shaped face, but paler and with greyish (not brown) back.* **JUV** *Like adult.* **STATUS** Widespread br resident and present in or around many restcamps. **HABITAT** Prefers open habitats (not forest), but commonly associated with man-made structures, which are used for roosting and br. **FOOD** Mainly small rodents, also birds, insects, lizards, frogs and termites. **CALL** Long hissing screech *schreeeee*. Most vocal in late summer and autumn when it breeds. **BR** Monogamous; frequently in man-made structures, including buildings, old Hamerkop nests or natural cavities in trees and cliffs. (**Nonnetjie-uil**)

AFRICAN SCOPS-OWL *Otus senegalensis* (396) 16 cm; 65 g
SEXES *Alike. Most like Southern White-faced Scops-Owl but only about half the size with grey (not white) face. Not much longer than a sparrow but about same weight as a wood-dove. Two colour forms: grey or grey-brown (uncommon).* **JUV** *Like adult, with wisps of down on underparts.* **STATUS** Common br resident occurring throughout the Park. May be heard or seen in all restcamps, and roosts in trees in picnic and camp sites, particularly Satara and the Afsaal and Muzanzeni picnic spots. **HABITAT** Savanna woodland, especially Mopane and *Acacia*. **FOOD** Mainly insects; also small rodents, birds, geckos and frogs. **CALL** Repetitive loud single *kruup* note every 5–8 seconds. **BR** Monogamous; nests in natural tree hole, also woodpecker nests. (**Skopsuil**)

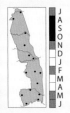

SOUTHERN WHITE-FACED SCOPS-OWL *Ptilopsis granti* (397) 27 cm; 210 g
SEXES *Similar, ♀>♂. Twice the size of African Scops-Owl and with a white facial disc, orange (not yellow) eyes and overall paler grey.* **JUV** *Resembles adult.* **STATUS** Generally uncommon but widespread in suitable dry regions. Resident. **HABITAT** Wide range of woodland habitats, favouring dry watercourses. Usually absent from riparian zones. **FOOD** Wide range of invertebrate and vertebrate prey (up to Laughing Dove size), mostly small mammals. **CALL** Series of bubbling notes, *popopopopopopreeo*. **BR** Monogamous. Uses stick nests built by other species, frequently old nests of sparrowhawks; occasionally nests in natural hollows in trees. (**Witwanguil**) **ALT NAME** White-faced Owl

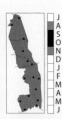

PEARL-SPOTTED OWLET *Glaucidium perlatum* (398) 19 cm; 76 g
SEXES *Similar, ♀>♂. Resembles African Barred Owlet but is smaller, has spotted head, streaked not barred chest and two dark patches at back of head known as false 'eyes'.* **JUV** *Reduced spotting on head.* **STATUS** Common and conspicuous resident. Sometimes calls (and hunts) during the day and is more likely to be found in restcamps than African Barred Owlet. **HABITAT** Mostly dry woodland habitats incl Mopane. **FOOD** Insects; also a bold hunter of rodents, bats, lizards, snakes, birds, etc., sometimes during daylight hours. **CALL** Series of loud shrill whistles reaching crescendo *peu peu peu peu peeu peeeu*. **BR** Monogamous. Most nest holes used are those of woodpeckers; occasionally uses natural hole. (**Witkoluil**) **ALT NAME** Pearl-spotted Owl

AFRICAN BARRED OWLET *Glaucidium capense* (399) 21 cm; 120 g
SEXES *Similar. Similar to Pearl-spotted Owlet but with barred (not spotted) head, neck and upper breast. Lacks false eyes at back of head.* **JUV** *Less barred above and spotted below.* **STATUS** Locally common to uncommon resident. Common on the hills and ridges around Punda Maria and in some restcamps (e.g. Letaba), also along the major seasonal rivers with well-developed riparian fringes. **HABITAT** Tall open woodland and broad-leaved bushveld. **FOOD** Mostly invertebrates but some birds, lizards and frogs recorded. **CALL** Main call a series of purring *wow wow wow ...* then *purr purr purr ...* **BR** Monogamous. Nests in natural tree cavities (woodpecker and barbet entrance holes usually too small). (**Gebande Uil**) **ALT NAME** Barred Owl

PLATE 21 Owls p211

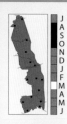

SPOTTED EAGLE-OWL *Bubo africanus* **(401) 45 cm; 700 g**
SEXES *Alike, ♂>♀. Mostly grey/brown above, finely barred grey and white below with pale grey blotches on chest. Eyes yellow. Uncommon rufous form has rufous brown (not grey) plumage and yellow-orange eyes.* **JUV** *Resembles adult with shorter 'ear' tufts.* **STATUS** Resident. Most common large 'eared' owl in region. **HABITAT** Tolerant of wide variety of habitats so occurs almost throughout the Park. Remains well concealed during the day but can usually be seen on night drives. **FOOD** Small insects to rodents, birds and mammal prey up to hare and bushbaby size. **CALL** Mellow 2- or 3-syllable hoots *hu whoooooo* or *whoo are you*, rising and falling in pitch, sometimes in duet. **BR** Monogamous. Wide range of nest sites: on the ground, in trees and on buildings. (**Gevlekte Ooruil**)

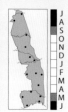

VERREAUX'S EAGLE-OWL *Bubo lacteus* **(402) 62 cm; ♂ 1,7 kg, ♀ 2,3 kg**
SEXES *Alike, but ♀ about 10% larger and 30% heavier than ♂. Pale, finely barred milky grey underparts, grey-brown back, pinkish eyelids and very large size make this owl unmistakable. Much larger and greyer than Spotted Eagle-Owl.* **JUV** *Fledges with grey-brown down, grey bars on head and body and with smaller 'ears'.* **STATUS** Widespread and fairly common. Resident and sedentary. **HABITAT** Dry savanna woodland, especially riparian with large trees in which to roost. **FOOD** Wide range of animal prey up to half-grown monkey and warthog piglet size. Large birds, including diurnal raptors, Pel's Fishing-Owl, herons and flamingos also taken as prey. **CALL** Deep grunting *hok-hok-hok,* sometimes in uneven duet, ♀ at lower pitch. Juveniles' characteristic begging call sometimes can be heard in daylight hours. **BR** Monogamous. Nests on large twig platform of other raptor species, or on Red-billed Buffalo-Weaver and Hamerkop nests. (**Reuse-ooruil**) **ALT NAME** *Giant Eagle Owl*

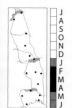

PEL'S FISHING-OWL *Scotopelia peli* **(403) 63 cm; 2,1 kg**
SEXES *Alike. Rufous colour, 'earless' head, dark brown eyes and large size distinguish it from all other owls in region. Legs unfeathered.* **JUV** *Fledges with white or pale rufous down, which it retains for about 6 months.* **STATUS** Listed as Vulnerable throughout its range. Resident and largely confined to the major river systems, especially the Luvuvhu in the north, also a few regular localities on the Olifants and Sabie rivers; threatened by disturbance, river siltation and pollution. Historically more widespread and common; current population estimate approximately 60 birds in the KNP. **HABITAT** Densely wooded riparian forest; with large shady trees overhanging water. **FOOD** Variety of small fish, usually 100–250 g, sometimes up to 2 kg; small crocodiles and frogs also recorded. **CALL** Low deep *hoot* repeated every 10–20 sec, audible 3 km away; also cat-like wail. **BR** Monogamous. Nests in large tree cavities; br on average every second year. (**Visuil**)

AFRICAN WOOD-OWL *Strix woodfordii* **(394) 33 cm; 290 g**
SEXES *Alike, ♀>♂. Only dark 'earless' medium-sized owl found in thickly wooded habitats. Some variation in rufous or rufous-brown coloration.* **JUV** *Fledges with pale rufous down which is retained for up to 5 months.* **STATUS** Distribution is patchy, but occurs throughout KNP. Fairly common resident in well-wooded and riverine habitats but uncommon or absent in drier woodland. **HABITAT** Dense woodland and riparian thickets. **FOOD** Mainly insects and small birds; also rodents, frogs, small snakes and mammals. **CALL** Rhythmic series of hoots: *who who, who who who-are-you,* given alone or answered by mate. Most vocal prior to br in the autumn and winter months. **BR** Monogamous. Nests in natural tree holes. (**Bosuil**) **ALT NAME** *Wood Owl*

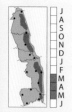

MARSH OWL *Asio capensis* **(395) 37 cm; 315 g**
SEXES *Alike. Dark brown above and variable buff (not white) facial disc that is more rounded than in African Grass-Owl.* **JUV** *Darker brown facial disc.* **STATUS** Locally common resident with fragmented range. Numbers in KNP fluctuate with rainfall and grass cover, most common on the eastern basaltic grasslands in wet summers. Frequently feeds in the late afternoon, at dusk and in the early morning. Probably most diurnal in late summer and autumn when feeding chicks. Often emerges to perch, preen and bask in the late afternoon sun in the winter months. **HABITAT** Moist grassland, and adjacent marshland. **FOOD** Mostly small rodents and birds, but also other vertebrates and invertebrates. **CALL** Grating croak, like tearing canvas. **BR** Monogamous. Nest a shallow pad concealed under grass tuft. (**Vlei-uil**)

PLATE 22 Nightjars p211

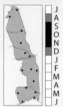

FRECKLED NIGHTJAR *Caprimulgus tristigma* **(408) 28 cm; 80 g**
SEXES *Similar. Large dark nightjar usually linked to rocky habitats.* **M** *White outer tail spots distinguish it from Pennant-winged Nightjar in flight.* **F** *Lacks white outer tail feathers.* **JUV** *White primary spots absent.* **STATUS** Resident: uncommon to locally common in suitable habitat; largely sedentary. **HABITAT** Strongly associated with bare granite and sandstone koppies, and also boulder-strewn hillsides, surrounded by broad-leaved woodland. **FOOD** Insectivorous. **CALL** Characteristic 2–3 syllabled dog-like yelp *pow-wow* or *pow-wow-wow*. **BR** Monogamous. Eggs laid in hollow on rock surface. (**Donkernaguil**)

FIERY-NECKED NIGHTJAR *Caprimulgus pectoralis* **(405) 24 cm; 52 g**
SEXES *Similar. Cryptic, with few distinctive features apart from broad rufous collar on hindneck.* **M** *Wing and tail spots white.* **F** *Less white on tail feathers.* **JUV** *Wing and tail spots buffish.* **STATUS** The most common nightjar in the Park; resident, perhaps with partial migrant populations. **HABITAT** Favours well-developed savanna woodland. **FOOD** Mainly nocturnal flying insects; often hunts from arboreal perch. **CALL** Very characteristic and distinctive *good lord deliver us* call; also rapid *wook, wook, wook ...* notes. **BR** Monogamous. Has strong site fidelity, returning annually to same locality to breed. Eggs laid on the ground in a shallow scrape, usually among leaf litter. (**Afrikaanse Naguil**)

SQUARE-TAILED NIGHTJAR *Caprimulgus fossii* **(409) 25 cm; 60 g**
SEXES *Similar.* **M** *With no distinctive field characters, all-white outer web of outer tail feather is best feature in flight.* **F** *Outer web of outer tail feather buff.* **JUV** *Similar to* ♀. **STATUS** Common resident; mostly sedentary in KNP, but with partial migratory population. A somewhat misnamed bird because tail tip not square, but rounded! **HABITAT** Occurs throughout the Park in open woodland, also on pan and floodplain margins. **FOOD** Insectivorous. **CALL** Continuous churring that sounds like a small engine running, changing speed and pitch of song (changing gear!). **BR** Monogamous. Eggs laid in natural depression in ground. (**Laeveldnaguil**) ALT NAME *Mozambique Nightjar*

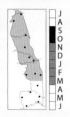

RUFOUS-CHEEKED NIGHTJAR *Caprimulgus rufigena* **(406) 24 cm; 60g**
SEXES *Similar. Difficult to identify when perched; narrow rufous hindneck collar. Call distinctive.* **M** *White wing spots and outer 2 tail feathers tipped white.* **F** *Wing and tail spots buff.* **JUV** *Similarly marked to* ♀. **STATUS** In KNP occurs patchily as a fairly uncommon br intra-African migrant mainly in the northern parts of KNP present Aug–Apr. **HABITAT** Dry open woodland and clearings. Unlikely to be found in areas of long grass, especially in years of good rainfall. **FOOD** Insects: mainly beetles and moths. **CALL** Sustained engine-like churring, given at set speed (unlike Square-tailed Nightjar). **BR** Monogamous. Eggs laid in depression in ground. (**Rooiwangnaguil**)

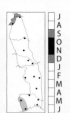

PENNANT-WINGED NIGHTJAR *Macrodipteryx vexillarius* **(410) 27 cm; 75 g**
SEXES M (**BR**) *Two long trailing pennants very conspicuous but only present for duration of br season.* **M** (**NON-BR**) *lacks pennants but broad white stripe across wing diagnostic.* **F** *Lacks pennants; primaries brown and rufous (not black and white as in* ♂). **JUV** *Similar to* ♀ *but more rufous.* **STATUS** Uncommon localised br intra-African migrant, present Sept–Feb. Found regularly only in the Punda Maria area but sporadic records also from the Pretoriuskop and Berg-en-Dal region. Earlier records in the Pretoriuskop (Napi) area have not been repeated in the recent past. **HABITAT** Favours broad-leaved woodland. **FOOD** Flying insects: mainly beetles and termites. **CALL** Insect-like high-pitched *tseeet-tseeet-tseeet-tseeet*. **BR** Polygynous. Nest scrape in leaf litter. (**Wimpelvlerknaguil**)

EUROPEAN NIGHTJAR *Caprimulgus europaeus* **(404) 27 cm; 67 g**
SEXES *Similar. Large dark nightjar with cryptic plumage with no distinctive field characters, but its direct and fast hawk-like flight, and habit of regularly roosting lengthwise on branches (unlike other nightjars in region), help aid identification.* **F** *Lacks white wing and tail spots.* **JUV** *Similar to* ♀. **STATUS** Fairly common throughout the Park. Non-br migrant, present Sept–Apr (most birds arrive late Nov). **HABITAT** Most woodland habitats, incl *Acacia* and broad-leaved such as Mopane and riparian. **FOOD** Insectivorous. **CALL** Continuous *churr* (similar to but faster than Square-tailed Nightjar). Mostly silent in s Africa. (**Europese Naguil**)

IN-HAND IDENTIFICATION OF SOUTHERN AFRICAN NIGHTJARS

Nightjars are nocturnal, highly cryptic and difficult birds to identify in the field. They are best identified by their far-carrying nocturnal whistles, or mechanical-sounding churrs. All six KNP species have characteristic and easily identifiable songs. Sadly, nightjars are frequently killed on roads, but these dead birds are useful as museum specimens and for distribution records. In spite of their mottled and similar plumages, they can be distinguished by selecting and comparing the following three feathers: the P9 primary wing feather (second to front or leading wing feather), and the two outer tail feathers (rectrices) R4 and R5 (see Fig.1). The combined characters of these three feathers are unique and diagnostic for adults of each of the seven species that occur in s Africa. The position of the white (♂) or buff (♀) wing spot on the trailing edge of the primary feather (P9) in relation to the emargination (kink or narrowing on the leading edge) varies with each species. For example, the centre of the wing spot is almost opposite the emargination in Fiery-necked Nightjar, slightly inwards for Rufous-cheeked and furthest in Square-tailed Nightjar. The white terminal tail spots are present in all males except Pennant-winged Nightjar. The extent and variation of these tail spots, when compared and used in combination with the primary feather (P9), should solve identification problems for all adult birds. In most cases juvenile males show some initial buffiness to the white spotting. The long pennants (P2) on Pennant-winged Nightjars in early spring are soon lost at the end of the breeding season, but males of this species can be identified in the non-breeding season by the broad white wing stripe. Additional mensural data is given by Jackson, *Ostrich* 71: 371–379, which also provides a dichotomous key to the identification of the nightjars.

FRECKLED NIGHTJAR *Apical patch longest on outer tail feather of ♂. Length of P9 > 175 mm, emargination > 65 mm, apical patch on R5 of ♂ 33–55 mm.*

FIERY-NECKED NIGHTJAR *♂ and ♀ very similar. White tail spots of ♂ longer (larger) than ♀. Length of P9 < 176 mm, emargination > 38% of length, apical patch on ♂ outer tail feather (R5) > 37 mm, and ♀ < 37 mm.*

SQUARE-TAILED NIGHTJAR *Outer web of outer tail feather (R5) white (♂), and buff (♀). No white on R4. Length of P9 < 175 mm, and apical patch on R5 > 80 mm (♂), and > 70 mm (♀).*

RUFOUS-CHEEKED NIGHTJAR *Tail patches on outer tail feathers of ♂ almost equal in length, barely visible in ♀. Length of P9 < 175 mm, emargination < 38%, and apical patch on R5 < 35 mm.*

PENNANT-WINGED NIGHTJAR *Females and immature males have P9 > 175 mm, emargination < 58 mm and R1 < R5.*

EUROPEAN NIGHTJAR *Tail and wing spots of ♂ similar to ♂ Rufous-cheeked Nightjar. Length of P9, however, diagnostic. P9 > 175 mm, emargination < 60 mm (♂), and < 63 mm (♀).*

Data acknowledgement: H.D. Jackson

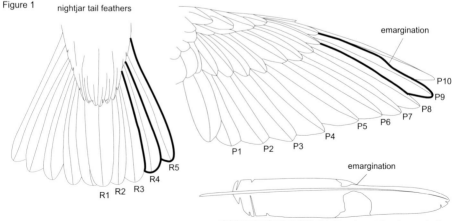

Figure 1 nightjar tail feathers

nightjar wing feather (P9) showing wing spot and emargination

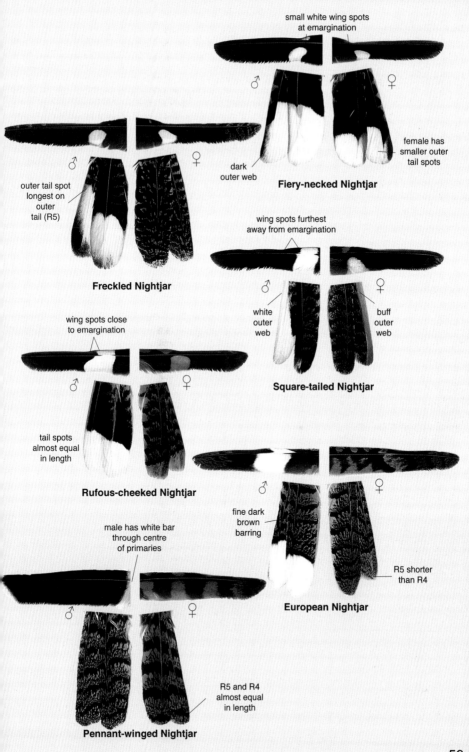

PLATE 24 Doves p211

LAUGHING DOVE *Streptopelia senegalensis*   **(355) 25 cm; 100 g**
SEXES *Very similar, ♀ paler than ♂. Only dove in region with pinkish-grey head and rufous chest with black spotting; lacks black collar on hindneck. White outer tail feathers conspicuous in flight.* **JUV** *Breast grey, lacks black speckling. Overall paler and browner than adult.* **STATUS** Common to abundant resident occurring throughout the Park. **HABITAT** Occurs in all KNP woodland habitats, especially *Acacia*. **FOOD** Mainly seeds from grasses, sedges, shrubs and trees. Some fruit taken as well as rhizomes and bulbs. **CALL** 6–8 notes in soft laughing tone *koo kuKUkuru-koo*. **BR** Monogamous. Nest a flimsy flat twig platform, usually 1–4 m high. (**Rooiborsduifie**)

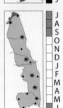

AFRICAN MOURNING DOVE *Streptopelia decipiens*   **(353) 30 cm; 160 g**
SEXES *Alike. Could be confused with both Red-eyed Dove and Cape Turtle-Dove, but combination of yellow eyes and red skin around eyes diagnostic. Usually gives presence away by bubbling call.* **JUV** *Browner than adult; wing coverts tipped buff.* **STATUS** Locally common in some areas, particularly in the Luvuvhu/Limpopo area. Strongly associated with human developments so common in restcamps such as Punda Maria, Shingwedzi, Letaba and Satara (and Tshokwane picnic place). **HABITAT** Restcamps and riparian and adjacent *Acacia* woodland. **FOOD** Seeds, also some small fruit and termite alates. **CALL** Loud bubbling notes *ookrroooooo, ookrroooooo*, followed by a series of *coos*. **BR** Monogamous. Flimsy twig platform nest placed in tree or bush. (**Rooioogtortelduif**)

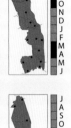

CAPE TURTLE-DOVE *Streptopelia capicola*   **(354) 27 cm; 153 g**
SEXES *Alike, ♂>♀. Similar to, but greyer than, larger Red-eyed Dove, and without bare red skin around eyes. Lacks chestnut colouring on upperparts as in Laughing Dove; white outer tail feathers conspicuous in flight. Black collar on hindneck narrowly rimmed with white.* **JUV** *Lacks black hindneck collar; upperparts and breast fringed buff.* **STATUS** Very common to abundant resident throughout the Park. **HABITAT** Woodland, open savanna. **FOOD** Seeds of grasses, cereals, shrubs and trees; also fruit, earthworms, termites and some insects. **CALL** Loud *kuk-koorr-ko* or 'work harder, work harder', repeated 10–40x; calls throughout the day. **BR** Monogamous. Nest flimsy platform 3–4 m high in tree or bush. (**Gewone Tortelduif**)

RED-EYED DOVE *Streptopelia semitorquata*   **(352) 35 cm; 252 g**
SEXES *Alike, ♂>♀. Largest 'grey' dove in region; pinkish-grey throat, neck and breast diagnostic. Dark red (not yellow) eyes distinguish it from African Mourning Dove. Deep purple-pink eye-ring distinguishes it from greyer-plumaged Cape Turtle-Dove. In flight, light buffy-grey rim to tail a key distinguishing feature.* **JUV** *Black collar partly developed; buffy feathers give it a light mottled appearance.* **STATUS** Common to fairly common resident. **HABITAT** Various well-wooded habitats, especially riparian. **FOOD** Mostly grass and tree seeds. **CALL** A 6-note *KOO KOO, ku-ku KOO-koo*, with emphasis on first two notes. **BR** Monogamous. Twig platform 3–18 m high. (**Grootringduif**)

EMERALD-SPOTTED WOOD-DOVE *Turtur chalcospilos*   **(358) 20 cm; 64 g**
SEXES *Alike, ♂>♀. Emerald green wing spots appear black if not reflected in good light. Base of bill may show dark red if seen from close.* **JUV** *Upperparts flecked and barred with buff or rufous, wing spots smaller and duller.* **STATUS** Locally very common resident, found singly or in pairs. **HABITAT** Occurs in all KNP habitats, especially *Acacia* woodland and riparian margins. **FOOD** Mostly seeds of herbs and grasses; also termites and some fruit. **CALL** Slow *hoo wuhoo hoo whoo* speeding up to faster *do do do dodododo ...* **BR** Monogamous. Nest a small flimsy platform of twigs and rootlets. (**Groenvlekduifie**) ALT NAME *Green-spotted Dove*

TAMBOURINE DOVE *Turtur tympanistria*   **(359) 22 cm; 71 g**
SEXES M *White face, throat and underparts diagnostic; rufous wings very evident in flight.* **F** *Face and underparts slightly buffy and not as white as ♂.* **JUV** *Breast grey in both sexes, barred rufous/brown upperparts.* **STATUS** A fairly common local resident, with a patchy distribution in KNP. Occurs mainly at Pafuri and the well-developed riverine woodlands on the Sabie and Crocodile rivers. Also in the Olifants Gorge area. **HABITAT** Riparian thickets, and riverine woodland. **FOOD** Seeds of grasses, shrubs and trees. Very partial to toxic seeds of alien Castor Oil *Ricinus communis*. **CALL** Slow *coos*, speeding up then ending abruptly. **BR** Monogamous. Nest a frail saucer made mainly of fine rootlets. (**Witborsduifie**)

PLATE 25 Pigeons, Doves, Green-Pigeon, Go-away-bird, Turaco p212

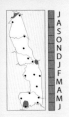

SPECKLED PIGEON *Columba guinea*   **(349) 33 cm; 344 g**
SEXES *Alike. Unlikely to be confused with any other pigeon in region; rusty brown plumage with white spotted wings, bare red skin around eyes, and red legs diagnostic. In flight, black terminal tail bar distinctive.* **JUV** *Browner than adult, and bare skin around eyes chocolate brown.* **STATUS** An uncommon bird in KNP restricted to areas of suitable habitat. Probably resident only in the Luvuvhu Gorges, but has been recorded in the Olifants Gorge and in the mountainous areas around Berg-en-dal. **HABITAT** Mountains, cliffs and rocky gorges. **FOOD** Mainly seeds, also green shoots. **CALL** Deep, echoing coos *doo-doo-doo-doo*, repeated 10–20x. **BR** Monogamous. On rock ledges. Usual clutch of 2 eggs laid on saucer-shaped twig platform. (**Kransduif**) **ALT NAME** *Rock Pigeon*

NAMAQUA DOVE *Oena capensis*   **(356) 26 cm; 40 g**
SEXES M *Conspicuous black face, throat and upper breast distinctive. Smallest-bodied dove in Africa and only one with a long tail. Yellow to yellow-orange bill with purple base very evident from close.* **F** *Lacks black face and throat of male and has blackish-brown bill; overall lighter in colour.* **JUV** *Lacks black face, and upperparts heavily spotted.* **STATUS** Typically a bird of the drier west, highly nomadic and can be abundant or fairly rare depending on previous season's rainfall, more common in drier years. Found singly, in pairs or small loose flocks. **HABITAT** Most common in more arid areas particularly *Acacia* savanna, and dry shrublands; occasionally found in broad-leaved woodland. **FOOD** Almost exclusively tiny seeds of grasses, sedges and weeds. **CALL** Mournful flufftail-like *kuh-whoo*, repeated frequently. **BR** Monogamous. Nest a frail platform of twigs and rootlets built by both sexes, usually low down in shrub or *Acacia* bush. Incubation period 13–14 days by both sexes. (**Namakwaduifie**)

AFRICAN GREEN-PIGEON *Treron calvus*   **(361) 29 cm; 235 g**
SEXES *Alike. Only colourful green-grey pigeon in region; when manoeuvring around branches in search of fruit has parrot-like appearance.* **JUV** *Similar to adult but duller and lacks lilac carpal patches.* **STATUS** Common resident particularly in the riparian zones of the major rivers: sedentary, moving only in response to food supplies. Gregarious in non-br season when usually in small flocks; flight fast and direct. **HABITAT** Dense woodland, evergreen forests, riparian thickets and bush clumps in savanna, especially where there are fruiting figs, Jackal-berry *Diospyros mespiliformis* and White-berry Bush *Flueggia virosa*. **FOOD** Almost entirely fruit (particularly figs). **CALL** High-pitched, fluted, whistling trills, followed by low, croaking, frog-like grunts. **BR** Monogamous. Flimsy twig platform built by both adults and placed 2–21 m high; usual clutch of 2 eggs incubated by both adults. (**Papegaaiduif**) **ALT NAME** *Green Pigeon*

GREY GO-AWAY-BIRD *Corythaixoides concolor*   **(373) 49 cm; 270 g**
SEXES *Alike. Very distinctive all-grey bird that is common in the bushveld and unlikely to be confused with any other species.* **JUV** *Paler than adult with shorter crest and buffy tinge to feathers.* **STATUS** Common and conspicuous resident occurring throughout the Park, mostly sedentary, forming flocks in non-br season. Roosts in groups of 3–5 birds at night. Alarm calls usually reveal the presence of raptors such as African Goshawk, African Hawk-Eagle, etc. **HABITAT** Dry savanna woodland (especially *Acacia*). **FOOD** Mostly fruit; also flowers, buds, invertebrates and nectar. **CALL** Highly vocal, especially when disturbed. Call a loud, drawn-out, nasal *go-away* or *g'way*. **BR** Monogamous, and occasionally breeds with nest helpers. Nest a shallow flimsy and unlined platform of twigs. (**Kwêvoël**) **ALT NAME** *Grey Lourie*

PURPLE-CRESTED TURACO *Gallirex porphyreolophus*   **(371) 42 cm; 285 g**
SEXES *Alike, ♂>♀. Only colourful turaco in KNP. Purple crest appears black unless seen in good light. Bright red primary feathers very conspicuous in flight.* **JUV** *Duller than adult and with less red in wings.* **STATUS** Common resident and mostly sedentary, particularly along the major rivers. Usually in pairs. Very agile; runs along branches and leaps quickly from branch to branch through canopy. **HABITAT** Dense woodland, forest, riparian margins and frequently in restcamp gardens such as Skukuza and Pretoriuskop. **FOOD** Frugivorous; also takes leaf buds. **CALL** Series of loud deep *kor-kor-kor-kor* ... notes, 10–15x. **BR** Monogamous. Nest a flimsy unlined platform. (**Bloukuifloerie**) **ALT NAME** *Purple-crested Lourie*

PLATE 26 Bustards, Korhaan, Sandgrouse p212

KORI BUSTARD *Ardeotis kori* (230) ♂ 135 cm, ♀ 112 cm; ♂ 12,4 kg, ♀ 5,7 kg
SEXES Similar, ♂ much larger than ♀. The largest bustard and one of the world's heaviest flying birds. Large size and all-grey (finely barred grey and white) neck distinctive. ♀ has black on crown reduced and less prominent eyebrow. **JUV** Crown paler than adult. **STATUS** Uncommon to locally common resident; mostly sedentary, local movements associated with patchy rainfall. Listed as Vulnerable in s Africa. Found singly or in loose groups (often single-sex groups) in non-br season. **HABITAT** In KNP mainly on eastern basaltic grasslands but also open savanna woodland and dwarf shrublands. Usually absent from dense vegetation. Vulnerable to collisions with power lines. KNP power lines over grassland areas fitted with 'bird flappers' to make them more visible. **FOOD** Wide range of insects, especially dung beetles, also lizards, chameleons, snakes and carrion. Vegetable matter includes seeds, berries, bulbs, wild melons and *Acacia* gum. **CALL** Far-carrying booming *wum-wum-wum* ... notes. **BR** Polygynous. Males display from regularly used sites with crest erect, neck and tail feathers fanned, wings drooped and booming call. Eggs laid on bare ground in shallow scrape; incubation and care of young by ♀ only. (**Gompou**)

RED-CRESTED KORHAAN *Lophotis ruficrista* (237) 50 cm; 675 g
SEXES M Elongated rufous crest feathers usually only visible in display during br season and seldom seen. Separated from ♂ Black-bellied Bustard by lack of black line down foreneck, greyish (not brownish) crown, and buff (not white) cheeks. Red crest seldom seen as crest only raised in rare displays. **F** Belly black as in ♂, crown brown flecked white. ♀ Black-bellied Bustard has white belly. **JUV** Like ♀ but with buff tips to flight feathers. **STATUS** Common near-endemic, occurring throughout Park, sedentary and territorial; usually found singly. **HABITAT** Dry, fairly dense woodland. **FOOD** Insectivorous; mainly termites and beetles, also tree seeds, berries and gum. **CALL** Main call by ♂ at regularly used call-sites is series of clicks followed by ventriloquial *kyip-kyip-kyip* ... notes. In spectacular display flight, ♂ flies up 10–30 m, throws itself backwards and falls (as though shot!) to ground, spreading wings at last moment. **BR** Polygynous. Eggs laid in shallow scrape, usually under shrub. (**Boskorhaan**)

BLACK-BELLIED BUSTARD *Lissotis melanogaster* (238) 62 cm; ♂ 2,3 kg, ♀ 1,2 kg
SEXES ♂>♀. Long legs and long slender neck differentiate it from all other s African bustards. **M** Black stripe down foreneck and white cheeks diagnostic. **F** Long legs, buff throat and white belly. **JUV** Like ♀ but cheeks, throat and neck rufous brown. **STATUS** Uncommon to locally common resident occurring throughout the Park but mainly on eastern basaltic grasslands, regarded as Near-threatened. Largely sedentary but influx into the southern half of the Park recorded in the summer months. **HABITAT** Tall grassland and open woodland with long grass. **FOOD** Omnivorous. Mainly insects such as locusts, grasshoppers, beetles and termites, but also berries, seeds and green leaves. **CALL** Males display from regularly used call-sites. Call a loud frog-like *quark,* followed by *kw-ick* (like a cork popping out of a bottle). Spectacular display flight in which male flies high with exaggerated wingbeats and then 'parachutes' slowly back to earth with wings held high. **BR** Polygynous. Nest scrape placed in tall grass. (**Langbeenkorhaan**) **ALT NAME** Black-bellied Korhaan

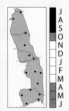

DOUBLE-BANDED SANDGROUSE *Pterocles bicinctus* (347) 25 cm; 230 g
SEXES M Only sandgrouse occurring in KNP. Black and white forehead band and narrow black and white breast band distinctive. **F** Cryptically plumaged with barred upper breast. **JUV** Similar to ♀, upperparts more pinkish fawn. **STATUS** Scarce to locally common near-endemic. An inconspicuous species that is easily overlooked. Commonly encountered along the road edges where it has the habit of freezing when disturbed; found in pairs or small groups. Undertakes some local movements, mainly in response to water availability, especially in the dry season. Unlike other sandgrouse in the western half of s Africa, drinks in the evenings at dusk when it often congregates in large numbers. Birds fly to waterholes just after sunset and socialise before approaching the water to drink, leaving about 40 min after sunset. Inactive in the heat of the day when it tends to lie under shaded trees. **HABITAT** Savanna woodland with preference for Mopane, but also inhabits *Acacia.* **FOOD** Feeds mainly on the seeds of legumes such as *Acacia* and *Tephrosia,* but also Blackjack and Hairy Thorn-apple. **CALL** Soft, quick *weep-weeu, chuk-chukki weep-weeu.* Mostly silent in flight. **BR** Monogamous. 2–3 eggs laid in shallow scrape. (**Dubbelbandsandpatrys**)

PLATE 27 Crakes, Snipes, Painted-snipe, Jacanas p212

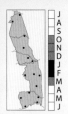

AFRICAN CRAKE *Crecopsis egregia* (212) 22 cm; 120 g
SEXES *Alike. Much shorter-billed than larger African Rail. Grey throat and barred belly distinctive.* **JUV** *Darker and duller than adult.* **STATUS** Uncommon and inconspicuous (may be more common than suspected) br intra-African migrant, present Nov–Apr. Occurrence and abundance determined by previous season's rainfall. Recorded at localities throughout KNP. **HABITAT** Favours seasonally flooded grassland, especially vlei margins, but also tall dry grassland. **FOOD** Earthworms, insects, small frogs and fish, seeds and plant matter. **CALL** ♂ advertising call a series of rapid grating *krrr* notes. **BR** Monogamous. Saucer nest placed in tuft of grass, usually over dry ground. (**Afrikaanse Riethaan**)

CORN CRAKE *Crex crex* (211) 28 cm; 155 g
SEXES *Similar,* ♀ *lacks grey face and neck. Seldom seen on ground, usually only when flushed, showing conspicuous chestnut wing coverts, and legs dangling well past tail.* **JUV** *Similar to adult* ♀. **STATUS** Uncommon to rare non-br Palearctic migrant, present end Nov to early Apr. Occurrence and abundance determined by previous season's rainfall and grass biomass. Recorded at grassy localities throughout KNP, but mainly on eastern basaltic grasslands. In s Africa, listed as Vulnerable. **HABITAT** Mostly in rank grassland adjacent to wet ground and marshes. **FOOD** Omnivorous. **CALL** Usually silent, but responds inquisitively to sound recordings. (**Kwartelkoning**)

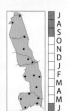

AFRICAN SNIPE *Gallinago nigripennis* (286) 28 cm; 113 g
SEXES *Alike. Bill longer, belly whiter and with less obvious white wing barring than Great Snipe. Blackish loral and eye-stripe. Well-known zigzagging take-off flight (showing whitish underwing).* **JUV** *Similar to adult.* **STATUS** A rare vagrant to KNP which may occur at any locality with suitable marshy habitat. Occurrence and abundance probably determined by previous season's rainfall. **HABITAT** Freshwater and brackish wetlands. **FOOD** Aquatic insects. **CALL** 'Mud-sucking' call *tchek* or *ghip* very characteristic. Gives impressive aerial drumming *wu-wu-wu ...* sounds. **BR** Monogamous. Nest in grass, just above mud. (**Afrikaanse Snip**) **ALT NAME** *Ethiopian Snipe*

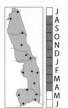

GREATER PAINTED-SNIPE *Rostratula benghalensis* (242) 25 cm; 122 g
SEXES ♀>♂. **M** *Distinguished from African Snipe by pale breast harness and white eye-patch.* **F** *More brightly coloured than* ♂; *dark chestnut neck and breast diagnostic.* **JUV** *Similar to adult* ♂. **STATUS** Throughout KNP but mostly rare; only fairly common in a few localities where suitable habitat exists. Highly nomadic and probably partially migratory. Listed as Near-threatened. **HABITAT** Favours vegetated waterside habitats with exposed mud. **FOOD** Worms, insects and crustaceans. **CALL** ♀ has series of low *vuk-oo* and hooting *boooo* notes; ♂ has less diverse repertoire. **BR** Usually polyandrous. Incubation and parental care by ♂. (**Goudsnip**) **ALT NAME** *Painted Snipe*

AFRICAN JACANA *Actophilornis africanus* (240) 23–32 cm; ♂ 140 g, ♀ 232 g
SEXES *Alike,* ♀ *considerably larger than* ♂. *Adult distinctive with chestnut plumage and pale blue bill and frontal shield.* **JUV** *Easily confused with adult Lesser Jacana, but larger, has light grey bill and frontal shield (rufous forecrown in adult Lesser Jacana) and, in flight, both adult and juv show no white trailing edge to secondaries.* **STATUS** Widespread resident occurring on almost all water bodies. **HABITAT** Most water bodies with low emergent vegetation especially if covered by water-lilies. **FOOD** Mostly aquatic insects and their larvae; also small fish, crustaceans and some plant matter. **CALL** Screeching, scolding and yapping *kyorrr, kyorrr* and *kraaaa, kraaaa ...* notes. **BR** Polyandrous. Nest a pad on floating vegetation. ♂ incubates and takes over all parental duties once last egg is laid. (**Grootlangtoon**)

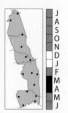

LESSER JACANA *Microparra capensis* (241) 16 cm; 41 g
SEXES *Similar,* ♀>♂ *Smallest member of jacana family. Confusion possible with juv African Jacana, which also has white supercilium and bold dark line through eye, but has rufous crown and forecrown, and darker back. White trailing edge to secondaries evident in flight.* **JUV** *Like adult but with buff-edged feathers to upperparts.* **STATUS** Very rare bird in KNP but recorded at widespread localities; regarded as Near-threatened in s Africa. Resident, but nomadic in response to water levels and availability of floating vegetation. **HABITAT** Permanent and seasonal shallow freshwaters with floating vegetation. **FOOD** Aquatic insects. **CALL** Low-pitched hoot *woot-woot-woot ...* and quieter *ksk-ksk-ksk*. **BR** Monogamous. Nest a small floating platform; incubation shared by both sexes. (**Dwerglangtoon**)

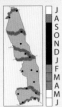

AFRICAN FINFOOT *Podica senegalensis* (229) 55 cm; 600 g
SEXES Similar, ♂>♀. *Easily distinguished from similar-looking cormorants by orange-red bill, and out of water by bright orange legs and feet. Dark head and neck of ♂ distinctive.* **JUV** *Similar to adult but with less red on bill.* **STATUS** Uncommon, highly secretive and easily overlooked, mostly sedentary; usually found singly. **HABITAT** Favours slow-flowing permanent rivers and streams (especially the Sabie River) with overhanging vegetation. **FOOD** Mainly insects; also crabs, snails, frogs and molluscs. **CALL** Mostly silent; loud *kak, kak, kak* ... **BR** Monogamous. Nest a deep untidy bowl usually overhanging water. (**Watertrapper**)

BLACK CRAKE *Amaurornis flavirostra* (213) 21 cm; 90 g
SEXES Alike, ♂>♀. *Unmistakable with jet-black body, lemon yellow bill and bright red legs.* **JUV** *Olive-brown above, dark grey to blackish below, and with black bill.* **STATUS** Very common resident occurring throughout Park on almost all water bodies; usually in pairs. A noisy active species seen most often in the early morning or evening on aquatic fringes. **HABITAT** Most water bodies fringed with emergent vegetation. **FOOD** Earthworms, insects, crustaceans, molluscs, small fish, plant matter and seeds. **CALL** Series of loud and fast clucking notes, given as duet. **BR** Monogamous. Deep cup nest usually placed in emergent vegetation just above water. (**Swartriethaan**)

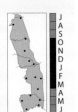

COMMON MOORHEN *Gallinula chloropus* (226) 34 cm; 260 g
SEXES Alike. *Combination of white flank stripe, red base to yellow-tipped bill and rounded red frontal shield diagnostic.* **JUV** *Initially black, upperparts become dark brown, bill and frontal shield greenish-brown.* **STATUS** Locally common resident, largely sedentary with local movements in response to water levels. Usually solitary or in pairs. **HABITAT** Most freshwater bodies with emergent vegetation. **FOOD** Mainly water plants, aquatic insects, tadpoles and molluscs. **CALL** Single high-pitched *krrruk* and range of clucking notes. **BR** Monogamous. Bowl nest placed in emergent vegetation. (**Grootwaterhoender**)

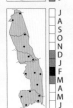

ALLEN'S GALLINULE *Porphyrio alleni* (224) 26 cm; 140 g
SEXES Similar, ♂>♀. *About half the size of African Purple Gallinule (vagrant to the Park) with blue or green frontal shield.* **BR** ♂*'s frontal shield blue,* ♀*'s apple green; eyes coral red, legs bright red.* **NON-BR** *Frontal shield brown, legs dull red.* **JUV** *No white flank stripe; body brown-buff.* **STATUS** Uncommon partial migrant to seasonally flooded areas, numbers dependent on availability of suitable habitat, a specialist colonising freshwater pans once sufficient rain has fallen. Best seen at Leeupan. Breeding recently recorded within view of the Lake Panic bird hide at Skukuza. Absent during droughts. **HABITAT** Freshwater marshes, seasonally flooded pans and wetlands, particularly with abundant emergent vegetation. **FOOD** Omnivorous. **CALL** Most frequent call repetitive deep *klip-klip* ... **BR** Monogamous. Cup nest placed in emergent vegetation. (**Kleinkoningriethaan**) **ALT NAME** *Lesser Gallinule*

LESSER MOORHEN *Gallinula angulata* (227) 23 cm; 135 g
SEXES Similar, ♂>♀. *Noticeably smaller than Common Moorhen; predominantly yellow bill has red ridge on culmen. Legs greenish (not yellow).* **JUV** *Bill pale brownish-yellow; paler, browner neck than juv Common Moorhen, and greyish-white underparts. Lacks white flank stripe.* **STATUS** Uncommon in much of its range in KNP, but may be locally common, especially in years of good rainfall. Best seen at Leeupan. Br intra-African migrant, present Nov–May (some overwintering in north). **HABITAT** Occupies similar habitats to Allen's Gallinule: temporary and permanent freshwater wetlands and pans with abundant emergent vegetation. **FOOD** Omnivorous. **CALL** Series of clucking notes. **BR** Monogamous. Small bowl nest placed in emergent vegetation. (**Kleinwaterhoender**)

RED-KNOBBED COOT *Fulica cristata* (228) 39 cm; 730 g
SEXES Alike, ♂>♀. *Only large all-black plumaged bird in aquatic habitats in the Park. White bill and frontal shield conspicuous. Two red knobs above shield more conspicuous during br season.* **JUV** *Upperparts dark brown, greyish below.* **STATUS** A rare nomad in KNP usually only occupying water bodies very temporarily – here today gone tomorrow. **HABITAT** Mostly (when recorded) on dams. **FOOD** Mainly plant material; also insects and seeds. **CALL** Sharp *kik* or *krik*. **BR** Monogamous. Nest a floating mound of vegetation, but no breeding records for KNP, the attached records reflect breeding outside the lowveld. (**Bleshoender**)

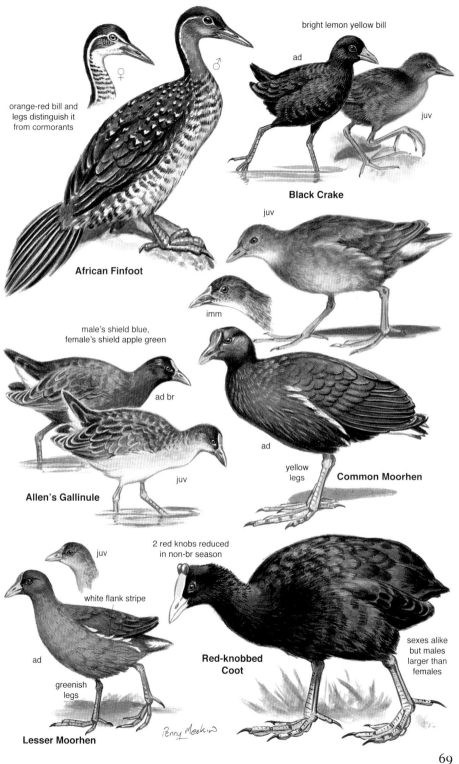

MARSH SANDPIPER *Tringa stagnatilis* (269) 24 cm; 70 g
SEXES *Alike*. **NON-BR** *Pale wader with light grey upperparts, very white underparts, long yellowish legs, slim body and needle-like straight black bill. Much smaller than Common Greenshank that has a thicker upturned bill. Rump and centre of back white.* **JUV** *Like non-br adult but darker and browner above.* **STATUS** A non-br Palearctic migrant, fairly common soon after arrival in Oct/Nov but a passage migrant as numbers decline as the summer progresses. Northward migration starts in Feb, with last birds leaving by Apr. Often found singly, also though in association with other waders. **HABITAT** Favours wetlands, ephemeral pools and river floodplains. Also found on the edge of dams and less commonly river margins. **FOOD** Aquatic insects, small molluscs and crustaceans. **CALL** Sharp *tew-tew-tew* on take-off. (**Moerasruiter**)

COMMON GREENSHANK *Tringa nebularia* (270) 32 cm; 185 g
SEXES *Alike*. **NON-BR** *Largest of the Tringa waders, much larger than Marsh Sandpiper, pale olive green legs (rarely yellowish) and slightly upturned heavier bill (with more extensive pale base) are key features. White on rump extends up centre of back.* **JUV** *Like non-br adult, upperparts browner.* **STATUS** Common non-br Palearctic summer visitor, main arrivals Aug/Sept, departing Mar/Apr. Usually found singly but sometimes feeds with other waders. **HABITAT** Margins of almost all water bodies in Park. **FOOD** Forages mainly by day; mainly fish fry, tadpoles and aquatic insects. Also crustaceans, molluscs and small fish. **CALL** Conspicuous strident 3- or 4-syllabled '*tew-tew-tew*'. (**Groenpootruiter**) **ALT NAME** Greenshank

GREEN SANDPIPER *Tringa ochropus* (265) 23 cm; 80 g
SEXES *Alike*. **NON-BR** *Key features distinguishing it from Wood Sandpiper are darker brown upperparts and breast, shorter greenish (not yellowish) legs, and supercilium that extends from eye to bill (inconspicuous or absent behind eye). In flight has blackish (not white) underwing and darkish upperparts contrasting with mainly white tail, rump and belly (tail has 3–4 incomplete dark bars at tip). White spotting (especially on wings) more conspicuous at onset of br season during late summer, but never as boldly spotted as Wood Sandpiper.* **JUV** *Buff spotting on upperparts.* **STATUS** Scarce but regular non-br Palearctic migrant, present Sept–Apr/May. Usually secretive and solitary when feeding. **HABITAT** In KNP found mainly at smaller quieter pans less favoured by other waders. Probably a passage migrant in KNP as most sightings are in early summer. **FOOD** Both aquatic and terrestrial invertebrates, also molluscs, small fish and some plant material. **CALL** Take-off call a musical *weet* or *klu-eet*. (**Witgatruiter**)

WOOD SANDPIPER *Tringa glareola* (266) 20 cm; 60 g
SEXES *Alike*. **NON-BR** *Upperparts finely spotted; this easily distinguishes it from Common Sandpiper. Conspicuous white supercilium extends behind eye (inconspicuous behind eye in Common Sandpiper). Legs yellowish (not greenish) and longer than Green Sandpiper.* **JUV** *Upperparts warmer brown, spotted buff.* **STATUS** Very common non-br Palearctic migrant. Most abundant wader in KNP with main arrivals from Aug and departure late Feb–May; some 1-year birds overwinter. Generally solitary or in loose groups. **HABITAT** Shorelines of almost all freshwater habitats. **FOOD** Wide range of aquatic and terrestrial insects, as well as molluscs, worms, small fish and frogs. **CALL** Take-off call *chiff-iff-iff*. (**Bosruiter**)

COMMON SANDPIPER *Actitis hypoleucos* (264) 20 cm; 47 g
SEXES *Alike. Darkish bronzy-brown upperparts, intricately lined (not white spotted) easily distinguishes it from Wood Sandpiper. Confusion most likely with Green Sandpiper, which lacks distinctive white shoulder crescent. Common Sandpiper shorter-legged and longer-tailed with wing tips falling well short of end of tail. Bill straight and about equal to head length; habit of bobbing its rear half distinctive.* **JUV** *Upperpart feathers slightly buff-tipped.* **STATUS** Very common non-br Palearctic migrant, main arrivals Sept/Oct, most departing by Mar; usually solitary. **HABITAT** Shorelines of most aquatic habitats. **FOOD** Aquatic and terrestrial insects, also molluscs, small frogs and tadpoles and small fish. Gleans insects from backs of Hippopotamus. **CALL** High-pitched shrill *seep-seep-seep* ... (**Gewone Ruiter**)

SANDERLING *Calidris alba* (281) 21 cm; 55 g
SEXES *Alike.* **NON-BR** *Palest sandpiper in region with white underparts and conspicuous black shoulder patch. In flight has grey rump and white wingbar. Birds in partial br plumage show partial chestnut upperparts.* **JUV** *Like adult; sides of breast washed buff.* **STATUS** A rare non-br migrant from Arctic tundra arriving Sept/Oct, probably a passage migrant through KNP as most records made soon after the initial arrival in early summer. Usually in small active flocks on shorelines of man-made dams. **HABITAT** Most KNP records from bare shore lines. **FOOD** Shrimps, small molluscs, insects, and washed-up fish and insect scraps. **CALL** Flight call soft *twick, twick*. (**Drietoonstrandloper**)

LITTLE STINT *Calidris minuta* (274) 13 cm; 23 g
SEXES *Alike, ♀>♂. Easily the smallest wader in the Park. Short bill, and wings that project beyond tail are key features but short-legged appearance and hunched stance while feeding good characteristics.* **NON-BR** *Back mottled greyish brown. In flight, centre of tail and rump black, sides of rump white and with narrow white wingbar. Bill and legs black.* **BR** *On arrival, and just before departure, some birds can be seen in partial br dress: upperparts washed rufous (including upper breast) but throat whitish.* **JUV** *Centre of crown buffy contrasting with whitish supercilium; feathers of upperparts fringed rufous.* **STATUS** Common non-br Palearctic migrant; peak arrivals Oct–Nov, probably a passage migrant as numbers decline as summer progresses. Departs Feb–Mar. Typically in small flocks but also singly along shorelines. **HABITAT** Most mud-lined water bodies in Park. **FOOD** Mainly small aquatic and terrestrial invertebrates. Because of small size and short legs, feeds only in shallow margins and mudflats. **CALL** Short *stit-it* notes, creating twittering chorus in flocks. (**Kleinstrandloper**)

CURLEW SANDPIPER *Calidris ferruginea* (272) 19 m; 55 g
SEXES *Alike in non-br dress, ♀ bill slightly longer.* **NON-BR** *Only sandpiper in region with distinctive decurved bill. Pale eye-stripe and white rump.* **BR** *On arrival in region, or prior to departure, some birds may show signs of breeding plumage; chestnut flecking or blotching to plumage on head, neck and breast.* **JUV** *Upperparts browner and with scaled appearance.* **STATUS** Common non-br Palearctic migrant, main arrivals Aug–Nov, probably a passage migrant as numbers decline in KNP as the summer progresses. Departs Mar–Apr. Gregarious, usually in small flocks and often feeds with Little Stints. **HABITAT** All water bodies in KNP with muddy fringes. **FOOD** Mainly aquatic insects, small crabs and molluscs. **CALL** Contact call a trilled *chirrup*. (**Krombekstrandloper**)

RUFF (♀=REEVE) *Philomachus pugnax* (284) ♂ 28 cm, ♀ 22 cm; ♂ 170 g, ♀ 100 g
SEXES *Similar in non-br plumage, ♂ much larger than ♀.* **NON-BR** *Distinctive scaled or scalloped upperparts, thickset appearance and shortish, dark stout bill (pink to orange-red in br). Legs orange-coloured, though this sometimes varies in shades of yellow, pink, vermilion or brown. Birds with bright orange/red legs could be mistaken for Common Redshank (not recorded to date in Park), but lack streaked underparts, and in flight the white secondaries. Small proportion of ♂♂ whitish on head, upperparts and breast.* **JUV** *Similar to non-br adult but buffier.* **STATUS** Common non-br Palearctic migrant, main arrivals Aug/Sept, probably also a passage migrant as numbers decline as the summer progresses. Departs Feb–Mar. Gregarious, usually in small flocks. **HABITAT** Most inland water bodies. **FOOD** Aquatic and terrestrial insects and their larvae that incl beetles, flies, waterbugs, moths, ants, grasshoppers and crickets. Also spiders, molluscs, small frogs, fish and fish fry. **CALL** Mostly silent in winter quarters. (**Kemphaan**)

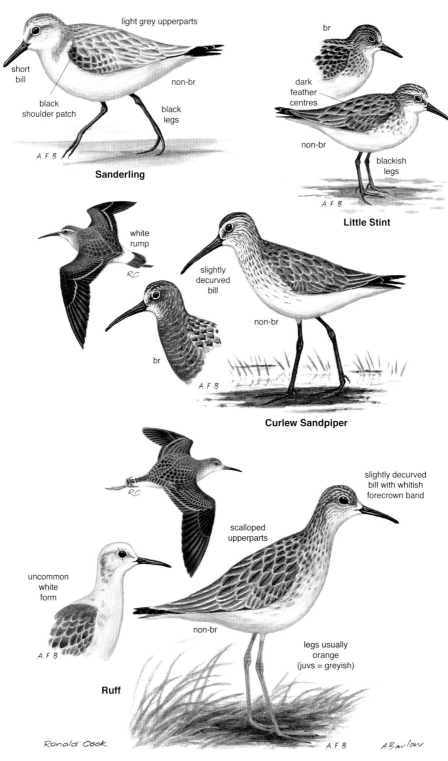

PLATE 31 Thick-knees, Stilt, Avocet p213

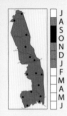

WATER THICK-KNEE *Burhinus vermiculatus* **(298) 40 cm; 303 g**
SEXES *Alike. Pale grey wing panel and accompanying narrow white bar disinguish it from similar Spotted Thick-knee. Upperparts streaked (not spotted). In flight, pale (grey) wing panel conspicuous.* **JUV** *More vermiculated upperparts and with buff freckling on wing coverts.* **STATUS** Locally common resident; undertakes local movements in response to water fluctuations. Crepuscular and nocturnal. Roosts most of the day, in summer usually under shaded tree or bush close to water, and in winter in the full sun often within metres of the water's edge. Usually in pairs, and in non-br season in loose flocks. Often seen on roads in KNP after dark. **HABITAT** Banks of rivers and dams. **FOOD** Wide range of prey, including crabs, aquatic beetles, frogs, insects, molluscs and small fish. **CALL** Loud far-carrying *ti-ti-ti-ti-tee-tee-teee-teeee*, fading towards the end. Although nocturnal, often calls in the evening and early morning. **BR** Monogamous. Nest a shallow scrape in sand, often among debris. (**Waterdikkop**) **ALT NAME** *Water Dikkop*

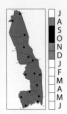

SPOTTED THICK-KNEE *Burhinus capensis* **(297) 43 cm; 465 g**
SEXES *Alike. Overall spotted appearance, and large yellow eyes distinctive. Likely only to be confused with Water Thick-knee, but lacks greyish wing panel, thin white wingbar, and has yellowish (not pale greenish) legs.* **JUV** *Similar to adult but more streaked above.* **STATUS** Fairly common resident but largely nocturnal and difficult to see during the day (most active on overcast days). Usually found singly or in pairs but gregarious in loose flocks of up to 40–50 birds in non-br season. Often seen on roads after dark. **HABITAT** Favours stony open savanna and grassland. **FOOD** Mostly insects, especially beetles and termites. **CALL** Mournful *ti-ti-ti-teeeteeeteee ti ti ti ...* **BR** Monogamous. Nest a simple scrape in almost bare ground, usually adjacent to debris or plants. Incubation by both adults, but mostly by ♀ during the day. (**Gewone Dikkop**) **ALT NAME** *Spotted Dikkop*

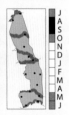

BLACK-WINGED STILT *Himantopus himantopus* **(295) 38 cm; 165 g**
SEXES *Similar, ♂>♀. Unmistakable black and white wader with the longest (red) legs relative to body size of any bird. During the non-br season, ♂ duskier on crown and nape. In flight, legs project way beyond tail.* **F** *Dusky crown and nape and duller (browner) upperparts.* **JUV** *Similar to ♀, legs greyish-pink.* **STATUS** Fairly common resident, nomadic and a partial migrant. Usually in small groups. **HABITAT** Most of the KNP's more permanent rivers and dams. **FOOD** Mainly larvae and adults of aquatic insects such as beetles, water-bugs and dragonflies. Also tadpoles, small fish and their eggs as well as amphibian eggs. Forages both during the day and at night (even moonless nights). **CALL** Sharp repetitive *kik-kik-kik ...* **BR** Monogamous. Nest usually surrounded by shallow water. (**Rooipootelsie**)

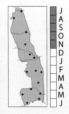

PIED AVOCET *Recurvirostra avosetta* **(294) 44 cm; 350 g**
SEXES *Alike, ♂>♀. Unmistakable with pied plumage and long thin upturned bill. In flight, pied plumage and pale blue-grey legs that project well beyond tail distinctive.* **JUV** *Like adult but black plumage replaced with dusky brown; white upperparts edged pale buffy brown.* **STATUS** A rare vagrant in KNP that may appear on any still water body. **HABITAT** Favours shallow pans, ephemeral wetlands and dams; rare along rivers. **FOOD** Highly specialised upcurved bill has internal lamellae which act as a filter during the bill scything action. This feeding technique enables the bird to filter small food items in the water or mud. Mostly aquatic insects and their larvae, crustaceans, and worms. Also occasionally molluscs, small fish and plant material. Forages both by day and at night. **CALL** Contact call fluty *clute, clute ...* **BR** Monogamous. Nest a scrape in damp or dry soil, frequently on an island not far from water's edge. (**Bontelsie**) **ALT NAME** *Avocet*

PLATE 32 Plovers p213

COMMON RINGED PLOVER *Charadrius hiaticula* **(245) 19 cm; 51 g**
SEXES *Alike. Broad white collar above single blackish breast band and orange legs distinguish it from Three-banded Plover. Non-br plumage duller.* **JUV** *Key features orange legs and dark upperparts.* **STATUS** A very rare visitor to the KNP. Probably occurs only as a passage migrant; could be seen on any of the KNP's rivers and dams. A migrant of almost circumpolar br origin. **HABITAT** Primarily coastal, especially muddy lagoons; inland water bodies used mainly during migration. **FOOD** Small crustaceans, molluscs, aquatic insects. **CALL** Contact call *tooLEE*. (**Ringnekstrandkiewiet**) **ALT NAME** *Ringed Plover*

KITTLITZ'S PLOVER *Charadrius pecuarius* **(248) 13 cm; 36 g**
SEXES *Similar.* **BR** *Combination of buffy breast, black line through eye and ear coverts, white neck collar, and greenish-black legs distinctive.* **NON-BR** *Black head markings usually brown (variable), eyebrow and collar buffy.* **JUV** *Like non-br adult, lacks black head markings; upperparts with scaled appearance.* **STATUS** Locally common resident, usually found on the margins of some of the KNP's man-made dams, but also on river sandbanks. **HABITAT** Favours muddy margins along open water bodies. **FOOD** Mostly terrestrial and aquatic invertebrates. Feeds both by day and at night. **CALL** Calls include *pipip* or *towhit*, and plaintive contact *tee peep* note. **BR** Monogamous. Nest a simple scrape lined with plant debris on bare ground. Eggs usually covered by nest debris and sand when incubating bird is disturbed. (**Geelborsstrandkiewiet**)

THREE-BANDED PLOVER *Charadrius tricollaris* **(249) 18 cm; 34 g**
SEXES *Alike. Only plover in region with double black breast band. Bright red eye-ring and red bill tip distinctive.* **JUV** *Duller than adult and lacks red eye-ring.* **STATUS** Very common resident occurring on most of the KNP's permanent and semi-permanent water bodies, mostly sedentary but undergoes large-scale movements in response to water levels. **HABITAT** Open shorelines at wide range of aquatic habitats. **FOOD** Mainly terrestrial and aquatic invertebrates and their larvae. Also crustaceans and small worms. Foot-trembles water by standing on one foot and vibrating toes of other foot on mud to attract prey to surface of water. **CALL** Plaintive *wik, wik*; flight call a shrill *weee-weet*. **BR** Monogamous. Nest a shallow scrape in river gravels or patch of pebbles, usually, but not necessarily, near water's edge. (**Driebandstrandkiewiet**)

WHITE-FRONTED PLOVER *Charadrius marginatus* **(246) 18 cm; 46 g**
SEXES *Similar. Overall light colour and very thin black eye-stripe distinguishes it from Kittlitz's Plover that has a broad bold eye-stripe and darker back. Underparts whitish, not buff. Legs greenish-grey;* ♀ *lacks dark crown band.* **JUV** *Lacks black on head.* **STATUS** A rather uncommon resident, largely sedentary and usually found in pairs. **HABITAT** Sandbanks on many of the KNP's larger rivers and verges of man-made dams. Seldom on muddy shorelines. **FOOD** Insects, crustaceans and worms. Foot-trembles by vibrating toes on substratum to disturb prey to surface. Forages by day or night. **CALL** *Wit, tirit-tirit* notes. **BR** Monogamous. Nest a scrape in sand. Covers eggs with sand prior to leaving nest when disturbed. (**Vaalstrandkiewiet**)

CASPIAN PLOVER *Charadrius asiaticus* **(252) 19 cm; 71 g**
SEXES *Alike in non-br plumage.* **NON-BR** *Long-legged plover with broad brown breast band, conspicuous broad light buffy-coloured eyebrow, and lores that are mostly white.* **JUV** *Upperparts with buffy scaled appearance.* **STATUS** Rare Palearctic migrant; more common in northern KNP. Numbers seem to have declined in KNP with closure of waterholes and resultant improvement of grass cover in previously overgrazed areas around such waterholes. Rarely remains in one locality for long. Arrives Sept–Nov and departs Feb–Mar. Usually in small flocks. **HABITAT** Mostly in short overgrazed dryland and sparsely vegetated plains, often far from water. Drinks frequently but does not remain long on water's edge. **FOOD** Small insects and larvae, especially beetles, termites and grasshoppers. **CALL** Sharp *tyup* notes. (**Asiatiese Strandkiewiet**)

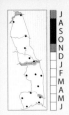

WHITE-CROWNED LAPWING *Vanellus albiceps* (259) 30 cm; 193 g
SEXES *Alike, ♂>♀. Best distinguishing features are black and white wing pattern, white breast and belly, and longer yellow wattles (lacking red at base). Forehead and central crown white, and long sharp carpal spurs usually visible.* **JUV** *Like adult but with less white on crown and throat, upperparts lightly mottled buff; spurs and wattles smaller.* **STATUS** Mostly an uncommon resident regarded as Near-threatened. In KNP restricted to certain major river systems with extensive sandbanks. Common on the Olifants and Luvuvhu/Limpopo rivers and on a stretch of the Sabie River, but paradoxically absent from the other rivers. Generally sedentary but moves when water levels become too low. **HABITAT** Sand and mudbanks of large rivers. **FOOD** Mainly insects; also frogs and even small fish. **CALL** Sharp high-pitched piping *peek, peeuw* or *kyip* and similar notes. **BR** Monogamous. Nest a shallow scrape in sand, usually not far from water, and vulnerable to unseasonal early flooding. (**Witkopkiewiet**) **ALT NAME** *White-crowned Plover*

BLACKSMITH LAPWING *Vanellus armatus* (258) 30 cm; 166 g
SEXES *Alike. Unlikely to be confused with any other species in region, with distinctive black, white and grey plumage, and black legs. ♂ has longer wing spurs than ♀: black with sharp ivory tip.* **JUV** *Black plumage browner, and with buff tips to feathers.* **STATUS** Common resident, mostly sedentary. Usually in pairs or small family groups in non-br season. **HABITAT** Shorelines of most water systems, short grassland, wetlands and mudflats. **FOOD** Aquatic and terrestrial invertebrates such as molluscs, crustaceans, worms and insects. Sometimes forages on water's edge by foot-trembling. **CALL** Piercing metallic *klink klink klink ...* like a blacksmith's hammer on an anvil. **BR** Monogamous. Nest a shallow scrape lined with vegetation, stones or mud flakes, usually close to water. (**Bontkiewiet**) **ALT NAME** *Blacksmith Plover*

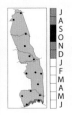

AFRICAN WATTLED LAPWING *Vanellus senegallus* (260) 35 cm; 254 g
SEXES *Alike. Largest lapwing in region. Red bases to wattles, streaked neck, grey breast, lack of white collar across hindneck, and in flight much less white in wings all features that distinguish it from White-crowned Lapwing. Black carpal spurs usually not visible.* **JUV** *Less white on forehead, small wattles, greyish bill and pale yellow legs.* **STATUS** Fairly rare resident, with some local migratory and nomadic movements recorded after br. Usually in pairs. Common only along the Crocodile River, but a rare vagrant to the rest of KNP. **HABITAT** Waterlogged or moist short grassland, floodplains, and burnt grass. **FOOD** Mainly insects such as grasshoppers, beetles, caterpillars, crickets and termites. **CALL** High-pitched *keep-keep-keep ...* , rising in pitch. Often calls at night during br season **BR** Monogamous. Nest scrape usually placed on bare ground or in short grass. (**Lelkiewiet**) **ALT NAME** *Wattled Plover*

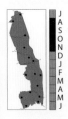

CROWNED LAPWING *Vanellus coronatus* (255) 31 m; 185 g
SEXES *Alike, ♂>♀. Sandy brown upperparts, bright red bill and legs, and distinctive head pattern unlike any other* Vanellus *plover. Legs and bill duller, more pinkish-red in non-br season.* **JUV** *Forehead buff-brown, crown band buffy. Upperparts buff-fringed.* **STATUS** Fairly common resident and mostly sedentary. Numbers apparently decline in wetter years when short grass habitats are limited; non-br flocks 10–40 birds. **HABITAT** Short grassland and burnt or other bare open areas. **FOOD** Mainly termites; also other invertebrates. **CALL** Strident *shreeek, shreeek* calls, accelerating to excited *kreeip-kreeip-kreeip ...* notes in flight. **BR** Monogamous. Nest a shallow scrape lined with dry grass, small stones and dried dung. (**Kroonkiewiet**) **ALT NAME** *Crowned Plover*

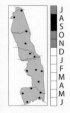

SENEGAL LAPWING *Vanellus lugubris* (256) 24 cm; 115 g
SEXES *Alike. The only lapwing in the Park with brown back, white forehead and blackish legs.* **JUV** *Head and breast pattern less well defined, forehead buffier and feathers on upperparts buff-tipped.* **STATUS** Fairly scarce resident and intra-African migrant mostly recorded in the southern half of KNP. Nomadic, undertaking local movements in search of short or burnt grass. Gregarious, usually in flocks of 5–10 birds; often active at night. **HABITAT** Favours dry open *Acacia* savanna with short or burnt grass. **FOOD** Mainly termites, also other terrestrial invertebrates. **CALL** Clear clarinet-like *chi-whoo* or *chi-hi-whoo ...* , lower-pitched on the last note; often active and calling high on the wing on moonlit nights. **BR** Monogamous. Nest a shallow scrape lined with grass stubble, bits of dung and pebbles. (**Kleinswartvlerkkiewiet**) **ALT NAME** *Lesser Black-winged Plover*

PLATE 34 Coursers, Pratincole, Terns p214

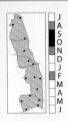

BRONZE-WINGED COURSER *Rhinoptilus chalcopterus* (303) 27 cm; 155 g
SEXES *Alike. Largest African courser. Key distinguishing features are narrow blackish breast band, bold facial markings and broad brown neck band. Outer 2 primary feathers uniquely tipped with metallic bronze sheen. Legs purplish-red.* **JUV** *Mottled buff on underparts and with less distinct facial markings and narrower breast band.* **STATUS** Usually an uncommon resident, but numbers apparently supplemented by an influx of birds in Nov that depart northwards again by May/June. Usually solitary or in pairs. Nocturnal so difficult to see during daylight hours. **HABITAT** Mopane, *Acacia* and broad-leaved woodland. **FOOD** Insects. **CALL** Mournful drawn-out *ji-ku-it*. **BR** Monogamous. Nest a shallow scrape. (**Bronsvlerkdrawwertjie**)

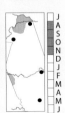

THREE-BANDED COURSER *Rhinoptilus cinctus* (302) 26 cm; 130 g
SEXES *Alike. Distinctive and boldly marked head pattern and three conspicuous neck and chest bands should easily distinguish it from Bronze-winged Courser. Other diagnostic features are yellowish tail (tipped black), and pale dusky yellow legs.* **JUV** *Resembles adult but duller; chestnut band faint.* **STATUS** Very uncommon resident, mostly nocturnal, usually only recorded in the Pafuri area, one record from Shangoni (where the Shingwedzi River enters KNP). **HABITAT** Favours dry woodland areas, especially Mopane. **FOOD** Insectivorous. **CALL** Series of high-pitched piping notes. **BR** Probably monogamous; 1–2 eggs usually laid in a shallow scrape. (**Driebanddrawwertjie**)

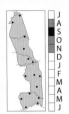

COLLARED PRATINCOLE *Glareola pratincola* (304) 25 cm; 75 g
SEXES *Alike. Appears as a large swallow in flight, showing white tips to secondaries and chestnut underwing coverts.* **BR** *Appears mainly in br dress in KNP; buff breast narrowly outlined with thin neat black line.* **NON-BR** *Thin black collar almost absent; throat whitish-cream.* **JUV** *Similar to adult in non-br dress, without collar.* **STATUS** Uncommon br intra-African migrant present Jul–Feb. Has been recorded throughout KNP but mostly on the eastern basaltic substrates; fairly regular at the Matambeni bird hide (Letaba). Regarded as Near-threatened in s Africa. **HABITAT** Mostly on floodplains and at edges of dams. **FOOD** Mainly insects taken in flight. **CALL** Various shrill harsh chattering notes. **BR** Monogamous and colonial br. Nests are shallow scrapes and loosely scattered. (**Rooivlerksprinkaanvoël**) **ALT NAME** *Red-winged Pratincole*

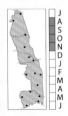

TEMMINCK'S COURSER *Cursorius temminckii* (300) 20 cm; 70 g
SEXES *Alike. Smallest African courser. Has broad black line extending behind the eye, and a rusty-brown crown. Trailing flight feathers completely black.* **JUV** *Upperparts mottled; breast pale reddish and belly blackish.* **STATUS** KNP population partially resident with some birds nomadic in search of favorable habitat, and others thought to be intra-African migrants. Nowhere common but probably the commonest courser in the region. Usually in small flocks; a diurnal species. **HABITAT** Short, open and burnt grasslands, especially in Mopane regions. **FOOD** Mostly insects. **CALL** High-pitched staccato *err-err-errrr* call likened to a squeaky gate. **BR** Monogamous; 2 eggs usually laid on bare open ground and incubated by both sexes. (**Trekdrawwertjie**)

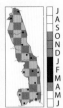

WHISKERED TERN *Chlidonias hybrida* (338) 24 cm; 100 g
SEXES *Alike.* **BR** *Dark cap and lead grey underparts distinctive, confusion only likely with White-winged Tern in non-br dress (only two terns in the Park).* **NON-BR** *Black line from eye to nape does not extend below eye (ear-patch) as in White-winged Tern.* **JUV** *Mottled above but similar to non-br adult.* **STATUS** Uncommon resident in the region, but nomadic in KNP, its occurrence dependent on above-average rainfall. **HABITAT** Favours man-made wetlands in KNP. **FOOD** Small fish, invertebrates and small frogs. **CALL** Harsh chattering notes. **BR** Monogamous and usually colonial. Nest built on floating vegetation, no records for KNP. (**Witbaardsterretjie**)

WHITE-WINGED TERN *Chlidonias leucopterus* (339) 21 cm; 54 g
SEXES *Alike. Recorded mainly in non-br dress in the summer months so confusion unlikely as most Whiskered Terns in full br dress during this period. Smaller than Whiskered Tern and with shorter bill.* **BR** *Occasionally recorded in partial breeding dress shortly after arrival and just prior to northward migration at the end of summer. Body and underwings partially black.* **NON-BR** *Black ear-patch that extends behind eye and over white crown gives a black 'headphone' appearance.* **JUV** *Upperparts mottled brownish-black; similar to non-br adult.* **STATUS** Uncommon non-br Palearctic migrant, present Sept–Apr, usually in flocks. More common than Whiskered Tern. **HABITAT** Man-made water bodies and pans. **FOOD** Fish, invertebrates and small crabs. **CALL** Hoarse *kirsch* or *kreek*. (**Witvlerksterretjie**)

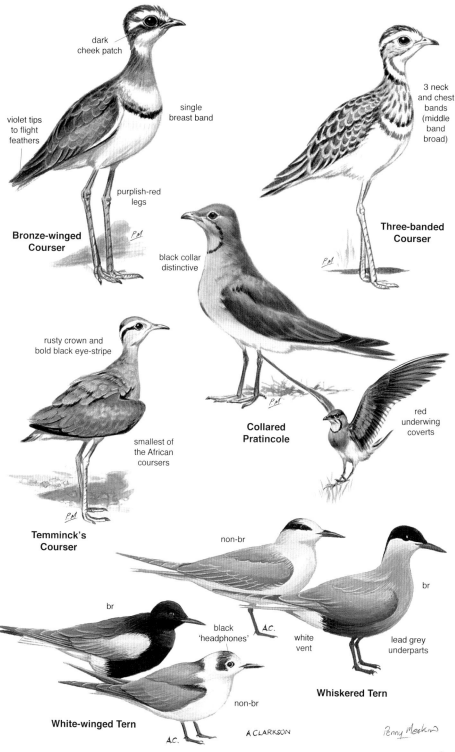

EUROPEAN HONEY-BUZZARD *Pernis apivorus* (130) 55 cm; 750 g
SEXES *Similar. Confusion most likely with Steppe Buzzard, but pigeon-like appearance with small round yellowish eyes and longer tail distinctive. In flight key features are broad terminal tail bar and dark carpal patches; 2 dark bars at base of tail sometimes difficult to see.* **M** *Head greyer and eyes orange-yellow.* **F** *Browner (less grey) than ♂ and with yellowish eyes.* **JUV** *Variable; lacks distinct tail band; eyes brown.* **STATUS** Rare non-br Eurasian migrant, present Nov–Mar/Apr. **HABITAT** Mostly well-developed riparian woodland. **FOOD** In s Africa, predominantly wasps, bees and their broods. **CALL** Silent in s Africa. (**Wespedief**) **ALT NAME** Honey Buzzard

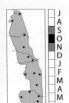

AFRICAN CUCKOO HAWK *Aviceda cuculoides* (128) 40 cm; 250 g
SEXES *Similar, ♀ slightly larger and less boldly marked. At a distance may resemble African Goshawk but tawny brown barring on underparts broader and bolder, flight slower. Small crest not always visible.* **M** *Dark eyes.* **F** *Eyes yellow; barring broader and paler.* **JUV** *Dark brown above; underparts white with red-brown tear-shaped spots.* **STATUS** Rare br resident. **HABITAT** Riparian bush and open woodland. **FOOD** Mostly chameleons and large (often green) insects. **CALL** Mostly 4-note *tickey-to-you* call; also plaintive *pleou*. **BR** Monogamous. Small flimsy nest usually placed high in tall tree. (**Koekoekvalk**) **ALT NAME** *Cuckoo Hawk*

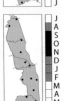

BAT HAWK *Macheiramphus alcinus* (129) 45 cm; 625 g
SEXES *Alike. Dark sooty brown with varying amounts of white below, pale eyelids and false white eye spots on nape. In flight has long pointed wings (key feature) and shortish square tail. At rest, wing tips protrude past tail. Often confused with smaller Eurasian Hobby which also hunts bats crepuscularly.* **JUV** *Underparts mostly white with brown chest band.* **STATUS** Rare resident; regarded as Near-threatened in s Africa. **HABITAT** Recorded in various woodland habitats, especially near large bat populations and their roosting sites in restcamps and large bridges. **FOOD** Mostly insectivorous bats taken between sunset and last light; also birds such as swifts, swallows and martins. Prey swallowed whole in flight. **CALL** High-pitched *kik-kik-kik*. **BR** Monogamous. Large platform nest usually placed on lateral branch of tall tree. In KNP br recorded once only in a Baobab tree in Olifants Gorge. (**Vlermuisvalk**)

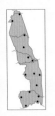

STEPPE BUZZARD *Buteo vulpinus* (149) 48 cm; ♂ 640 g, ♀ 820 g
SEXES *Alike, ♀ considerably larger than ♂. Thickset appearance, upright stance and habit of hunting from top of exposed conspicuous perch helps aid identification. Highly variable plumage, with 3 distinct colour phases (and variations within these groups): rufous, grey-brown and dark brown. Intensity of barring on belly, whitish breast patch and blotching on underparts all so variable that almost no 2 birds look alike. Most birds show pale to rufous barring on belly, a distinct whitish lower breast patch and pale rufous upper tail (whitish at base).* **JUV** *Streaked on underparts, not barred like adult.* **STATUS** A fairly uncommon Palearctic migrant; most arrive late Oct/Nov and depart Feb–Apr. **HABITAT** Open woodland, and grassland. **FOOD** Mainly insects, also rodents, reptiles and occasionally birds. **CALL** Mostly silent in Africa. (**Bruinjakkalsvoël**)

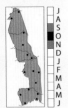

YELLOW-BILLED KITE *Milvus aegyptius* (126b) 55 cm; 674 g
SEXES *Alike, ♀>♂. Large brown kite with yellow bill, forked tail and, in flight, angled wings.* **JUV** *Bill black, eyes dark, with blackish facial mask and pale streaks on head, neck and underparts.* **STATUS** Common and widespread intra-African br migrant, present Jul–Feb/Mar. Until recently regarded as a subspecies of Black Kite (a non-br Palearctic migrant). Fairly common especially around restcamps. **HABITAT** Almost all habitats. **FOOD** Wide range of small vertebrates and insects, much of it scavenged. Also feeds at carcasses. **CALL** Plaintive *kleeeeu-errrr* or *kleeeu-ki-ki*. **BR** Monogamous; nest usually in thickly foliaged tree. (**Geelbekwou**)

BLACK KITE *Milvus migrans* (126) 55 cm; 780 g
SEXES *Alike, ♀>♂. Until recently Black and Yellow-billed kites were viewed as conspecific but current molecular evidence demonstrates that they are separate species. Distinguished from Yellow-billed Kite by pale greyish head, dark bill, pale eyes, less deeply forked tail and, in flight, more angled wings.* **JUV** *Bill black, eyes dark, blackish facial mask and pale streaks on head and underparts. Kites recorded with black bills during mid-summer are likely to be juv Yellow-billed Kites and not Black Kites.* **STATUS** Uncommon non-br migrant, present Oct/Nov–Mar. Recorded throughout KNP, but nowhere common. **HABITAT** Almost all habitats. **FOOD** Small vertebrates and insects, much of it scavenged, also at road kills. **CALL** Similar to Yellow-billed Kite but largely silent in s Africa. **BR** Europe, n-w Africa and Asia. (**Swartwou**)

OSPREY *Pandion haliaetus* (170) 59 cm; 1,49 kg
SEXES *Similar, ♀ heavier and with bolder breast streaking. Long wings and tail, gull-like flight action, white underparts, yellow eyes and broad blackish band through eye distinguish it from juv African Fish-Eagle (which lacks black face mask, and has broad wings and short tail).* **JUV** *Paler than adult, with white edges to dark upperparts.* **STATUS** Rare non-br Palearctic migrant; arrives Aug/Sept, departs by May with some overwinter records (mostly juvs). **HABITAT** Aquatic habitats throughout KNP, especially along rivers and man-made dams. **FOOD** Almost entirely fish. Recorded taking fish up to 3 kg but most prey 200–500 g. **CALL** Melodious whistle; usually silent in s Africa. **BR** Monogamous; no br records for KNP and no recent breeding records for s Africa, with only 2 historical nesting attempts. (**Visvalk**)

BLACK-SHOULDERED KITE *Elanus caeruleus* (127) 30 cm; 248 g
SEXES *Alike, ♀>♂. Distinctive smallish grey and white raptor with black shoulder patch and bright red eyes. In flight, black primaries and white underparts distinctive. Frequently hovers in search of prey, and tail often wagged up and down apparently as a threat to conspecifics.* **JUV** *Eyes yellow, turning orange, then red. Mottled brownish above and with rusty neck and breast.*
STATUS Usually uncommon, but numbers increase in KNP during rodent irruptions which can take place in wet years that follow a drought period. Nomadic in response to prey availability. May be absent in drought years. **HABITAT** Mainly in open woodland and grassland habitats. **FOOD** Primarily diurnal rodents, to a lesser extent birds, lizards and insects. **CALL** High-pitched whistling *peee-oo*. **BR** Monogamous. Nest usually placed just below tree canopy. (**Blouvalk**)

AFRICAN FISH-EAGLE *Haliaeetus vocifer* (148) 68 cm; 2,5 kg
SEXES *Similar, ♀>♂ and has broader white breast patch. Plumage distinctive, unlikely to be confused with any other species except Palm-nut Vulture, facial skin yellow (not pink or pink-orange) and shoulder patch dark rufous (not whitish).* **JUV** *Takes 4–5 years to reach adult plumage; scruffy brown and white plumage easy to confuse with Osprey. Dark brown eyes and lack of eye-stripe distinguish it at all ages from Osprey.* **STATUS** Locally common resident, largely sedentary and highly territorial, year-round. **HABITAT** Usually associated with the larger rivers and man-made water bodies. **FOOD** Predominantly fish; capable of lifting prey up to 2 kg from the water. Known, however, to kill prey over 3 kg but large fish fed on at water's edge. Specialises on fish caught just below water surface in graceful sweeping dive, with feet thrown forward to grasp fish. Also preys on large waterside birds, reptiles, and small mammals. Known to steal fish from other birds such as herons, storks and even kingfishers! **CALL** Loud ringing *weee-ah, hyo-hyo-hyo*, often in duet; ♀ voice deeper. Head is flung back while calling, both while perched and in flight. **BR** Monogamous. Breeds during winter when water levels are low and prey most vulnerable. Large stick nest usually placed in tall tree close to water. (**Visarend**)

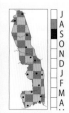

PALM-NUT VULTURE *Gypohierax angolensis* (147) 60 cm; 1,6 kg
SEXES *Alike, ♀>♂ and with individual variation in plumage. Most likely to be confused with African Fish-Eagle but has conspicuous pink or pink-orange facial skin and white wing coverts (pale orange-brown in br season), giving it an overall whiter appearance.* **JUV** *Dark brown with yellow facial skin; lacks the barred tails of juv snake-eagles.* **STATUS** Rare vagrant to KNP recorded only at a few widespread localities. Very nomadic (both adults and imm birds) that wander widely sometimes throughout s Africa, especially during the non-br season. **HABITAT** Usually linked to presence of Kosi Palms *Raphia australis* on which it feeds and breeds, but these do not occur in KNP which accounts for its rarity. **FOOD** Mostly flesh from palm fruits, but in the Park, carrion, crabs, frogs and fish. **CALL** Usually silent away from br sites; gives grunting *grog, grog-grog* notes. **BR** Monogamous; stick nest placed high in palm fronds. (**Witaasvoël**)

PLATE 37 Vultures pp214–5

HOODED VULTURE *Necrosyrtes monachus* **(121) 70 cm; 2,1 kg**
SEXES *Alike. The smallest of the vultures in the region; very slender bill and small head distinctive.*
JUV *Darker plumage with dark brown down at back of head.* **STATUS** Fairly common resident, regarded as Vulnerable in S Africa. Usually found singly but may gather in numbers at carcasses. **HABITAT** Woodland throughout the Park. **FOOD** Primarily a scavenger at carcasses, its slender bill enabling it to extract scraps not accessible by the larger vultures. If first to arrive at unopened prey, feeds on eyes, but soon displaced by the arrival of larger vultures. Also feeds on dung of larger predators such as lions and sometimes on termite alates. **CALL** Usually silent. **BR** Monogamous. Large stick platform usually placed under canopy of well-foliaged tree. (**Monnikaasvoël**)

WHITE-BACKED VULTURE *Gyps africanus* **(123) 95 cm; 5,5 kg**
SEXES *Alike. Smaller and darker than Cape Vulture, though plumage tends to become paler with age. In flight, key feature is the primary and secondary feathers that are uniform dark brown, and contrast with much paler wing coverts. Characteristic off-white rump and lower back visible only in flight or when wings are spread. Neck dark grey, covered sparsely with down. Face dark grey or blackish, eyes dark brown.* **JUV** *Darker than adult, plumage with streaked appearance. Lower back and rump spotted dark brown; attains adult plumage after sixth year.* **STATUS** Most common vulture in KNP, gathering in considerable numbers at large carcasses. Regarded as Vulnerable in s Africa owing to decrease in numbers. Resident, with long-distance movements (especially juvs). **HABITAT** Savanna woodland and bushveld. **FOOD** Primarily a scavenger at carcasses. **CALL** Usually silent; squeals, hisses and grunts at carcasses. **BR** Monogamous. Stick platform nest usually placed on crown of thorn tree. (**Witrugaasvoël**)

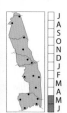

CAPE VULTURE *Gyps coprotheres* **(122) 101 cm; 8,5 kg**
SEXES *Alike. Paler than White-backed Vulture, eyes yellow, face bluish and neck thicker. Key feature in flight is darker primaries that contrast with lighter-coloured, black-tipped secondaries (unlike White-backed Vulture with uniform flight feathers).* **JUV** *Slightly darker (warmer brown) than adult, with dark eyes, neck and throat patches pink to magenta; adult plumage reached after 6 years.* **STATUS** Uncommon near-endemic resident, numbers limited in KNP as there are no suitable cliffs for nesting. KNP birds mainly from Manoutsa br colony on the Drakensberg escarpment to the west of KNP (Hoedspruit). Listed as Vulnerable in S Africa. **HABITAT** Linked to cliff br sites in mountainous areas but ranges widely in search of food. **FOOD** Aggressive scavenger at carcasses. **CALL** Cackles, grunts and hisses at carcasses. **BR** Monogamous and colonial. Platform of sticks and grass placed high on cliff ledges. No breeding records for KNP. (**Kransaasvoël**)

LAPPET-FACED VULTURE *Aegypius tracheliotos* **(124) 102 cm; 6,7 kg**
SEXES *Alike. Largest and most dominant of the vultures. Main features apart from large size dark plumage, bare reddish head with conspicuous skin folds, and enormous yellowish-horn bill with blue-grey base. In flight, underwings mostly dark with narrow white bar.* **JUV** *Browner and paler than adult, bill horn-coloured and leggings initially brown. Age to first br at least 6 years.* **STATUS** Uncommon resident, listed as Vulnerable. Extensive movements recorded in non-br birds. **HABITAT** Favours semi-arid open woodland. **FOOD** Feeds on carcasses, tackling skin, tendons and ligaments too tough for other vultures; seldom takes meat. Also stranded fish, and terrapin nests. **CALL** Generally silent. **BR** Monogamous. Large stick nest placed on crown of isolated flat-topped tree. (**Swartaasvoël**)

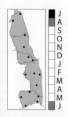

WHITE-HEADED VULTURE *Aegypius occipitalis* **(125) 85 cm; 4,2 kg**
SEXES *Similar, ♀>♂. In flight, ♀ secondaries conspicuously white, ♂ greyish-black. Very distinctive vulture with dark plumage, white crown, pink facial skin, red bill and white leggings. Pale tips to greater wing coverts give appearance of a pale wing bar.* **JUV** *Lacks white plumage, woolly cap brown, face and neck duller. Overall plumage dark brown.* **STATUS** Uncommon resident, locally nomadic, and listed as Vulnerable owing to population declines elsewhere. **HABITAT** Woodland throughout the Park. **FOOD** Scavenger at carcasses; also kills small mammals such as porcupines, Bat-eared Fox, African Wild Cat, genets, hyraxes, hares and monkeys. **CALL** Usually silent; shrill whistle given in aggression. **BR** Monogamous and probably pairs for life. Stick nest placed on top of emergent tree such as Baobab or large *Acacia*. (**Witkopaasvoël**)

PLATE 38 Snake-Eagles, Bateleur, Harriers p215

BROWN SNAKE-EAGLE *Circaetus cinereus* **(142) 74 cm; 2,0 kg**
SEXES Alike. The largest snake-eagle. Dark brown plumage, large yellow owl-like eyes, upright posture and round head very distinctive. Legs scaled and featherless, cere light grey. In flight, all-brown except for pale or silvery-grey flight feathers (diagnostic); tail with 3 narrow white (sometimes indistinct) bars. **JUV** Similar to adult; overall darkish brown, but highly variable, some flecked white above, others mottled white below; tail barred. **STATUS** Locally common resident, may be nomadic in non-br season. **HABITAT** Most dry woodland habitats. **FOOD** Mainly snakes (including large poisonous species), also monitor lizards, lizards and chameleons. **CALL** Hoarse *kok-kok-kok-kaw* ... , usually given in flight. **BR** Monogamous. Stick nest usually placed on top of flat-topped tree such as an *Acacia*. (**Bruinslangarend**)

BLACK-CHESTED SNAKE-EAGLE *Circaetus pectoralis* **(143) 66 cm; 1,5 kg**
SEXES Alike. Smaller than Martial Eagle with unspotted white underparts, scaled (not feathered) lower legs, round head and conspicuous large yellow eyes. In flight easily differentiated by white (not brown) underwings with narrow black bars across secondaries. **JUV** Upperparts variable from pale to darkish brown, sometimes flecked with white. Underparts more rufous brown than juv Brown Snake-Eagle, becoming whiter with age. In flight, underwing coverts brown, flight feathers pale with 3 distinct narrow bars. Eyes yellow. **STATUS** Uncommon resident, nomadic in non-br season. **HABITAT** Occurs throughout Park, often seen on Caborra Bassa powerlines near Punda Maria. **FOOD** Mainly snakes, also lizards, water monitors, frogs, rodents and insects. **CALL** Ringing *kwo-kwo-kwo* ... *kweeu-kweeu* ... **BR** Monogamous. Nest platform placed on top of tree or pylon. (**Swartborsslangarend**) **ALT NAME** Black-breasted Snake Eagle

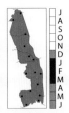

BATELEUR *Terathopius ecaudatus* **(146) 63 cm; 2,25 kg**
SEXES Similar. Very distinctive with red facial skin, grey shoulder patches and long wings that extend past the very short tail. **M** When perched, flight feathers appear black; in flight, black trailing edge to white underwing considerably broader than ♀. **F** When perched, primaries show pale grey wing panel (below grey-brown shoulder patch). In flight underwing white with narrow black trailing edge. **CREAM MORPH** Uncommon to rare. Back, rump and tail creamy buff. **JUV** Uniform brown plumage; cere and facial skin grey-green, and short tail distinctive. **STATUS** Common and widespread in KNP, regarded as Vulnerable in s Africa. Resident and largely sedentary. **HABITAT** Open and closed woodland. **FOOD** Mainly small mammals, birds and reptiles. Scavenges carrion. **CALL** Generally silent; gives guttural *kow-aw* or *ko-waaaa*. **BR** Monogamous. Stick platform nest placed below canopy in large tree. (**Berghaan**)

MONTAGU'S HARRIER *Circus pygargus* **(166) 44 cm; ♂ 260 g, ♀ 370 g**
SEXES ♀ larger than ♂. Flight pattern more buoyant and slower than Pallid Harrier. **M** In flight shows more black in primaries than ♂ Pallid Harrier, has black wingbar, and is streaked rufous on underwing coverts, belly, flanks and thighs. **F** Rump white, hence referred to as 'ringtails'. Best distinguished from similar ♀ Pallid Harrier by smaller dark face patch that is confined more to rear of eye. White neck collar indistinct or absent. **JUV** Similar to adult ♀, but upperparts darker, and with rufous underparts (unstreaked) and underwing coverts; eyes brown. **STATUS** Uncommon non-br Palearctic migrant, present Oct–Apr. More common than Pallid Harrier. **HABITAT** Almost exclusively limited to the eastern grasslands but also associated with open pans and floodplains. **FOOD** Insects and mice. **CALL** Silent in Africa. (**Blouvleivalk**)

PALLID HARRIER *Circus macrourus* **(167) 44 cm; ♂ 310 g, ♀ 440 g**
SEXES ♀ larger than the ♂. **M** Upperparts pale grey, whitish below. In flight from below, the 4/5 black outer primaries form a wedge at tip of white underwing. From above, pale grey black-tipped upperwing lacks black wingbar of ♂ Montagu's Harrier. **F** Larger dark face patch than ♀ Montagu's Harrier that extends forward below eye to base of bill; also more distinct thin white collar below dark face patch. **JUV** Like adult ♀, but with unstreaked rufous underparts; eyes brown. **STATUS** Rare non-br Palearctic migrant, present Nov–Apr. Listed globally as Near-threatened. **HABITAT** Almost exclusively limited to the eastern grasslands, but also associated with open pans and floodplains. **FOOD** Mainly insects, also small mammals, birds and reptiles. **CALL** Silent in Africa. (**Witborsvleivalk**)

PLATE 39 Buzzard, Goshawks, Shikra, Sparrowhawks p215

LIZARD BUZZARD *Kaupifalco monogrammicus* **(154) 36 cm; 273 g**
SEXES *Alike, ♀>♂. Most similar to Gabar Goshawk but stockier in appearance, and black throat stripe through white chin diagnostic. Single white tail band conspicuous in flight (also uncommon double tail band recorded in some birds).* **JUV** *Upperparts have pale brown wash.* **STATUS** Uncommon to locally common resident, largely sedentary and usually found singly. **HABITAT** Favours broad-leaved woodland, thus more common in vegetation on the western granitic soils. **FOOD** Mainly lizards, small snakes, rodents, frogs and invertebrates. **CALL** Main call loud and distinctive *kli-ooo-klukluklu.* **BR** Monogamous. Stick nest built by both adults, usually 6–10 m high in main fork of tree. (**Akkedisvalk**)

GABAR GOSHAWK *Melierax gabar* **(161) 32 cm; 164 g**
SEXES *Alike, ♀ considerably larger than ♂.* **PALE MORPH** *Plain dove-grey upperparts, throat and breast distinguish it from other similar-sized raptors.* **DARK MORPH** *Very uncommon; entirely black except for barred white flight feathers and 3 narrow white tail bars.* **JUV** *Upperparts brown, head and breast streaked brown, belly barred and rump white; eyes yellow.* **STATUS** Fairly common resident; largely sedentary. **HABITAT** Acacia and broad-leaved woodland. **FOOD** Mainly birds (up to young francolin size); also rodents, bats, reptiles and insects. **CALL** Piping *twi-twi-twi-twi* ... and other similar notes. **BR** Monogamous. Stick nest usually has social spider colony (*Stegodyphus* sp.) incorporated that surrounds nest with web. (**Kleinsingvalk**)

AFRICAN GOSHAWK *Accipiter tachiro* **(160) ♂ 38 cm, ♀ 45 cm; ♂ 220 g, ♀ 360 g**
SEXES *♀ considerably larger than ♂. Larger than Little Sparrowhawk with greenish-grey (not yellow) cere, and lack of white on rump.* **M** *Upperparts grey, tail with white spots in centre of 2–3 pale grey bars.* **F** *Upperparts brownish; tail lacks central white tail spots on 3 broad dark bars.* **JUV** *Brown upperparts like ♀, white underparts heavily blotched dark brown. Median throat stripe and pale eyebrow distinguishes it from smaller juv Little Sparrowhawk.* **STATUS** Locally common resident. **HABITAT** Well-developed woodlands. **FOOD** Mainly birds, also bats, reptiles and insects. **CALL** Repetitive *tchick, tchick, tchick* ... at 2–3 sec intervals (mostly in flight). **BR** Monogamous. Nest built in well-foliaged tree. (**Afrikaanse Sperwer**)

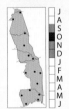

SHIKRA *Accipiter badius* **(159) 29 cm; 133 g**
SEXES *Similar, ♀ larger than ♂ and upperparts tinged browner. Key features are the bright red (♂) or deep orange (♀) eyes and all-grey central tail feathers. The 4 light grey tail bars difficult to see unless in flight.* **JUV** *Eyes yellow. Upperparts dark brown, white below with rusty-brown spots and barring on breast and belly.* **STATUS** Fairly common resident in the drier regions. **HABITAT** Favours open woodland with some tall trees. **FOOD** Mainly lizards (unique among the accipiters); also insects, small birds, bats, rodents and frogs. **CALL** Main call a loud piercing *kli-vit* ... **BR** Monogamous. Small platform nest in tall tree. (**Gebande Sperwer**) **ALT NAME** Little Banded Goshawk

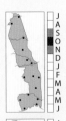

LITTLE SPARROWHAWK *Accipiter minullus* **(157) 25 cm; ♂ 74 g, ♀ 106 g**
SEXES *Similar, ♀ larger and slightly browner on upperparts than ♂. Laughing Dove size. Similar to larger ♂ African Goshawk but with yellow cere, and, in flight, white rump (above 2 central white eye spots on tail).* **JUV** *Upperparts dark brown, white underparts heavily blotched brown. Lacks dark throat stripe juv African Goshawk.* **STATUS** Generally an uncommon resident; sedentary. **HABITAT** Forests, dense riverine woodland and wooded areas. **FOOD** Mainly small birds up to dove size, also small rodents, bats, lizards and insects. **CALL** Rapid high-pitched piping notes *ke-ke-ke* ... , ♂ lower pitched. **BR** Monogamous. Small stick platform built in tall tree. (**Kleinsperwer**)

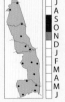

OVAMBO SPARROWHAWK *Accipiter ovampensis* **(156) 36 cm; ♂ 140 g, ♀ 260 g**
SEXES *Alike, ♀ considerably larger than ♂. Key features separating it from smaller Shikra are dark (not red or orange) eyes and yellow-orange to pale red (not yellow) cere and legs. Tail pattern unique: 3 pale grey bars each with white feather shafts. Rare melanistic form occurs.* **JUV** *Two colour forms: underparts either white or rufous, both, however, streaked and barred; upperparts brown (including rump), and both have light eyebrow and dark ear-patch.* **STATUS** Very uncommon resident in KNP. **HABITAT** Tall open woodland. **FOOD** Almost entirely birds. **CALL** Fast *kwee-kwee-kwee* ... ascending in tone. **BR** Monogamous. Nests in tall trees. (**Ovambosperwer**)

PLATE 40 Sparrowhawk, Chanting Goshawk, Hawk-Eagles p215

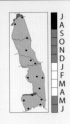

BLACK SPARROWHAWK *Accipiter melanoleucus* (158) 52 cm; ♂ 540 g, ♀ 900 g
SEXES *Alike, ♀ considerably larger than ♂. Upperparts black, underparts vary from white (most common) to completely black with small white throat patch. Very distinctive and unlikely to be confused with any other raptor in region except African and Ayres's hawk-eagles. Smaller, however, and lacks streaking on underparts (adult) and has unfeathered legs.* **JUV** *Two colour phases, rufous (most common phase) or white. Both have dark brown upperparts; underparts streaked (more heavily in ♀).* **STATUS** Very rare br resident in KNP, secretive and seldom seen. Recorded mainly in well-developed riparian woodlands on the Sabie, Luvuvhu and Limpopo rivers and also at scattered localities elsewhere in suitable well-wooded habitats. Sedentary. **HABITAT** Riparian forests and well-developed woodland. **FOOD** Predominantly birds, especially pigeons and doves. **CALL** Long-drawn-out *kweeeee-uw*, or *chip-chip* and *ky-ip* notes. **BR** Monogamous. Large stick nest most frequently placed in tall trees. (**Swartsperwer**)

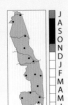

DARK CHANTING GOSHAWK *Melierax metabates* (163) 45 cm; 682 g
SEXES *Alike, ♀>♂. Considerably larger than Gabar Goshawk (which has a white rump). Upright stance very characteristic and with red cere and legs.* **JUV** *Upperparts brown-grey, breast mottled or streaked brown, rest of underparts below gorget barred brown and white. Legs and cere orange and uppertail coverts barred brown.* **STATUS** Rather uncommon resident, and generally sedentary. Found singly or in pairs. **HABITAT** Favours tall well-developed broad-leaved woodland. **FOOD** Small mammals, birds up to size of young guineafowl, francolin and hornbills, also snakes, lizards and insects. **CALL** Flute-like *whee-o, whew, whew ...* notes. **BR** Monogamous. Stick nest often festooned with spider web. (**Donkersingvalk**)

AYRES'S HAWK-EAGLE *Aquila ayresii* (138) 51 cm; ♂ 655 g, ♀ 1 kg
SEXES *Alike, ♀>♂. Smaller, more compact and more heavily streaked than African Hawk-Eagle, though this streaking is highly variable, ranging from dark to fairly light. Streaking always extends onto thighs, a key distinguishing feature. Small crest on hind crown (nuchal crest) not a good field character, as it is not often raised. In flight, from above upperwings uniform blackish brown, lacking white windows on primaries; from below flight feathers and tail heavily barred at all ages. Lacks prominent black trailing edge to wing and broad subterminal tail band.* **JUV** *Upperparts grey-brown, underparts variable, usually buffy rufous with dark brown streaking. Distinguished in flight from juv African Hawk-Eagle by heavily barred flight feathers and tail, and from juv Booted Eagle (rare vagrant to Park) by barred (not blackish) flight feathers.* **STATUS** Very rare in KNP. A few records exist for the Pafuri area, and also for the s-e Lebombo Mountains. Regarded as Near-threatened in s Africa. Both resident and intra-African migrant populations occur in s Africa; the KNP birds are probably all migrant visitors, mainly present Jan–Apr. **HABITAT** Favours dense woodland and forest edge in hilly country. **FOOD** Almost entirely birds, especially doves and pigeons, but wide range of bird species taken (including small raptors); also squirrels and fruit bats. **CALL** In display high-pitched *hueeeep-hip-hip-hip-hueeep*. **BR** Monogamous. Nest concealed in large well-foliaged tree but no br records for KNP. (**Kleinjagarend**) ALT NAME *Ayres's Eagle*

AFRICAN HAWK-EAGLE *Aquila spilogaster* (137) 63 cm; ♂ 1,3 kg, ♀ 1,6 kg
SEXES *Similar, ♀ larger than ♂, with browner upperparts and more heavily streaked below. Larger and longer-winged than Ayres's Hawk-Eagle, less heavily streaked below with white (not black-streaked) thighs. In flight from above, has diagnostic pale panels (windows) at base of primaries (unlike Ayres's Hawk-Eagle, which is uniform blackish-brown). From below, flight feathers pale (not barred), with contrasting black tips. Tail finely barred with broad subterminal band.* **JUV** *Upperparts dark brown, underparts pale rufous brown, tail lacks subterminal bar and eyes are brown.* **STATUS** Fairly common; resident and largely sedentary. Usually in pairs; territorial year-round. **HABITAT** Throughout KNP in sparse savanna to tall broad-leaved woodland. **FOOD** Mainly birds, particularly gamebirds such as francolin and guineafowl, but wide range of smaller species also taken. Mammal prey includes squirrels, hares, hyraxes, mongooses, bushbabies, fruit bats and rodents. **CALL** Musical *klu-klu-klu-kluee*. **BR** Monogamous. Nest a large platform usually placed below canopy of tall tree; also recorded br on pylons and cliff faces. (**Grootjagarend**)

PLATE 41 Harrier-hawk, Eagles pp215–6

AFRICAN HARRIER-HAWK *Polyboroides typus*   **(169) 63 cm; 750 g**
SEXES *Alike. Largest grey hawk in region. Head small with yellow face and long thin yellow legs. Facial skin flushes pink or reddish during social interactions, especially in br season.* **JUV** *Dark brown with small head; in flight has broad wings (characteristic of species).* **STATUS** Fairly common resident, largely sedentary. **HABITAT** Wide range of woodland; most habitats in KNP except grasslands. **FOOD** Eggs and nestlings of birds ranging in size from waxbills to herons (specialises in robbing weaver nests); extracts bats from crevices in trees; also small mammals, reptiles, frogs and insects. **CALL** Plaintive high-pitched *peeee*, and in flight *sueeeee-oh*. **BR** Monogamous. Large stick nest in tall tree or cliff face. (**Kaalwangvalk**) **ALT NAME** *Gymnogene*

TAWNY EAGLE *Aquila rapax*   **(132) 71 cm; 1,94 kg**
SEXES *Alike, ♀ larger, and usually darker and more streaked. Plumage variable, from dark rufous brown to light brown (most common) and pale buff. Usually paler than Steppe Eagle and always more rufous. Eyes yellowish; yellow gape extends only as far back as to below centre of eye (diagnostic). Legs fully feathered and more shaggy than Lesser Spotted Eagle. Larger bill than Steppe Eagle.* **JUV** *Like adult but dark brown eyes and generally paler plumage.* **STATUS** Fairly common in protected areas, but because of population decrease elsewhere, regarded as Vulnerable in S. Africa. Resident and territorial. Usually found singly or in pairs (unlike Steppe Eagle). **HABITAT** Favours open savanna woodland. **FOOD** Wide range of prey including mammals (up to young antelope size), birds, reptiles, frogs, fish and insects. Scavenges at carcasses and also kleptoparasitises other eagles, storks and hornbills. **CALL** Guttural *kioh* or *kowk*. **BR** Monogamous. Nest placed on top of tree canopy. (**Roofarend**)

STEPPE EAGLE *Aquila nipalensis*   **(133) 77 cm; 2,75 kg**
SEXES *Alike, ♀>♂. Largest of 'brown' eagles in region; plumage dark brown, eyes brown (yellowish in adult Tawny and Lesser Spotted eagles); most birds have pale rufous nape patch. Leggings broad, right down to feet (not tapered as in Lesser Spotted Eagle). Yellow gape extends to below rear of eye (middle of eye in smaller Tawny Eagle). In flight, wings long and broad, trailing edge to wings conspicuously 'S' shaped. Flight feathers and tail narrowly barred.* **JUV** *Plumage paler; prominent broad white bar across upper- and underwings diagnostic.* **STATUS** Locally common non-br Palearctic migrant, arrives Oct/Nov and departs Mar/Apr. Usually very localised, often gregarious in flocks where good rains have fallen, particularly at termite alate emergences and quelea colonies. **HABITAT** Favours open savanna woodland. **FOOD** Mainly termites; also Red-billed Quelea nestlings. **CALL** Usually silent in Africa. (**Steppe-arend**)

LESSER SPOTTED EAGLE *Aquila pomarina*   **(134) 62 cm; 1,35 kg**
SEXES *Alike, ♀>♂. Overall brownish colour, similar to larger Steppe Eagle and smaller Wahlberg's Eagle. Smaller and weaker-billed than Tawny Eagle. Best identified by 'stove-pipe' lower leggings (not as baggy as in smaller Wahlberg's Eagle) and yellowish eyes. Head shape distinctive with long loose (floppy) nape feathers. In flight, tail shortish and rounded, and base of primaries white, giving distinct window appearance from above. From below, wing coverts usually paler than dark flight feathers.* **JUV** *Upperwing coverts flecked white, nape streaked buff or rufous, eyes brown. In flight from above, white crescent at base of tail, and narrow white line separating flight feathers from wing coverts.* **STATUS** Fairly common but localised non-br Palearctic migrant; gregarious, often in large numbers at quelea breeding colonies. Present Oct–Mar. **HABITAT** Open woodland. **FOOD** Termites (flocks with Steppe Eagles at alate emergences and at quelea colonies), nestlings, rodents and frogs. **CALL** Silent in Africa. (**Gevlekte Arend**)

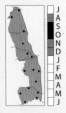

WAHLBERG'S EAGLE *Aquila wahlbergi*   **(135) 58 cm; ♂ 1,0 kg, ♀ 1,3 kg**
SEXES *Alike, ♀ considerably larger than ♂. Relatively small eagle with colour variations from pale, intermediate (most common) to dark. Distinctive features include small face, dark eyes and dark patch in front of eye, small crest (not always visible), baggy leggings and long narrow tail. In flight, long narrow ruler-like (square-tipped) tail and long square-ended wings (also ruler-like).* **JUV** *Like respective adult, but with pale fringes to feathers on upperparts.* **STATUS** Common br intra-African migrant, present Aug–Apr; some overwintering. **HABITAT** Well-wooded savanna. **FOOD** Predominantly birds, reptiles and small mammals; also frogs and insects. Also recorded catching bats at Letaba high-water bridge at dusk. **CALL** Fast *quee-quee-quee ...* and mournful *kleeee-eee*. **BR** Monogamous. Nest placed below canopy of tall tree, especially Knobthorn *Acacia nigrescens*. Selective removal of this tree species by elephants is a cause for concern for future breeding in KNP; same site used year after year. (**Bruinarend**)

PLATE 42 Eagles, Secretarybird p216

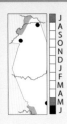

VERREAUXS' EAGLE *Aquila verreauxii*   **(131) 88 cm; ♂ 3,7 kg, ♀ 4,5 kg**
SEXES *Alike, ♀ larger and with more white on back. Only very large black eagle in region. In flight shows distinctive white 'Y' pattern on back and a white rump. Pale 'windows' in wings similar to Long-crested Eagle, but very much larger and without tail bands.* **JUV** *Overall mottled brown and black plumage, with blond crown and chestnut nape.* **STATUS** Recorded regularly only in the Luvuvhu Gorges at Pafuri where it is a br resident (± 7 pairs). Occurs sporadically around the granite koppies north of Phalaborwa. **HABITAT** Mountainous and rocky areas with large cliffs. **FOOD** Main prey is rock hyrax; wide range of other mammalian prey includes monkeys, young baboons and antelope, squirrels and hares. Also preys on birds (particularly game birds) and reptiles. **CALL** Clucking *pyuck*; in display *weeeee-oh*. **BR** Monogamous. Most nest on inaccessible cliffs. (**Witkruisarend**) **ALT NAME** *Black Eagle*

MARTIAL EAGLE *Polemaetus bellicosus*   **(140) 81 cm; 4 kg**
SEXES *Similar, ♀ larger, on average darker and more spotted below. Spotted underparts, and, in flight, dark brown underwings key features distinguishing it from smaller Black-chested Snake-Eagle.* **JUV** *Neck grey flecked (not plain as in African Crowned Eagle), upperparts grey-brown.* **STATUS** Fairly common resident in KNP; territorial. Listed as Vulnerable in S. Africa. **HABITAT** Mostly open savanna woodland. **FOOD** Diet varies regionally, mainly small mammals such as hares, small antelope including impala lambs, jackal, mongooses and young baboons. Bird prey includes francolin, guineafowl, ibises and small bustards; also reptiles. In KNP preys especially on Monitor Lizards. **CALL** Not very vocal; in display gives rapid *kwi-kwi-kwi-klooee-klooee* ... **BR** Monogamous. Nest a large platform of sticks placed in tall tree or on pylon. (**Breëkoparend**)

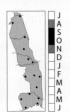

LONG-CRESTED EAGLE *Lophaetus occipitalis*   **(139) 56 cm; 1,06 kg**
SEXES *Similar, ♀ usually has browner leggings, but differences not consistent. Dark plumage and long crest very distinctive. In flight, combination of large white wing patches and barred tail distinguishes it from much larger Verreauxs' Eagle (which has all-black tail).* **JUV** *Like adult with white leggings, upperparts speckled pale brown, crest shorter and eyes initially greyish, becoming yellow at 1 year.* **STATUS** Very rare vagrant in KNP with records from widely scattered localities. It is suspected that these birds originate from the escarpment to the west dispersing eastwards down riparian corridors. **HABITAT** Favours moist open woodland. **FOOD** Mainly rodents, especially vlei rats (*Otomys* species), also frogs, reptiles and insects; rarely eats birds. **CALL** Loud *keee-ee, keee-ay* in aerial display. **BR** Monogamous. Nest usually placed in mid-canopy of large tree, but no br records for KNP. (**Langkuifarend**)

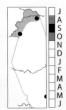

AFRICAN CROWNED EAGLE *Stephanoaetus coronatus*   **(141) 85 cm; 3,6 kg**
SEXES *Similar, ♀ considerably larger than ♂ and with 2 (not 3) narrow bars on flight feathers (excluding terminal bar), and usually darker underparts. Most powerful eagle on the continent; large size, dark plumage and blotched and barred underparts distinctive. Crest (crown) at back of head not often seen held upright.* **JUV** *White (not grey-flecked) neck separates it from juv Martial Eagle.* **STATUS** Very rare bird in KNP, resident only in the Pafuri/Punda Maria area with possibly only 4 br pairs; records of birds along the Sabie River are probably young dispersing birds from w of the Park; regarded as Near-threatened in S Africa. **HABITAT** In KNP favours tall riparian forest. **FOOD** Mainly mammals such as monkeys, antelope, hares, mongooses, hyraxes and genets. Avian prey mainly gamebirds and Hadeda Ibis. **CALL** Most vocal in aerial displays, ♂ calling repetitively *kewick-kewick-kewick* ... , rising and falling in pitch; ♀ calls lower-pitched *kooi-kooi-kooi* ... **BR** Monogamous. Nest built in tall canopy tree. (**Kroonarend**) **ALT NAME** *Crowned Eagle*

SECRETARYBIRD *Sagittarius serpentarius*   **(118) 1,38 m; 4 kg**
SEXES *Similar, ♂ larger and with slightly longer crest and tail. Orange facial skin, grey-black plumage and characteristic gait striding with long legs through grass distinctive.* **JUV** *Plumage browner, facial skin yellow and with shorter tail.* **STATUS** Uncommon in KNP and listed as Near-threatened in S. Africa. Resident but generally not sedentary. Usually in pairs, sometimes solitary. **HABITAT** Favours open grasslands on the eastern side of KNP with scattered trees or shrubs, but recorded throughout the Park. **FOOD** Wide variety of prey including insects, amphibians, reptiles, birds and their eggs, small mammals and rodents. Most prey (including snakes) caught on ground and killed by hard downward blows with the feet. Large prey, which may include francolin, mongoose and large snakes, torn up and swallowed, small prey such as tortoises swallowed whole. **CALL** Mostly silent away from nest. **BR** Monogamous. Nest a stick platform usually placed at top of flat thorn tree. (**Sekretarisvoël**)

DICKINSON'S KESTREL *Falco dickinsoni* (185) 29 cm; 210 g
SEXES *Alike, ♀ larger and slightly darker grey. The palest kestrel in the Park; ♂ is overall slightly paler than the ♀. Key features pale grey square-shaped head contrasting with dark grey body, pale grey rump and boldly barred upper tail.* **JUV** *Pale brown wash and less contrast in head and body colours. Eye-ring and cere blue-green or pale yellow.* **STATUS** Uncommon resident, generally sedentary; recorded mainly in the northern parts of KNP particularly on the eastern grasslands. More conspicuous during the winter months probably because of its habit of perching conspicuously in the open and in the sun. Perches upright, generally hunts from perch but may also hover-hunt. **HABITAT** Low-lying open savanna woodland regions, especially where there are tall palms and Baobab trees; also found in open Mopane woodland. **FOOD** Mainly insects such as grasshoppers and crickets, also birds, frogs, snakes, small mammals (including bats) and crabs. Attracted to, and hunts at the edge of, veld fires where it may take aerial prey. **CALL** Strident *keh-keh-keh* and whistle-like *kill-koo*. **BR** Monogamous. Nest usually placed on top of dead palm stump, dead tree stump or even on top of a Hamerkop nest. (**Dickinsonse Grysvalk**)

AMUR FALCON *Falco amurensis* (180) 29 cm; 142 g
SEXES ♀>♂. **M** *White (not grey) underwing coverts very conspicuous in flight. Uniform dark grey plumage darkest on back and lightest on cheeks, throat and breast. Indistinct blackish malar stripe.* **F** *Distinguished from Eurasian Hobby by grey (not black) crown, barred (not streaked) flanks, less prominent malar stripe, and lighter chestnut to lower belly.* **JUV** *Similar to ♀, feathers on upperparts tinged buff; eye-ring and cere orange-yellow.* **STATUS** Common non-br Palearctic migrant, traveling an enormous distance from e Asia where it br in e Siberia, n-e Mongolia, n Korea and e China. Present mainly in the eastern regions of s Africa; first arrivals in Nov and departs Mar/Apr, some as late as May. Highly gregarious and nomadic; undertakes local movements in response to rainfall and insect emergences. Well known for its communal roosting habits. The regular use of communal roost sites not recorded in KNP (usually uses eucalyptus trees close to human habitation outside the Park). **HABITAT** Most common on the highveld grasslands, but in the Park found mainly over eastern grasslands and lightly wooded grassland. **FOOD** Predominantly insects such as locusts, grasshoppers, termite alates and armoured crickets, also to a lesser extent beetles and bees. Occasionally small birds. **CALL** Generally silent; gives high-pitched *kew-kew-kew …* at communal roosts. (**Oostelike Rooipootvalk**) **ALT NAME** *Eastern Red-footed Kestrel*

EURASIAN HOBBY *Falco subbuteo* (173) 32 cm; 215 g
SEXES *Alike, ♀>♂. Densely streaked underparts, deep rufous thighs and lower belly, white cheeks and throat (with bold moustachial stripe) distinguish it from other similar-looking falcons. Most likely to be confused with female Amur Falcon, but has black (not grey) crown, streaked (not barred) flanks, yellow (not orange-red) cere and far more prominent malar stripe. Often confused with Bat Hawks because of their habit of hunting bats.* **JUV** *Underparts more boldly streaked or blotched; upperparts browner and crown paler.* **STATUS** Fairly uncommon non-br Palearctic migrant, present Oct–Mar/Apr. Found singly, occasionally in small groups moving in loose association. **HABITAT** Favours low-altitude open woodland and riparian margins. **FOOD** Hunts mainly at dusk, but also in the early morning. Mostly large insects such as locusts, dragonflies and beetles. Also small birds (even fast-flying species such as swifts and swallows) and bats caught at dawn and dusk, behavior that can cause confusion with that of Bat Hawks. Regularly feeds in association with other falcons on termite alate emergences and at quelea aggregations. **CALL** Generally silent in region; gives falcon-like screams. (**Europese Boomvalk**) **ALT NAME** *Hobby Falcon*

LESSER KESTREL *Falco naumanni* (183) 29 cm; 148 g
SEXES ♀>♂. *Slim, long-tailed and short-legged raptor that is normally highly gregarious but in KNP is rather shy and normally found singly. In flight has long slim wings and frequently hovers.* **M** *Distinctive with blue-grey and chestnut upperparts (unspotted); buff below. Confusion only likely with Rock Kestrel, but wings grey (not brown) when perched.* **F** *Lacks grey head and wings, duller with heavy barring above and streaking below. Distinguished from ♀ Rock Kestrel by brown (not grey) head and barred tail.* **JUV** *Like ♀ but paler and duller with less distinct malar stripe.* **STATUS** Uncommon non-br Palearctic migrant in KNP, recorded throughout the Park. Globally now regarded as Vulnerable with enormous reduction in population in past few decades. Major threats include habitat modification and poisoning by pesticides (particularly those applied to swarming locusts). Present Oct/Nov–Mar/Apr. Usually gregarious and roosts communally, but this behaviour not recorded in KNP. **HABITAT** Open savanna, shrublands and grassland, particularly in the eastern half of the Park. **FOOD** Mainly insects, occasionally small rodents, birds and reptiles. Often feeds in association with Amur Falcons. **CALL** High-pitched chattering and screaming at communal roosts. (**Kleinrooivalk**)

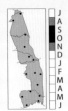

ROCK KESTREL *Falco rupicolus* (181) 32 cm; 215 g
SEXES *Similar, ♀>♂. Black spotting on rufous upperparts distinctive. In flight white underwing lightly barred.* **M** *Brown (not grey) wings distinguishes it from ♂ Lesser Kestrel. Tail grey with broad black subterminal tail band (white-tipped).* **F** *Browner with narrow black bars to grey tail (in addition to broad black subterminal band), and more pronounced streaking to crown.* **JUV** *Head brownish and more heavily streaked; tail and flight feathers tipped buff.* **STATUS** Very uncommon resident, generally sedentary with some local movements. **HABITAT** Wide variety of habitats, usually associated with rocky outcrops (br sites), especially during the br season. Probably only permanently resident in the gorges of the Luvuvhu River. **FOOD** Diet varies regionally; mainly small mammals, birds up to dove size, reptiles and insects. Also recorded taking bats. **CALL** Metallic high-pitched *kik-kik-kik* ... and soliciting *kreee-kreee* ... notes. **BR** Monogamous. Nest a simple scrape on cliff ledge but also uses crows' nests and man-made structures where no suitable natural sites. (**Kransvalk**)

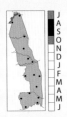

PEREGRINE FALCON *Falco peregrinus* (171) 39 cm; 650 g
SEXES *Alike, ♀ considerably larger than ♂. Black 'hangman's hood' appearance, compact build and densely barred underparts diagnostic. In flight appears stockier with shorter wings and tail than slightly larger Lanner Falcon.* **JUV** *Browner above, underparts pale to beige with conspicuous narrow streaks (not bars).* **STATUS** Scarce to uncommon with both resident and migratory (rare) populations; regarded as Near-threatened. Br population generally sedentary particularly in the gorges of the Luvuvhu and Olifants rivers; birds recorded elsewhere probably migrants or dispersing juveniles. **HABITAT** Largely restricted to areas near high cliffs (resident birds). **FOOD** Mainly birds, especially doves, swifts and starlings; also bats and occasionally insects. When hunting over woodland, may also take species such as hornbills, hoopoes and mousebirds. **CALL** Harsh *kek-kek* ... screams and *krrChuck-krrChuck* ... notes. **BR** Monogamous. Usually nests on highest cliff site available. (**Swerfvalk**)

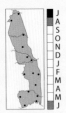

LANNER FALCON *Falco biarmicus* (172) 42 cm; 595 g
SEXES *Alike, ♀ considerably larger than ♂. Rufous crown, greyer upperparts, buff (almost unmarked) underparts and larger size distinguish it from Peregrine Falcon.* **JUV** *Underparts buff, heavily streaked brown; upperparts browner.* **STATUS** Rare in KNP and most records probably migrant post-br birds as breeding not recorded in Park; listed as Near-threatened in South Africa. Found singly or in pairs. **HABITAT** Favours open grassland or woodland near cliffs. **FOOD** Birds make up >80% of prey; also small mammals (including bats), reptiles and insects. An opportunistic feeder that often hunts at water holes where birds aggregate to drink. Pigeons and doves make up the bulk of bird prey, but also feeds heavily on smaller seedeaters. Recorded preying on Crowned Plover in KNP. **CALL** Wailing *weeah-weeah* ... and other piercing notes. **BR** Monogamous. Nests mostly on cliffs; uses stick nests of other species when br in trees or on pylons (elsewhere in country). (**Edelvalk**)

PLATE 45 Grebe, Darter, Cormorants p216

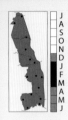

LITTLE GREBE *Tachybaptus ruficollis* **(8) 20 cm; 170 g**
SEXES Alike. **BR** Not likely to be confused with any other species, with no other grebes recorded in the Park. Sides and front of neck dark rufous-chestnut; pale gape-spot and dark red eyes distinctive. **NON-BR** Overall paler, chestnut on neck becoming brownish-white. **JUV** Like non-br adult with black streaking on head and neck. **STATUS** Common and widespread in wetter years but almost absent during droughts. Resident with local movements in response to open water availability; usually in pairs or small groups. **HABITAT** Found on most water bodies, especially pans and dams with emergent vegetation but rare on rivers. **FOOD** Aquatic animals including fish, frogs, tadpoles and insects. Most prey taken from under the surface of the water, but insects such as falling termite alates taken from the surface. Sometimes follows Hippopotamuses to feed on disturbed aquatic insects. Like other grebes, eats own feathers to wrap fish bones so later on can be regurgitated in pellets. **CALL** Long, high-pitched descending trill. **BR** Monogamous. Nest a floating heap of plant material usually anchored to submerged vegetation. Eggs always covered by nesting material when incubating bird is disturbed and leaves the nest. (**Kleindobbertjie**) **ALT NAME** *Dabchick*

AFRICAN DARTER *Anhinga rufa* **(60) 90 cm; 1,4 kg**
SEXES Similar. Sits low in water; this, with its long thin neck (diagnostic) and straight dagger-like (not hooked) bill, has given it the colloquial name 'Snakebird'. Lacks red bill and legs of adult African Finfoot. Br ♂ has rufous neck, blacker plumage, white cheek stripe and prominent mantle plumes. Non-br ♂ and ♀ paler and browner. **JUV** Underparts usually cream or pale buffy-brown. **STATUS** Fairly common resident, sedentary; local movements linked to water levels and food availability. Some birds have complete post-br moult, simultaneously replacing flight feathers leaving the bird flightless till these feathers have grown sufficiently. **HABITAT** Most of the rivers and dams in KNP. **FOOD** Mainly fish. Typically spears fish with sharp bill, large prey with both mandibles, small prey usually with upper mandible only. Prey is brought to the surface, tossed in the air, caught and swallowed head first. Also takes frogs. **CALL** Usually silent away from nest; occasional series of *chack, chack, chack* notes. **BR** Monogamous and colonial. Nest a platform of twigs placed in tree or reedbed. Incubates with feet, undertaken by both sexes. (**Slanghalsvoël**) **ALT NAME** *Darter*

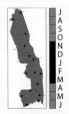

REED CORMORANT *Phalacrocorax africanus* **(58) 55 cm; 550 g**
SEXES Alike, ♂>♀. Only small cormorant to occur in the Park. **BR** Plumage glossy black; bare yellow-orange skin turns red during courtship. **NON-BR** Upperparts dark brown, underparts usually buffy or whitish, eyes not as bright red. **JUV** Browner above, throat grey and underparts whiter; eyes pale brown. **STATUS** The most common cormorant in the park; a br resident. Generally sedentary but subject to local movements that are dependent on food supplies and water levels. Typically rests with wings spread which helps dry feathers, but probably more likely for body warming. **HABITAT** Most rivers and dams in KNP; often roosts communally in reedbeds. **FOOD** Mainly fish, also frogs, some insects and plants. **CALL** Mostly silent away from colonies. **BR** Monogamous; in KNP usually nests colonially in reedbeds on islands in major rivers, or in a stick nest placed in tree over water. Incubation by both sexes. (**Rietduiker**)

WHITE-BREASTED CORMORANT *Phalacrocorax lucidus* **(55) 90 cm; 3 kg**
SEXES Alike, ♂>♀. Large size and white breast should eliminate confusion with other cormorants in region. Non-br birds duller and browner above, facial skin duller yellow-orange; eyes remain green. **JUV** Browner above, underparts all-white and eyes brown. **STATUS** Uncommon resident, but may be seen on most of KNP's rivers and dams; nomadic in response to water levels. Usually gregarious (both at br and roost sites). **HABITAT** Mainly man-made dams but also larger rivers. **FOOD** Mostly fish, but also preys on frogs and crabs. Fish caught between mandibles and not speared as in the case of African Darter. May attempt to steal prey from smaller Reed Cormorant and also African Darter. **CALL** Usually silent away from nest. **BR** Monogamous; in KNP mainly in small colonies, especially in dead trees in man-made dams. Nest a stick platform, incubation by both sexes. (**Witborsduiker**)

LITTLE EGRET *Egretta garzetta* **(67) 64 cm; 520 g**
SEXES *Alike. Black bill, black legs and bright yellow feet (non-br) distinctive.* **BR** *Eyes and legs orange to red. Develops long plumes from back and scapulars that may extend to tip of tail.* **JUV** *Base of bill and lower legs may be greenish.* **STATUS** Fairly common resident, with nomadic movements that may be triggered by food or water shortages. Usually solitary but roosts communally at night. **HABITAT** Shallow water bodies, mainly larger rivers but also dams. **FOOD** Mostly fish, but also amphibians, reptiles, insects, worms, molluscs and even small birds and mammals. Yellow feet supposedly attract prey within striking range of the bill. Sometimes perches on and feeds from the backs of Hippopotamus. **CALL** Harsh rattling *ggrow*. **BR** Monogamous and colonial. Stick nest usually placed in tree, bush or in reedbeds over water. Incubation by both sexes. (**Kleinwitreier**)

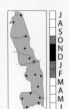

YELLOW-BILLED EGRET *Egretta intermedia* **(68) 69 cm; 415 g**
SEXES *Alike, ♂ > ♀. Slightly taller than Little Egret but with yellow bill and bicoloured legs. Considerably smaller than Great Egret (which has all-black legs).* **BR** *Bill red, tipped orange; eyes red, lores green; develops long elongated plumes at the onset of br season.* **JUV** *Similar to non-br adult.* **STATUS** Very uncommon vagrant; wanders in response to water levels. **HABITAT** Shallow water margins on dams and rivers; favours seasonally flooded marshes and grasslands during the wet season. May follow in the wake of Hippopotamus to take advantage of disturbed prey. **FOOD** Mostly fish, also frogs, reptiles, insects, worms and even small birds and mammals. Sometimes stamps and shuffles feet under water to flush hidden prey. **CALL** Usually silent away from nest. **BR** Monogamous and colonial. Nests over water in tree or reedbeds. (**Geelbekwitreier**)

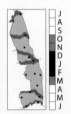

GREAT EGRET *Egretta alba* **(66) 95 cm; 1,1 kg**
SEXES *Alike. Largest white egret, as tall as Grey Heron. Legs extend beyond tail in flight.* **BR** *For short period at the onset of br season, bill becomes black, eyes red and lores and eye-ring emerald green; develops long white plumes.* **NON-BR** *Bill yellow.* **JUV** *Like non-br adult.* **STATUS** Fairly common and localised resident, partially nomadic in response to water levels. Usually shy and solitary. Roosts communally in trees, usually with other associated water birds. **HABITAT** Open water systems (mainly rivers but also dams). **FOOD** Mainly fish, also frogs, insects and reptiles. Hunting techniques include foot-stirring and following in the wake of Hippopotamus. **CALL** Raucous *croak-croak* given in flight. **BR** Monogamous and colonial. Nests above water in trees or in reedbeds. (**Grootwitreier**) **ALT NAME** *Great White Egret*

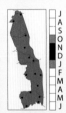

CATTLE EGRET *Bubulcus ibis* **(71) 54 cm; 365 g**
SEXES *Alike. Smallest of the white egrets. Historically confined to Africa, now has almost global distribution range.* **BR** *Develops rufous plumes and dorsal aigrettes; bill orange or red, and legs shades of red.* **NON-BR** *Lacks rufous plumes; bill, lores and eyes yellow.* **JUV** *Bill, legs and feet black.* **STATUS** Rather uncommon resident; numbers have increased in southern KNP over the past few decades, perhaps as a result of increasing buffalo and white rhino numbers with which they usually associate. Undertakes local nomadic movements; mostly gregarious, roosting in trees and reedbeds, individuals often occupying same perch each night. **HABITAT** Open grassland and woodlands, frequently on aquatic margins. **FOOD** Mainly insects, also frogs and small mammals. **CALL** Mostly silent; raucous croaking around br colonies. **BR** Monogamous and colonial; nests in trees or reedbeds. (**Veereier**)

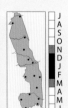

SQUACCO HERON *Ardeola ralloides* **(72) 43 cm; 300 g**
SEXES *Alike. Small squat buffy-brown heron with bittern-like appearance when hunting, but resembles Cattle Egret in flight when conspicuous white flight feathers can be seen.* **BR** *Feathers elongated and edged black on crest and neck, bill china blue with black tip.* **NON-BR** *Crest shorter, more heavily streaked on back and throat, bill greenish-yellow with black tip, legs yellow-green.* **JUV** *Back browner and breast more heavily streaked.* **STATUS** Rather rare resident, sedentary and usually found singly. **HABITAT** Found mainly on well-vegetated dams but also on river shorelines with dense vegetation. Follows rains taking advantage of ephemeral pans and flood plains. **FOOD** Fish, frogs, aquatic insects and crustaceans. **CALL** Usually silent, harsh squawks and clucks during br season. **BR** Monogamous and usually colonial. Nest a platform of sticks placed in tree, low bush or reedbed just above water level. (**Ralreier**)

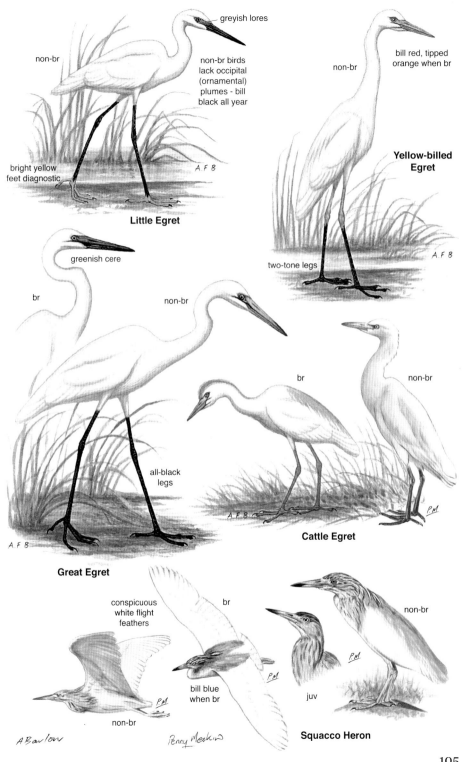

BLACK HERON *Egretta ardesiaca* (69) 55 cm; 310 g
SEXES Alike. Dark grey (almost black) plumage with black legs and yellow feet. **JUV** Duller and browner; lacks plumes. **STATUS** Very uncommon but widespread vagrant usually recorded in summer; usually gregarious, but often recorded singly in KNP. **HABITAT** In KNP mainly on stiller water bodies such as dams and ephemeral pans. **FOOD** Mostly fish, caught using unique foraging behaviour, under 'umbrella' formation of wings (illustrated), also crustaceans and insects. 'Umbrella' formation of wings usually only held for 2–3 sec. Reasons proposed for the use of this feeding tactic include elimination of reflection on the water and luring of fish to false refuge. Sometimes toe-trembles during canopy formation. Yellow feet supposedly also attract prey within striking range. **CALL** Low cluck. **BR** Probably monogamous; nests in mixed-species colonies, usually in reedbeds. **(Swartreier) ALT NAME** *Black Egret*

GREY HERON *Ardea cinerea* (62) 94 cm; 1,4 kg
SEXES Alike, ♂>♀. *Usually an 'aquatic' species found on water's edge (unlike Black-headed Heron usually found away from water bodies. White head and hindneck, with black eyebrow and plumes, distinguish it from Black-headed Heron, which has a black crown and hindneck. In flight entire underwing appears uniform dark grey. Bill becomes bright orange and legs pinkish-red at onset of br.* **JUV** *Paler grey body, with grey on forehead and crown. Distinguished from juv Black-headed Heron by white ear coverts and yellowish (not black) legs.* **STATUS** Common resident. **HABITAT** Shallow water bodies (rivers and dams). Active both diurnally and nocturnally. **FOOD** Mainly fish such as tilapia, but a wide range of aquatic animals such as frogs, molluscs, worms and insects, also small rodents and birds. **CALL** Mostly in flight: a raucous *kraank*. **BR** Monogamous. Nests sometimes singly, but usually colonially, either in monospecific or mixed-species colonies in trees or reedbeds. **(Bloureier)**

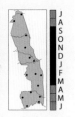

BLACK-HEADED HERON *Ardea melanocephala* (63) 92 cm; 1,1 kg
SEXES *Black crown and hindneck distinguish it from similar-sized Grey Heron. Legs black, and in flight has two-tone black-and-white underwing pattern. Eyes become red when br.* **JUV** *Top of head and hindneck grey to brownish-grey; underparts whitish.* **STATUS** A rather uncommon bird in KNP, probably vagrant, active both diurnally and nocturnally. **HABITAT** Favours open grassland; sometimes near, but not dependent on, water bodies (unlike Grey Heron). **FOOD** Primarily invertebrates but wide range of small mammals, reptiles and even birds (up to dove size). **CALL** Loud raucous call, *kuaark*. **BR** Br rarely recorded in KNP. Monogamous; usually colonial and in small mixed heronries. Nest a large stick platform usually placed in tall tree; also in reedbeds. **(Swartkopreier)**

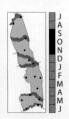

GOLIATH HERON *Ardea goliath* (64) 1,43 m; 4,3 kg
SEXES *Alike. World's largest heron, almost double the size of Purple Heron. Legs and feet black, not yellowish.* **JUV** *Similar to adult but duller. Grey upperparts edged rufous; underparts streaked white.* **STATUS** Widespread and reasonably common resident. **HABITAT** Mostly shallow margins of larger rivers but also on dams. **FOOD** Primarily fairly large fish but also frogs, reptiles and small mammals. Large, partly digested fish regurgitated on nest to feed the young chicks. **CALL** Usually silent; call a raucous *kowoork-kowoork* ... **BR** Monogamous. Normally solitary but does sometimes br in loose colonies. Large stick nest usually placed in tree over water, or in reedbeds, or on ground on rocky islands. Olifants River Gorge in Lebombos an important breeding area. **(Reusereier)**

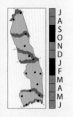

PURPLE HERON *Ardea purpurea* (65) 85 cm; 870 g
SEXES *Much smaller size, yellowish (not black) legs and black facial markings distinguish it from much larger Goliath Heron. Crown, hindneck and long plume feathers black. Plumage duller in non-br plumage, and lacks head plumes.* **JUV** *Tawny brown above, facial markings absent and generally duller than adult in appearance.* **STATUS** In KNP probably an uncommon resident; may be more common than suspected owing to its retiring nature. **HABITAT** Requires dense vegetation, especially *Phragmites*, fringing shallow wetlands; seldom hunts in the open, so probably occurs mainly on the larger rivers. **FOOD** Mainly fish; also insects, frogs, reptiles and occasionally small mammals and birds. **CALL** Seldom calls away from colonies; in flight a rasping *kraak*. **BR** Monogamous. Sometimes br singly, more commonly in mixed colonies, both in reedbeds and tree thickets. Nest a flat platform of sticks and reed stems constructed by both adults. **(Rooireier)**

PLATE 48 Heron, Night-Heron, Bittern p217

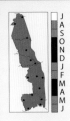

GREEN-BACKED HERON *Butorides striata* **(74) 41 cm; 215 g**
SEXES *Alike, ♀ slightly duller. Small size, squat appearance with hunched stance, blackish-green gloss to back and grey underparts distinctive. When disturbed sometimes adopts bittern stance with bill and head pointed vertically.* **BR** *Eyes orange; legs orange to orange-red.* **NON-BR** *Duller; eyes yellow, legs yellow-brown.* **JUV** *Dark brown above; pale below and heavily streaked.*
STATUS Common resident, mostly sedentary, and with solitary and skulking habits is generally overlooked; partly nocturnal **HABITAT** Favours vegetated margins on almost all KNP's rivers and dams. **FOOD** Fish, frogs and wide range of aquatic insects and animals. Has been observed using insects, bread and paper as bait to attract fish to the surface when 'fishing' from a branch low over water. **CALL** Varied harsh *kyah* and similar notes. **BR** Monogamous. Nest a platform of sticks, usually placed low over water. Incubation is shared by both sexes. (**Groenrugreier**)

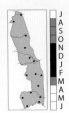

BLACK-CROWNED NIGHT-HERON *Nycticorax nycticorax* **(76) 56 cm; 630 g**
SEXES *Alike, ♂>♀. With thickset, short-legged appearance, and black, white and grey plumage, confusion unlikely with any other heron in region.* **JUV** *Larger than juv Green-backed Heron, white spotting on upperparts more prominent.* **STATUS** Fairly common resident, probably more common than suspected owing to its nocturnal, crepuscular and retiring habits. Locally nomadic in response to water levels; roosts communally by day. **HABITAT** Vegetated margins of slow-moving water bodies and dams and ponds. **FOOD** Fish, amphibians, reptiles, small birds up to Laughing Dove size, small mammals and insects. Known to rob bird colonies of eggs and chicks. **CALL** Harsh *quock* when flushed. **BR** Monogamous and colonial. Nests in both reedbeds and trees; incubation by both sexes. (**Gewone Nagreier**)

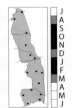

WHITE-BACKED NIGHT-HERON *Gorsachius leuconotus* **(77) 53 cm; ca 440 g**
SEXES *Alike. Blackish head, conspicuous white eye-ring and rufous upper chest and neck are key features. Narrow white back patch seldom visible in field.* **JUV** *Underparts streaked brown and white, wings spotted white.* **STATUS** Probably occurs through most of Park. Uncommon br resident but generally overlooked (owing to its retiring and nocturnal habits) and probably more common than thought; resident. Listed as Vulnerable in S Africa; mostly sedentary and found singly or in pairs. **HABITAT** Overhanging vegetation along quiet backwaters. **FOOD** Crustaceans, insects, small frogs and fish. **CALL** Harsh frog-like *krak-krak-krak*. **BR** Monogamous. Platform nest, usually low over water. (**Witrugnagreier**)

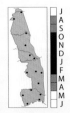

LITTLE BITTERN *Ixobrychus minutus* **(78) 36 cm; 110 g**
SEXES *Similar. Pale buff underparts and, in flight, wing coverts that contrast with the dark flight feathers distinguish it from Dwarf Bittern that is similar sized but an overall darker bird.* ♀ *browner and more heavily streaked.* **JUV** *Duller and more reddish than ♀, upperparts more mottled.* **STATUS** Generally uncommon, with 2 populations: non-br Palearctic migrants (*I. m. minutus*) present Dec–Apr, and br resident birds (*I. m. payesii*). In KNP resident race *payesii* may be present all year in some localities particularly after very wet summers. Local race distinguished from migratory birds by buff-rufous (not grey) cheeks and neck. **HABITAT** Mainly bulrushes and reedbeds, also on the margins of well-wooded streams and rivers and in rank vegetation around ponds. **FOOD** Mainly aquatic invertebrates, frogs and fish. **CALL** Mostly silent; deep *gak* or *ak*. **BR** Monogamous. Nests over water in thick reedbeds, incubation by both sexes. (**Kleinrietreier**)

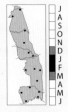

DWARF BITTERN *Ixobrychus sturmii* **(79) 30 cm; 140 g**
SEXES *Similar. Smallest bittern; upperparts greyish-black, underparts with broad vertical streaks.* ♀ *has paler tawny underparts. Lower mandible and lores greenish-yellow or bluish (especially during courtship).* **JUV** *Paler with buff-tipped feathers on upperparts.* **STATUS** Uncommon br intra-African migrant, present Oct–Apr, majority present Nov–Mar, particularly in years of above average summer rainfall. Probably absent from KNP during drought years. Usually solitary, crepuscular and partly nocturnal. Assumes bittern posture when threatened. **HABITAT** Favours seasonally inundated pans and floodplains with scattered trees. **FOOD** Mainly frogs, fish, aquatic insects, crabs and spiders. **CALL** Deep *hoot-hoot-hoot*. **BR** Monogamous. Usually nests low over water in temporarily flooded tree or bush, incubation by both sexes. (**Dwergrietreier**)

PLATE 49 Ibis, Hamerkop p217

GLOSSY IBIS *Plegadis falcinellus*   (93) 58 cm; 625 g
SEXES *Alike, ♂>♀. Small slender ibis; head, neck and shoulders dark chestnut, rest of plumage has metallic purple and green sheen. Non-br plumage duller and with white flecking to neck.* **JUV** *Head and neck sooty brown; duller, and lacks chestnut and purple gloss.* **STATUS** Very rare vagrant to KNP. Usually seen alone or in small groups. **HABITAT** Shallow pans, marshes, ephemeral pans, edges of dams and flooded grassland. Often the first birds to locate ephemeral habitats. Birds recorded in KNP seem to remain only for a day or two. **FOOD** Mainly aquatic insects and their larvae, crustaceans and worms. Also to a lesser extent, fish, frogs, lizards and small mammals. **CALL** Generally silent; gives harsh *graa-graa* notes. **BR** Monogamous and colonial. Usually nests in reedbeds, but no breeding records for KNP. (**Glansibis**)

HAMERKOP *Scopus umbretta*   (81) 56 cm; 500 g
SEXES *Alike, ♂>♀. Unmistakable with dull brown plumage and backward-projecting crest.* **JUV** *Similar to adult.* **STATUS** Common resident, mostly sedentary and found singly or in pairs. Diurnal, but often active at dusk. Unlike storks, does not sit or rest on tarsi, and does not excrete on its legs during hot weather. Often roosts in or near large nest. **HABITAT** Margins of most rivers, dams and seasonally flooded pans. **FOOD** Mainly frogs, tadpoles, small fish, aquatic insects, termite alates, small mammals and earthworms. Also recorded scavenging scraps from carcasses. Specialised bill structure with hook at tip allows bird to remove individual prey from among vegetation, and foot-trembles in water to flush aquatic prey. Opportunistic feeder, hunts from backs of Hippopotamus, follows groups of Banded Mongoose and feeds in association with Buffalo. Feeds to a great extent on Platanna (*Zenopus laevis*), frogs and tadpoles. **CALL** Repetitive slow yelping *kyip*, increasing in tempo to trill, often given in flight. **BR** Monogamous. Builds unique and enormous domed nest with side entrance; diameter approx 1–2 m. Usually constructed in large tree (which may take months to complete), but may also be placed on edge of cliff. Nest construction and incubation undertaken by both sexes. (**Hamerkop**)

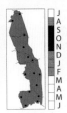

HADEDA IBIS *Bostrychia hagedash*   (94) 76 cm; 1,25 kg
SEXES *Alike, ♂>♀. Larger, heavier-billed and with overall paler plumage than Glossy Ibis. Both bronze and green metallic sheen to wing coverts, depending on angle of light.* **JUV** *Duller than adult, lacking metallic sheen.* **STATUS** Common and widespread breeding resident. Mainly sedentary and usually in pairs or small groups. Roosts communally at regularly used sites such as cliffs, forest patch or reedbed. These sites may often be used for years. **HABITAT** Favours riparian zones of larger rivers and often seen in the restcamps of the KNP. **FOOD** Largely invertebrates, earthworms and insects. Most prey obtained from under the surface of the ground, using long bill to probe into the earth. **CALL** Loud and raucous 4-note *ha-ha-HAaa*. **BR** Monogamous. Nest usually placed on horizontal branch overhanging water. Both sexes incubate; occasional records of double brooding. (**Hadeda**)

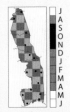

AFRICAN SACRED IBIS *Threskiornis aethiopicus*   (91) 77 cm; 1,5 kg
SEXES *Alike, ♂>♀ and with longer bill. Plumage distinctive; blue-black scapular plumes extend over lower back and tail.* **JUV** *Bare head and neck sparsely feathered.* **STATUS** A very rare vagrant to KNP, usually seen at the larger man-made dams. Usually seen singly in the Park but occasionally in small groups. **HABITAT** Margins of wetlands and dams. **FOOD** Invertebrates and their larvae, worms, frogs, reptiles and fish. At communal br sites (outside the Park) sometimes raids nests of other species for eggs and nestlings. **CALL** Generally silent; at nest harsh croaks and squeals. **BR** Monogamous and colonial. Usually nests in trees but sometimes on ground, but no breeding records in KNP. (**Skoorsteenveër**) **ALT NAME** *Sacred Ibis*

PLATE 50 Flamingos, Spoonbill, Pelican p217

GREATER FLAMINGO *Phoenicopterus ruber* **(96) 1.55 m; 2,7 kg**
SEXES *Alike, ♂>♀. Larger and whiter than Lesser Flamingo, with pale pinkish, black-tipped bill. In flight shows conspicuous crimson-red and black wings. Non-br birds have whiter plumage.* **JUV** *Dark grey-brown with grey bill.* **STATUS** Rare vagrant in KNP as a result of lack of suitable habitat – birds seen in KNP are probably on passage migrating to more suitable habitats, regarded as Near-threatened in S Africa. Highly nomadic and partially migratory, usually in large flocks at suitable habitats elsewhere in Africa. **HABITAT** Favours saline or brackish shallow water bodies (no suitable habitats in KNP, but regularly recorded at the Phalaborwa mines settling dams on the KNP's western boundary). **FOOD** Aquatic invertebrates such as brine shrimps and brine fly (larvae), also algae. Wades in water with bill upside down, filtering small invertebrates from mud. Large tongue pumps water in and out through lamellae on edge of bill to filter out prey. **CALL** Goose-like double *honk-honk* given in flight. **BR** Monogamous and colonial. Nest a cone of mud. No records for KNP. (**Grootflamink**)

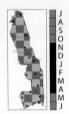

LESSER FLAMINGO *Phoenicopterus minor* **(97) 1,18 m; 1,7 kg**
SEXES *Alike, ♂>♀. Smaller than Greater Flamingo, usually richer pink in colour, and with a maroon (not pale pinkish) bill.* **JUV** *Brownish-grey and with a dark grey bill.* **STATUS** As with the Greater Flamingo, a rare vagrant in KNP as a result of lack of suitable habitat. Birds seen in KNP on passage migration to more suitable habitats. Listed as Near-threatened in S Africa. **HABITAT** Primarily eutrophic shallow wetlands, especially saltpans. No such habitats exist in KNP but regularly recorded at the Phalaborwa mines settling dams on the KNP's western boundary. **FOOD** Microscopic blue-green algae and diatoms. Feeds by both day and night, filtering water through bill in same manner as Greater Flamingo. **CALL** High-pitched honking. **BR** Monogamous; colonial, builds cone mud nest. No breeding records for KNP. (**Kleinflamink**)

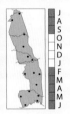

AFRICAN SPOONBILL *Platalea alba* **(95) 83 cm; 1,6 kg**
SEXES *Alike, ♂>♀ and with longer and slightly more decurved bill. Red face, white plumage and bill shape distinctive; unlikely to be confused with any other species.* **JUV** *Blackish streaks on crown, bill yellowish to horn-coloured.* **STATUS** Rare summer vagrant to KNP though some possibly resident; gregarious and nomadic in response to rainfall and habitat availability. Occasionally breeds in KNP. **HABITAT** Shallow aquatic margins of larger dams and in KNP's larger rivers. **FOOD** Small fish and aquatic invertebrates. Feeds by wading through water with bill partly submerged (and opened) sweeping side-to-side. As tactile feeders they have sensitive bills able to detect and snap up aquatic prey. **CALL** Usually silent; gives guttural croaks and grunts. **BR** Monogamous and colonial. Stick platform nest usually in reedbeds, sometimes in trees over water. (**Lepelaar**)

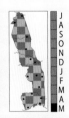

GREAT WHITE PELICAN *Pelecanus onocrotalus* **(49) 1,6 m; 9,5 kg**
SEXES *Similar, ♂>♀. Larger than Pink-backed Pelican, with whiter plumage and longer bill and neck. In flight distinguised by conspicuous black flight feathers. During br season forehead swells knob-like, plumage is tinged pink, and has a yellowish breast patch. Sexes differ slightly in br plumage with ♀ developing bright orange facial skin. Bright yellow pouch slightly duller during non-br season.* **JUV** *Head, neck and upperparts dull buff-brown.* **STATUS** Rare vagrant in KNP, probably on passage in search of higher water levels and better food sources. Usually recorded at the larger dams as water levels decline during winter. May stay in such localities for several days while fish stocks last, spending up to about 25% of day fishing. Gregarious. Listed as Near-threatened in S Africa. **HABITAT** In KNP mainly large pans and dams. **FOOD** Mostly fish, also shrimps. **CALL** Grunts and moos at colony, silent elsewhere. **BR** Monogamous and colonial. Nests on ground, usually on predator-free islands; nest a shallow scrape. No br records for KNP. (**Witpelikaan**)
ALT NAME *White Pelican*

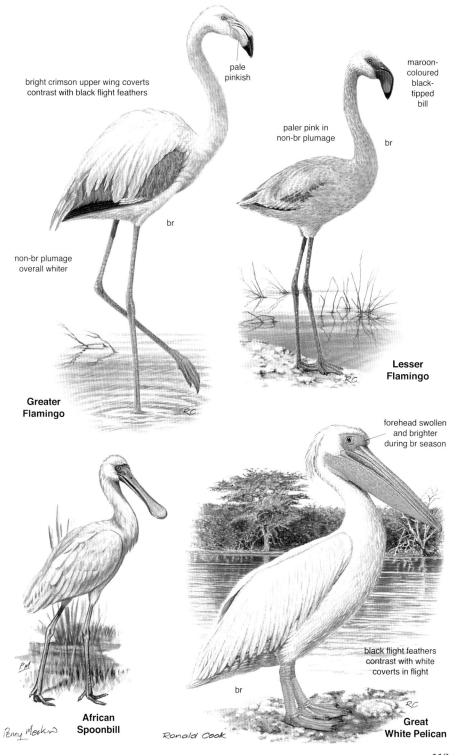

PLATE 51 Openbill, Storks pp217–8

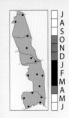

AFRICAN OPENBILL *Anastomus lamelligerus* **(87) 82 cm; 1,1 kg**
SEXES *Similar, ♂>♀ and with larger bill. Relatively small stork, blackish plumage glossed green, purple and brown.* **JUV** *Like adult but duller; bill initially straight and shorter.* **STATUS** Rather uncommon throughout KNP but has been recorded in some numbers at br colonies in fever tree forests along the Limpopo/Luvuvhu flood plain. May be resident or nomadic. Considered Near-threatened in s Africa. Usually singly or in small groups. Excretes on legs when overheated. **HABITAT** Wetlands, floodplains, shallow rivers, and dams. **FOOD** Almost entirely snails and mussels, occasionally frogs. Uniquely shaped bill has conspicuous gap specially adapted to facilitate removal of molluscs from their shells. Upper mandible is relatively straight and used to hold prey (usually under water), while sharp-tipped lower mandible is used to open and extract the flesh. **CALL** Raucous croaks and honks. **BR** Monogamous and colonial. Stick platform nest placed in tree above water. (**Oopbekooievaar**) **ALT NAME** *Openbilled Stork*

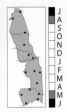

BLACK STORK *Ciconia nigra* **(84) 103 cm; 3 kg**
SEXES *Alike, ♂>♀ and with slighty longer, heavier and sometimes more recurved bill. Distinguished from smaller Abdim's Stork by red bill and legs, and in flight by black (not white) rump. Underparts white. Black plumage has glossier green and purple sheen during br season.* **JUV** *Browner than adult and lacking iridescence. Bill, facial skin and legs dull grey-green.* **STATUS** Uncommon br resident, nomadic in non-br season. Regarded as Near-threatened in S Africa. Usually found singly or in pairs but sometimes also in small groups in the non-br season. **HABITAT** Associated with mountainous regions, but not restricted to them. **FOOD** Mainly fish, also frogs, tadpoles, small mammals, nestling birds, tortoises, reptiles, insects and snails. **CALL** Generally silent when not br. At nest gives high-pitched whistling calls. **BR** Monogamous. Nest a platform of sticks placed on cliff ledge or in pothole and re-used in successive years. In KNP breeds in the gorges of the Luvuvhu, Mutale and Olifants rivers. (**Grootswartooievaar**)

ABDIM'S STORK *Ciconia abdimii* **(85) 76 cm; 1,3 kg**
SEXES *Alike, ♂>♀. Easily distinguished from Black Stork by smaller size, dull grey-green bill (tipped dull red) and greenish-olive legs (toes and 'knees' red). In flight shows white rump. Bare skin on face duller blue during non-br season.* **JUV** *Duller and browner than adult.* **STATUS** Rare to abundant non-br intra-African migrant. Most years almost absent from KNP, but has been recorded in abundance in others. Most arrive in Nov and depart by Apr (some overwintering). Sometimes in large flocks up to several thousand at insect outbreaks, although such numbers have not been recorded in KNP for several years. **HABITAT** Grassland and savanna woodland. **FOOD** Mainly insects, especially grasshoppers, locusts and crickets, and wide range of small vertebrates such as frogs, lizards, mice and small fish. Gathers in large numbers at insect irruptions, eg locusts, Armyworm, American Bollworm, caterpillars of Mopane Emperor Moth and termite alates. **CALL** Usually silent in s Africa. (**Kleinswartooievaar**)

WOOLLY-NECKED STORK *Ciconia episcopus* **(86) 84 cm; 1,8 kg**
SEXES *Alike. Only stork in region with black body and white (woolly-looking) neck. In flight, all-black wings and chest contrast with white neck, belly and undertail. Bill tip and ridge of upper mandible red. Upperparts glossed purple and eyes deep crimson-red during br season.* **JUV** *Upperparts duller and browner and face blacker; neck pale sooty brownish.* **STATUS** Generally uncommon, but sometimes congregates in some numbers in non br season, listed as Near-threatened in S Africa. Resident but migrant populations may also be present in non br season. **HABITAT** Wetlands and river margins. **FOOD** Mainly large insects, also frogs, crabs, molluscs, and some fish. Attends grass fires to feed on burnt insects and reptiles. Also feeds at termite alate irruptions, often with other storks and raptors. **CALL** Generally silent; at nest gives raucous calls. **BR** Monogamous. Nest placed high in well-foliaged tree, usually over swampy ground. (**Wolnekooievaar**)

PLATE 52 Storks p218

WHITE STORK *Ciconia ciconia* (83) 1,13 m; 3,5 kg
SEXES *Alike, ♂>♀. Large, mainly white with black flight feathers, bill and legs red. Unlikely to be confused with any other species.* **JUV** *Bill black with reddish base, legs dull brownish-red and wing coverts brown-tinged.* **STATUS** Common to abundant non-br Palearctic migrant. Numbers vary greatly between years; present Oct–May with a few old, injured or juv birds overwintering. In hot weather, cools down by defecating on legs. Travels in large flocks when temperatures rise, circling on thermals or up-currents to gain height, then glides to next thermal. **HABITAT** Grassland, wetland margins and open woodland. Usually in flocks of 10–50 birds but sometimes many hundreds congregate at insect irruptions. Attends veld fires to feed on flushed insects. **FOOD** Insects, especially caterpillars, locusts and crickets; also mice, small reptiles, frogs, tadpoles and termite alates. **CALL** Mostly silent. (**Witooievaar**)

YELLOW-BILLED STORK *Mycteria ibis* (90) 97 cm; 1,8 kg
SEXES *Alike. Large white stork with conspicuous yellow bill; unlikely to be confused with any other species. In flight distinguished from White Stork by black (not white) tail. At onset of br, naked facial skin becomes bright red, and back and upperwing covert feathers tinged pink. During non-br season bill duller yellow and legs dull red to pink.* **JUV** *Upperparts brownish, underparts washed grey-brown and bill greyish-yellow. Adult plumage attained after about 3 years.* **STATUS** Uncommon to locally common; regarded as Near-threatened in S Africa. Smaller resident KNP population may be augmented by non-br intra-African nomads and migrants between Oct and Apr. Found singly but more often in pairs. Nomadic in response to water levels and fish availability. **HABITAT** Shorelines of most KNP dams and larger rivers. **FOOD** Mainly fish, also frogs, aquatic insects, worms and crustaceans. Feeds with bill immersed and slightly open while stirring mud with foot. Bill intensely tactile, snapping closed if prey is detected. **CALL** Usually silent; during br season gives loud squeaks and other notes likened to a squeaking hinge. **BR** An occasional breeder in KNP in late winter. Monogamous and colonial in mixed-species heronries. Large stick nest placed in tree, usually over or near water. (**Nimmersat**)

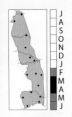

SADDLE-BILLED STORK *Ephippiorhynchus senegalensis* (88) 1,5 m; 6 kg
SEXES *Similar, ♀ has bright yellow eyes, ♂ has brown eyes and small pair of yellow pendant wattles at base of bill. Sometimes have round patch of dark red skin exposed in middle of breast.* **JUV** *Upperparts initially grey-brown, and bill (including saddle) black or dark brown, lacking any red till at least 6 months; legs greenish.* **STATUS** Uncommon resident occurring throughout KNP, with an estimated population of about 16 pairs, listed as Endangered in S Africa, but KNP population apparently stable. Sedentary but nomadic during drought. Found singly or in pairs. **HABITAT** Along large river systems, dam margins and wetlands. **FOOD** Mainly fish, also frogs, reptiles, small mammals, birds, termite alates, crustaceans and aquatic insects. Stabs at fish within reach and may impale large fish with one mandible. Attracted to Red-billed Quelea br colonies, presumably in search of young that may have fallen from nests. Drinks frequently, especially after feeding. **CALL** Adults apparently silent. **BR** Monogamous. Large platform nest built on top of tree in full sunlight, usually near water. (**Saalbekooievaar**)

MARABOU STORK *Leptoptilos crumeniferus* (89) 1,52 m; 6,4 kg
SEXES *Alike, ♂>♀. Large size and shape distinctive. The large bulbous pendant air sac may be up to 350 mm in size when inflated; a second orange-red air sac is partly hidden beneath the white ruff at base of hindneck, and only visible when inflated.* **JUV** *Wing feathers dark brown; first-year birds more feathered on the head.* **STATUS** Locally fairly common nomad; has been recorded in groups numbering several hundreds at locust or armoured cricket irruptions; birds in constant attendance at rubbish disposal sites in KNP. As regular breeding does not occur in KNP, it is suspected that these birds 'commute' between KNP and br sites further to the north, or even to a limited extent from Swaziland. Listed as Near-threatened in s Africa. On hot days may excrete urine onto legs for evaporative cooling effect. **HABITAT** Favours semi-arid areas; KNP population concentrated around rubbish tips where carrion and other scraps readily available. **FOOD** Primarily a scavenger but able to catch prey such as mice, birds, fish, frogs, newly hatched crocodiles and insects. **CALL** Mostly silent away from nest. **BR** Monogamous and loosely colonial. Only two breeding attempts recorded in KNP, one at Skukuza which failed, and one at Pafuri (outcome uncertain). Large stick platform built in tree. (**Maraboe**)

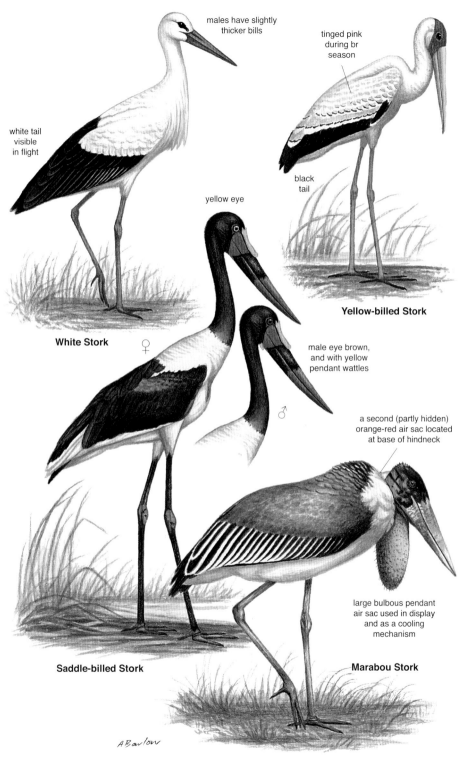

PLATE 53 Orioles, Paradies-Flycatcher, Drongo, Crested-Flycatcher p218

EURASIAN GOLDEN ORIOLE *Oriolus oriolus* **(543) 23 cm; 64 g**
SEXES M *Bright yellow body, black wings (also a feature in flight), and short yellow wing bar make the ♂ unmistakable.* Black loral streak hardly extends behind eye. **F** *Whitish lightly streaked underparts and more greenish upperparts distinguish it from ♀ African Golden Oriole.* **JUV** *Similar to ♀; more heavily streaked below.* **STATUS** Non-br Palearctic migrant, fairly common in some localities within the Park; shy nature seldom allows good views. Main arrivals in Nov, departing Feb–Mar. **HABITAT** Savanna and riparian woodland, favouring closed canopy woodlands. **FOOD** Both insects and fruit. Hawks termites aerially. **CALL** Mostly silent; call *who-are-you* or *weelawoo*. (**Europese Wielewaal**) **ALT NAME** *European Golden Oriole*

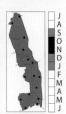

BLACK-HEADED ORIOLE *Oriolus larvatus* **(545) 25 cm; 65 g**
SEXES *Similar. Head and chest black, back greenish-yellow, belly bright yellow and bill coral pink.* **JUV** *Similar to adult but black head flecked yellow and breast streaked black. Bill blackish.* **STATUS** Common resident, with possibly local movements to regions outside the KNP during winter months. **HABITAT** Both sparse and dense woodland (particularly riparian) and savanna. **FOOD** Insects, fruit and nectar, particularly of flowering aloes esp. *A. marlothii*. **CALL** Loud liquid *kleeeu*; also other melodious notes. **BR** Monogamous. Male often calls in nest vicinity. Deep cup nest, often made of *Usnea* lichen, suspended hammock-like in thin horizontal fork in outer tree branch. (**Swartkopwielewaal**)

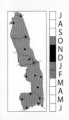

AFRICAN PARADISE-FLYCATCHER *Terpsiphone viridis* **(710) ♂ 37 cm, ♀ 17 cm; 14 g**
SEXES *Similar, ♂ larger than ♀ and with longer tail. Chestnut upperparts, dark head, bluish eye-ring and beak diagnostic.* **M** (**BR**) *Eye-wattle becomes larger and bluer during br; long central feathers temporarily lost during post-br moult (usually Jan–May).* Short-tailed males recorded (rare): may be recognised during br season by slightly heavier short tail and enlarged blue eye-wattles. **F** *Like ♂ but without long central tail feathers; eye-ring smaller.* Long-tailed females recorded (rare). **JUV** *Like ♀ but duller.* **STATUS** Common br intra-African migrant; some individuals overwinter in KNP. Most arrive Sept or early Oct and depart Mar/Apr; migratory birds return to same locality annually. Usually found singly or in pairs. **HABITAT** Favours riparian woodland; common in dense woodland, and well-wooded restcamp gardens. **FOOD** Mainly insects; spiders and small berries also recorded. **CALL** Very vocal. High-pitched *jig-see* or, when br, 5–6 liquid notes *tzzee switty-tsweep-sweepy-taweep*. **BR** Monogamous. Neat cup-shaped nest placed on exposed branch built by both sexes who also share incubation and feeding duties. Not a regular host of any of the cuckoos. (**Paradysvlieëvanger**) **ALT NAME** *Paradise Flycatcher*

FORK-TAILED DRONGO *Dicrurus adsimilis* **(541) 25 cm; 44 g**
SEXES *Similar, ♀>♂ and with less deeply forked tail. Most likely to be confused with Southern Black Flycatcher but larger; deeply forked tail, more robust bill and red (not brown) eye diagnostic.* Tail often appears double-forked during moult. In flight wings are pale and translucent. Mobs large raptors; aggressive towards other large bird species. **JUV** *Dark grey below, with buff and grey-tipped feathers. Yellow gape visible, but not prominent.* **STATUS** Very common widespread and conspicuous resident, generally sedentary; found singly or in pairs. **HABITAT** All woodland habitats, less common in evergreen forest; enters restcamp gardens. **FOOD** Insects (especially bees and termites), mostly caught in flight. Small birds and nectar also recorded, often follows elephants and other large herbivores hawking disturbed insects. **CALL** Loud repetitive *twik*; also jumble of 'unoiled wagonwheel' creaks and rasping notes. Mimics other bird calls. **BR** Monogamous. Nest a thinly made, wide shallow cup, suspended hammock-like between 2 horizontal branches. Only known regular host of African Cuckoo. (**Mikstertbyvanger**)

BLUE-MANTLED CRESTED-FLYCATCHER *Trochocercus cyanomelas* **(708) 15 cm; 10 g**
SEXES M *Glossy blue-black head and crest very distinctive; unlikely to be confused with any other species in the park.* **F** *Pale (not dark) grey head and crest, and narrow wingbar.* **JUV** *Like adult but more grey-brown above and with rufous-tipped wing coverts.* **STATUS** Very uncommon winter visitor resulting from altitudinal movements from the escarpment forests down riparian corridors of the major rivers; usually found singly. **HABITAT** In KNP occurs in undergrowth and lower canopy of riparian zones of the larger rivers. **FOOD** Insects. **CALL** Rasping *jig-zwee, jig-zwee*, similar to African Paradise-Flycatcher. Song more melodious *kew-ew-ew-ew-ew* or *say-say-say-say* ... and similar notes. **BR** Monogamous. Nest a neat thin-walled cup usually placed in upright fork, but no br records for KNP. (**Bloukuifvlieëvanger**) **ALT NAME** *Blue-mantled Flycatcher*

BRUBRU *Nilaus afer*   **(741) 14 cm; 24 g**
SEXES *Similar, ♀ duller with sooty brown (not black) upperparts and paler rufous flanks. Smallest of the shrikes in the region with typical shrike appearance.* Conspicuous with broad white eyebrow and conspicuous rufous flank stripe; dark eye and broad long eye-stripe distinguishes it from smaller and similar batises. **JUV** *Upperparts mottled brown and buff, wing stripes buff, underparts cream, barred brown.* **STATUS** Fairly common resident; sedentary. **HABITAT** Favours fairly dry woodland, especially *Acacia* but also occurs in Mopane and other broad-leaved woodland. **FOOD** Insectivorous. Sometimes joins mixed-species foraging flocks. **CALL** ♂ gives repetitive trilling *tippy-tip-prrrrreeeeeee* and variations thereof. **BR** Monogamous. Cup nest placed in fork of tree and extremely well camouflaged with lichen. Incubation by both sexes. (**Bontroklaksman**)

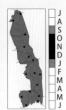

BLACK-BACKED PUFFBACK *Dryoscopus cubla*   **(740) 17,5 cm; 26 g**
SEXES *Similar, ♂>♀. ♀ duller and paler with greyish crown, back and wings. During display, ♂ erects back and rump feathers to create spectacular white 'powder puff'. ♀ may also raise back 'puff' but to lesser extent and less frequently.* **JUV** *Initially buff-brown below, later similar to ♀ but with horn-coloured bill.* **STATUS** Common widespread resident, sedentary and found singly or in pairs. Joins mixed species foraging flocks. Aggressive in defence of nest and against threats such as small snakes which it may attempt to displace or dislodge from tree or branch. **HABITAT** Moist well wooded and riparian forest, but also dry woodland; common in restcamp gardens. **FOOD** Insectivorous, also occasionally fruit and buds. **CALL** Loud 2-syllabled click-whistle *chick-wheeu, chick-wheeu ...*, mostly by ♂. **BR** Monogamous. Cup nest bound with spider web and placed in fork of branch. Incubation by both sexes. (**Sneeubal**) **ALT NAME** *Puffback*

BLACK-CROWNED TCHAGRA *Tchagra senegalus*   **(744) 21 cm; 53 g**
SEXES *Alike. Combination of black (not brown) crown and greyish underparts distinguishes it from Brown-crowned Tchagra. Bill black.* **JUV** *Bill horn-coloured, crown mottled and tail buff-tipped.* **STATUS** Common resident, found singly or in pairs. Usually shy and reclusive, retiring to centre of bush if alarmed; may run along the ground to escape attention while looking for cover. **HABITAT** Occupies wide range of habitats, from dry savanna to moist broad-leaved woodland, but prefers more open eastern mesic grasslands rather than the dense woodland favoured by Brown-crowned Tchagra. **FOOD** Mainly insects, also tadpoles, small snakes and lizards. Forages mainly on ground but sometimes joins mixed species feeding flocks for short spells. **CALL** Very characteristic loud warbling whistles: *wheeya, heeeea, hyoooee, whee-hoo ...* **BR** Monogamous. Shallow cup nest usually placed 1–2 m above ground in shrub, also sometimes on aloe leaf; incubation by both sexes. (**Swartkroontjagra**)

BROWN-CROWNED TCHAGRA *Tchagra australis*   **(743) 17,5 cm; 33 g**
SEXES *Alike. Combination of brown crown and buff or grey-buff underparts distinguishes it from other tchagras. Bold white eyebrow flanked on either side by conspicuous black stripes. Has brown chestnut wings and white tips to outer tail feathers. Bill may be dark, horn-brown or black.* **JUV** *Bill grey or horn-coloured, terminal tail spots buff and wings less rufous.* **STATUS** Common resident, usually occurring singly or in pairs. Less arboreal than Black-crowned Tchagra, more prone to skulking and creeping along the ground. Reluctant flier, preferring to dart for cover low down. **HABITAT** Semi-arid woodland, broad-leaved woodland, riverine thickets but favours denser habitats rather than more open mesic grasslands. **FOOD** Mainly insects, also small vertebrates. Forages mainly on the ground, especially beneath shrubs, but also gleans insects from branches and takes termite alates aerially. Frequent member of mixed-species foraging flocks. **CALL** In descent from characteristic aerial display flight gives loud descending trill. **BR** Monogamous. Shallow cup nest placed in tree, shrub or on aloe leaf; incubation by both sexes. (**Rooivlerktjagra**) **ALT NAME** *Three-streaked Tchagra*

PLATE 55 Boubous, Shrike, Bush-Shrike p218

TROPICAL BOUBOU *Laniarius aethiopicus* (737) 23 cm; 50 g
SEXES *Alike. Distinguished from Southern Boubou by creamy-white to pinkish underparts and very distinctive call. Southern Boubou has rufous flanks and belly.* **JUV** *Upperparts dull black, spotted tawny; bill greyish-brown.* **STATUS** Common but very localised resident; in KNP occurs only in the Luvuvhu and Limpopo riparian zones, range does not overlap with that of Southern Boubou. Sedentary and usually in pairs but may be in small family groups after br season. **HABITAT** Dense vegetation, found almost exclusively in riverine thickets. **FOOD** Mainly insects, but also small vertebrates such as small snakes, chameleons, skinks, frogs, rodents and snails; occasionally fruit. **CALL** Main song a synchronised duet, *haw-wheee-haw* or *wah-heeya-wah* ... , first and third notes usually by ♂, second by ♀. Also characteristic frog-like *ghwaaa*. **BR** Monogamous. Thin untidy cup nest well concealed in bush or among climbers. Parasitised by Black Cuckoo. (**Tropiese Waterfiskaal**)

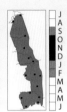

SOUTHERN BOUBOU *Laniarius ferrugineus* (736) 22 cm; 59 g
SEXES *Buffy belly and distinctive call help distinguish this species from Tropical Boubou, with which it is reported sometimes to hybridise in Limpopo River valley (outside the KNP).* **M** *Black upperparts; white throat and breast merge to rufous on flanks and belly.* **F** *Slate grey above, underparts more extensively rufous than ♂, giving overall duller appearance.* **JUV** *Paler grey above, mottled buff-brown above and lightly barred rufous on white underparts.* **STATUS** Common endemic except in the Luvuvhu/Limpopo valley where it is absent, and its range does not overlap with that of Tropical Boubou. Sedentary, secretive and found singly or in pairs. **HABITAT** Dense vegetation and riparian thickets, also in restcamp gardens; particularly common in the Nwambiya sandveld. **FOOD** Mainly insects and snails, also geckos, mice, earthworms, fruit and nectar. Snails banged repeatedly against rock or branch in attempt to break shell and reach flesh. Probes aloe flowers for nectar, and recorded eating the eggs of other birds. **CALL** Synchronised duets vary somewhat, giving each pair their own unique tone or combination. Main call a loud ringing *whee-whee-hohoho* or *hohoho-wheeeyoo* and similar combinations that may be initiated by either sex. **BR** Monogamous. Nest an untidy bowl usually well concealed low in leafy bush; incubation by both sexes. Host of Black Cuckoo (fairly common). (**Suidelike Waterfiskaal**)

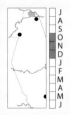

CRIMSON-BREASTED SHRIKE *Laniarius atrococcineus* (739) 22 cm; 48 g
SEXES *Alike. One of the most striking birds in the region; unlikely to be confused with any other species. Rare yellow form occurs (brilliant crimson underparts replaced by yellow).* **JUV** *Initially mottled and barred buff-brown, becoming blacker on upperparts, with varying amounts of red below.* **STATUS** Very rare vagrant in KNP, recorded only in the arid areas between the Limpopo and Luvuvhu rivers. May be resident but not recorded in all years. Near-endemic, sedentary and usually found singly or in pairs. **HABITAT** Semi-arid *Acacia* savanna. **FOOD** Invertebrates, especially beetles, ants and caterpillars; also small fruits. **CALL** Main call loud and fast *tyotyo* or *quipquip* and variations thereof, either solo or in duet. **BR** Monogamous, but br not recorded in KNP. Nest a neat cup of bark strips, bound with cobweb. Host of Black Cuckoo in s Africa. (**Rooiborslaksman**)
ALT NAME *Crimson-breasted Boubou*

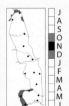

GORGEOUS BUSH-SHRIKE *Telophorus viridis* (747) 19 cm; 37 g
SEXES *One of the most strikingly beautiful birds on the continent, shy and difficult to see.* **SEXES** *Similar.* **M** *Black gorget extends up side of throat to lores.* **F** *Black gorget, lores and moustachial strip greatly reduced or almost absent, throat duller orange-red.* **JUV** *Lacks black gorget and red throat, underparts dull yellowish-green. Distinguished from juv Orange-breasted Bush-Shrike by greenish (not greyish) head.* **STATUS** Locally common resident, sedentary and found singly or in pairs. Disjunct distribution in KNP found only in the Limpopo/Luvuvhu valley, the Nwambiya sandveld and the hilly south-western Pretoriuskop/Berg-en-Dal area, and along the western parts of the Sabie River. Shy and skulking, seldom perching on exposed branches. In the winter months, may perch in the sun on semi-exposed branch. **HABITAT** Dense dry woodland, and riparian zones. **FOOD** Insectivorous. **CALL** Distinctive 2–4 note call, *kong-kong-koweet* or *kong-koweet-koweet*, and variations thereof. **BR** Monogamous. Nest a shallow cup of twigs and rootlets placed in dense low shrub, incubation by both sexes. (**Konkoit**)

PLATE 56 Bush-Shrikes, Helmet-Shrikes p219

ORANGE-BREASTED BUSH-SHRIKE *Telophorus sulfureopectus* (748) 19 cm; 27 g
SEXES *Alike, but ♀ duller and with narrower eyebrow. Much smaller than Grey-headed Bush-Shrike, distinguished by lighter bill, yellow forehead and eyebrow, and reddish-brown (not pale yellow) eyes. More often heard than seen.* **JUV** *Head and face pale grey, lightly barred white. Underparts pale yellow, with only a faint tinge of orange.* **STATUS** Common resident; sedentary, shy and usually seen hopping from branch to branch in upper tree canopy. Usually located by its distinct call. **HABITAT** Dense mixed *Acacia*, open riparian and broad-leaved woodlands. **FOOD** Insectivorous; most insects gleaned from upper tree canopy. Also hawks bees. **CALL** Loud, ringing 4-note call. Repetitive, with various versions of the phrase *kew-tee-tee-tee-tee* and *what-to-to-dooo*. **BR** Monogamous. Nest a shallow bowl of twigs, usually placed in thorn tree. (**Oranjeborsboslaksman**)

GREY-HEADED BUSH-SHRIKE *Malaconotus blanchoti* (751) 26 cm; 77 g
SEXES *Similar, ♀ slightly duller. Largest 'yellow' shrike in region. Distinguishable from much smaller Orange-breasted Bush-Shrike by enormous bill, pale yellow (not dark) eye and eye (not yellow) forehead and eyebrow.* **JUV** *Head mottled brown and grey, underparts with only a few orange feathers, and eyes brown becoming grey or whitish.* **STATUS** Uncommon to locally common resident, mostly sedentary. **HABITAT** Well-developed *Acacia* woodland, riverine forest and wooded restcamps. **FOOD** Mainly insects which it gleans from twigs, leaves and branches as it works its way along upper canopy branches. Also small vertebrates such as chameleons, lizards, small snakes, bats, birds, eggs and nestlings. Joins mixed-species feeding flocks during the non-br season; occasionally hawks flying insects; sometimes hunts for prey on the ground. **CALL** Repetitive and mournful drawn-out whistle, *whooooooooo ...* **BR** Monogamous. Nest a broad shallow cup placed in upper tree branches, incubation by both sexes. (**Spookvoël**)

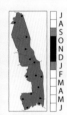

WHITE-CRESTED HELMET-SHRIKE *Prionops plumatus* (753) 18 cm; 34 g
SEXES *Alike, ♀>♂. Usually seen in flocks moving among trees. Grey crown, yellow eye-wattle and black and white plumage distinctive.* **JUV** *Duller than adult, upperparts greyish-brown, underparts buffy and lacks eye-wattle.* **STATUS** Locally common resident with considerable local movement in the non-br season and less common in KNP during the summer months. Breeding groups vary from 2–10 birds, non-br groups in excess of 20 at times when separate flocks may team together. Flocks move as cohesive unit with undulating flight from tree to tree. Groups roost huddled together in row on a branch. Aggressive towards predators which they mob as a group, dive-bombing and bill-snapping. Joins mixed-species feeding flocks (sometimes together with Retz's Helmet-Shrike). Wanders in response to environmental conditions, more common in KNP in winter months. **HABITAT** Broad-leaved woodland, but also mixed *Acacia*, particularly along drainage lines. **FOOD** Insects, spiders and small lizards. Forages both in the canopy and on the ground. **CALL** Usually given as a group chorus, a ratchety *krawow, krawow, kreee, kreepkraw ...* **BR** Co-operative br and probably monogamous. Cup nest constructed by all members of group and thickly bound with cobweb. Incubation shared by all members. (**Withelmlaksman**) ALT NAME *White Helmet-Shrike*

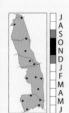

RETZ'S HELMET-SHRIKE *Prionops retzii* (754) 19,5 cm; 44 g
SEXES *Alike, ♀>♂. Dark plumage, bright orange-red eye-wattle and bill, and habit of moving through the trees in small flocks very distinctive.* **JUV** *Grey-brown above, buffy-white below, legs yellow or orange and lacks wattles.* **STATUS** Fairly common resident, nomadic in response to food supplies. Average group size about 5; up to 30 birds in non-br season when separate flocks may team together. Move among trees as a cohesive group. **HABITAT** Favours tall broad-leaved woodland, but moves into other mixed woodland habitats during the winter months. **FOOD** Insects, spiders and a small gecko also recorded. Joins mixed-species foraging flocks. **CALL** Wide variety of weird grating *tweeooh-tweew ...* notes. **BR** Co-operative br, and probably monogamous. Cup nest neatly bound with cobweb and placed high in upper branches of tree; built by dominant pair with help from group members; incubation mainly by br pair with help from other members. Parasitised by Thick-billed Cuckoo. (**Swarthelmlaksman**) ALT NAME *Red-billed Helmet-Shrike*

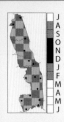

CAPE BATIS *Batis capensis* **(700) 13 cm; 12 g**
SEXES *Small, compact, shrike-like forest flycatcher.* **M** *Deep rufous flanks and wing panel diagnostic.* **F** *Lacks broad black chest band, rufous flanks and wing panel slightly paler than ♂.* **JUV** *Like ♀ but black face mask absent and eyes paler.* **STATUS** Very rare vagrant. An altitudinal migrant that moves in severe winters from escarpment forest into KNP down riparian corridors along major rivers. Near-endemic to s Africa. **HABITAT** Common in Afromontane forest to the west of KNP but occasionally also present in lowland evergreen riparian forest in winter. **FOOD** Insects gleaned from the lower or mid strata of riparian trees. **CALL** Most common call 3-note *whew-whew-whew*, also 'stone rubbing' sounds. **BR** Monogamous. Compact cup nest well camouflaged with lichen; not known to breed in the Park. Host of Klaas's Cuckoo. (**Kaapse Bosbontrokkie**)

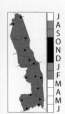

CHINSPOT BATIS *Batis molitor* **(701) 13 cm; 12 g**
SEXES ♀>♂. *Only resident batis in KNP so confusion with other species should not occur.* **M** *Broad black chest band, white wing and eye-stripe, and pale yellow eyes.* **F** *Dark chestnut breast band and throat patch easily distinguish ♀ from ♂.* **JUV** *Similar to ♀ but upperparts mottled buff.* **STATUS** Very common resident, sedentary and usually in pairs. **HABITAT** Savanna woodland, especially *Acacia*, also in broad-leaved woodland such as Mopane and in restcamp gardens. **FOOD** Insects. Most prey gleaned from leaves and twigs, but also hawks insects. Regularly joins mixed-species feeding flocks. **CALL** Usually 3 descending notes, *weep-woop-wurp* or 'three-blind-mice'; also gives 'stone rubbing' sounds. **BR** Monogamous. Nest a small compact cup, camouflaged with lichens and bark flakes, incubation by both sexes. Occasionally parasitised by Klaas's Cuckoo. (**Witliesbosbontrokkie**)

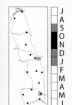

BLACK-THROATED WATTLE-EYE *Platysteira peltata* **(705) 13 cm; 13 g**
SEXES *Glossy black upperparts, mostly white underparts and conspicuous bright red wattle above eye diagnostic.* **M** *Throat white with narrow black chest band.* **F** *Similar to male but with all-black throat.* **JUV** *Initially smoky grey becoming light brown-grey; wattle dull red.* **STATUS** Locally common only in the Limpopo/Luvuvhu valley. Often seen at the Pafuri picnic site. Old records exist for the Olifants Gorge and Lower Sabie. Resident and sedentary and listed as Near-threatened in S Africa. Usually in pairs, occupying the undercanopy of riverine vegetation. **HABITAT** Dense riparian forest and thickets, seldom far from water. **FOOD** Insectivorous, most prey taken aerially; joins mixed-species foraging flocks along riverine vegetation. **CALL** Rasping and unmusical *djip-djip-djip-zipweet* ... **BR** Monogamous. Neat cup nest usually placed on low branch. (**Beloogbosbontrokkie**) **ALT NAME** *Wattle-eyed Flycatcher*

RED-BACKED SHRIKE *Lanius collurio* **(733) 18 cm; 29 g**
SEXES M *Male unlikely to be confused with any other shrike in region, with combination of brick-red back and wings and grey head.* **F** *Duller than ♂; lightly scalloped underparts and fine vermiculations on crown distinguish it from Marico Flycatcher.* **JUV** *Similar to ♀ but more heavily vermiculated above.* **STATUS** Very common non-br Palearctic migrant; majority arrive mid- to late Nov, most departing by Apr. **HABITAT** Primarily semi-arid open woodland, especially *Acacia*. Males generally prefer more open habitat with fewer and more shrubby trees, while females often favour denser, taller *Acacia* woodland. **FOOD** An insect specialist on non-br grounds, foraging mostly using a sit-and-wait technique, then pouncing on prey on the ground. Occasionally hawks insects such as termite alates. **CALL** Series of harsh warbles (including mimicry), especially in early morning and evening during the late summer months just prior to migration. (**Rooiruglaksman**)

LESSER GREY SHRIKE *Lanius minor* **(731) 21 cm; 46 g**
SEXES *Similar, ♀ slightly duller, mask narrower and with scattered pale feathers. Grey crown, nape and back easily distinguish it from Common Fiscal (which has prominent white scapular bar). Salmon pink underparts conspicuous in fresh plumage during late summer months just prior to northward migration; not always visible on non-br grounds.* **JUV** *Paler than adult, more buff above and yellowish below.* **STATUS** Fairly common non-br Palearctic migrant; most arrive second half Nov, and depart Mar/Apr. **HABITAT** Arid and semi-arid open *Acacia* savanna; favours the eastern basaltic grassland habitats but also found in dry broad-leaved woodland. **FOOD** Mainly insects, especially beetles; does not drink. **CALL** Harsh grating *geer-geer ... shrek-shrek-shrek-shrek* ... notes. (**Gryslaksman**)

PIED CROW *Corvus albus* (548) 49 cm; 550 g
SEXES *Alike.* ♀ *has slightly shorter and more slender bill. White breast extending to broad white collar on hindneck unmistakable.* **JUV** *White feathering tipped dusky and less glossy than adult.* **STATUS** Rare vagrant to KNP. Seldom seen in natural habitats, usually only associated with human settlements and KNP records reflect this. Recorded regularly only at Punda Maria which is fairly close to the Park boundary and has a considerable human presence close by. An unsuccessful breeding attempted was recorded on a radio mast at Skukuza. **HABITAT** In KNP has been recorded in a wide range of habitats; mostly associated with human settlements. **FOOD** Omnivorous but primarily plant material, including seeds, fruit and roots, and nectar; also invertebrates, such as termite alates. Vertebrate prey and scavenged food items includes lizards, small mammals, bats, snakes, birds, nestlings, eggs and fish. Regularly takes carrion. **CALL** Loud harsh *kraah*, and a snoring *khrrr*. **BR** Monogamous. Nest a large bowl of sticks and twigs built by both sexes, placed in isolated tree, telephone pole, pylon, windmill or occasionally on a building. Incubation by both sexes; successful br not recorded in KNP. Parasitised by Great-spotted Cuckoo elsewhere in s Africa. (**Witborskraai**)

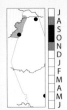

WHITE-NECKED RAVEN *Corvus albicollis* (550) 52 cm; 800 g
SEXES *Alike,* ♀>♂. *White collar on hindneck, heavy bill with whitish tip and, in flight, broad wings and short tail distinctive. Usually in mountainous regions or where there are cliffs and gorges on which it can br.* **JUV** *Browner than adult with narrow whitish breast band. White feathers on hindneck sometimes flecked black.* **STATUS** Uncommon vagrant recorded only in the Punda Maria area. **HABITAT** Mountains, gorges and cliffs, but hunts over adjacent plains and patrols roads for road kills. **FOOD** Wide variety of animal and insect prey including birds, tortoises (usually dropped onto rocks below), eggs; scavenges from carcasses and dump sites. Fruit and seeds also recorded. **CALL** Series of loud deep *kraak, kraak, kraak* notes, often given in flight. **BR** Monogamous. Nest a large bowl of sticks lined with grass, hair and wool, placed on inaccessible ledge or pothole in cliff. Br not recorded in KNP. Incubation by both sexes. (**Withalskraai**)

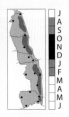

MAGPIE SHRIKE *Corvinella melanoleuca* (735) 45 cm; 82 g
SEXES *Similar. Striking long graduated tail and black and white plumage distinctive.* ♀ *can be distinguished by creamy white flanks.* **JUV** *Tail initially shorter and with dark brown (not black) plumage.* **STATUS** Locally common resident, sedentary but moves locally within KNP in response to drought and fires. Usually in groups of 3–12 birds perched conspicuously on the outer branches of trees or shrubs. Birds roost in loose groups usually about 1 m apart in regularly used trees. **HABITAT** Mainly on eastern grasslands on basalt, but elsewhere open *Acacia*-dominated woodland with scattered trees. **FOOD** Mainly insects, also small reptiles, mice, fresh and rotting meat and fruit. Most prey taken from ground but occasionally hawks insects. **CALL** 1–3 syllabled clear flute-like whistles *plee-teeooo* ... **BR** Monogamous and often co-operative. Cup nest usually placed 3–5 m up *Acacia* tree. Incubation by ♀ only, fed by ♂ and other members of the group. Nestlings fed by all group members. (**Langstertlaksman**) ALT NAME *Long-tailed Shrike*

SOUTHERN WHITE-CROWNED SHRIKE *Eurocephalus anguitimens* (756) 24 cm; 70 g
SEXES *Alike. White crown and dark line through eye diagnostic. Distinguished from similar-looking Northern White-crowned Shrike (distribution only in e Africa) by greyish-brown (not white) rump.* **JUV** *Crown off-white, mottled brown; bill yellowish.* **STATUS** Locally common near-endemic; sedentary, but with some nomadic movements as birds not always recorded at same localities. Recorded throughout the Park in pairs or small groups of 4–8 birds. Frequently squats down when perched, with only toes showing. **HABITAT** Favours dry deciduous woodland and park-like savanna with sparse ground cover; also occupies mixed *Acacia* woodland. **FOOD** Almost entirely invertebrates. Group may form nucleus of mixed-species feeding flock in the non-br season. Most prey taken from ground, but also gleans insects from foliage and hawks prey aerially. **CALL** Harsh, nasal *skwee-kwee-kwee* notes. **BR** Monogamous and co-operative. Nest built by alpha ♂ and ♀, assisted by 1–2 other group members; cup walls well bound with cobweb and placed in fork of outer branch of tree. Incubation mainly by alpha ♀, but sometimes by helpers for short periods. (**Kremetartlaksman**) ALT NAME *White-crowned Shrike*

PLATE 59 Cuckooshrikes, Penduline-Tit, Tit p219

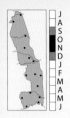

WHITE-BREASTED CUCKOOSHRIKE *Coracina pectoralis* **(539) 27 cm; 56 g**
SEXES M *White underparts, grey throat and upperparts and conspicuous black eye distinctive; slightly larger than Laughing Dove.* **F** *Chin and upper throat white.* **JUV** *Barred black and white on grey above, spotted black below.* **STATUS** Very uncommon but probably resident; territorial when br but with some nomadic movements in dry season. Sporadic records from throughout the Park. **HABITAT** Well-developed woodland, particularly Mopane; also riparian zones. **FOOD** Insectivorous, feeding mainly on caterpillars. Most insects gleaned from foliage and branches. Hawks termite alates aerially; may join mixed-species feeding flocks. **CALL** ♂ gives whistled *duid-duid* calls, ♀ a trilled *chreeeeee*. **BR** Monogamous. Shallow cup well camouflaged with lichen, placed 6–20 m high on thick branch or in horizontal fork. Incubation by both sexes. (**Witborskatakoeroe**)

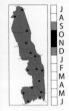

BLACK CUCKOOSHRIKE *Campephaga flava* **(538) 21 cm; 32 g**
SEXES *Bright orange gape (more conspicuous in ♂) diagnostic and a good field character especially when differentiating it from Fork-tailed Drongo and Southern Black Flycatcher.* **M** *Black with bluish gloss; rounded tip to tail and dark brown or blackish (not red) eyes distinguish it from Fork-tailed Drongo. Yellow shoulder patch (when present) separates ♂ from drongo and Black Flycatcher, but is absent in many birds (more common in south of range such as KZN and E Cape than in KNP).* **F** *Fine barring and yellow edges to wing and tail feathers give it similar appearance to* Chrysococcyx *cuckoo females, but larger.* **JUV** *Similar to ♀.* **STATUS** Fairly common resident, usually solitary and unobtrusive. **HABITAT** Well developed tall broad-leaved, mixed *Acacia* and riparian woodland margins. **FOOD** Insects, particularly caterpillars, also some fruit. Gleans prey from foliage and branches and frequently joins mixed-species feeding flocks. **CALL** High-pitched and prolonged cricket-like *kreeeee* ... that may be heard some distance away. **BR** Nest a beautifully made shallow cup camouflaged with lichen and placed in upright fork; incubation by ♀ only. (**Swartkatakoeroe**)

GREY PENDULINE-TIT *Anthoscopus caroli* **(558) 8–9 cm; 6,5 g**
SEXES *Alike. Forehead plain buffy (not black and white). Distinguished from slightly larger Yellow-bellied Eremomela by cinnamon-buff (not yellow) underparts, and shorter, sharper bill.* **JUV** *Like adult but belly paler buff.* **STATUS** Fairly common resident but generally an easily overlooked species. Sedentary, with little evidence of nomadic movements in KNP during winter months. Occurs in pairs or small family groups (usually 3–5 birds) that roost together in old nest or old weaver nest. **HABITAT** More common in broad-leaved woodland than *Acacia*-dominated habitats, but does occur in both. **FOOD** Insects; also feeds on aloe nectar. Forages in mixed-species flocks with other small insectivores such as eremomelas, white-eyes and Willow Warblers. **CALL** Calls range from buzzing *chi-chi-dzwizzz, ti-ti-sssweeee* or *tsi-tsi-eeee* notes to a rasping *chiZEE-chiZEE-chiZEE*. **BR** Monogamous. Nest structure unique to family and built by both sexes. An oval ball placed 5–10 m above the ground is constructed of woven spider web, plant down and other woolly material giving it a smooth and very strong felt-like finish. The side-top has a thin-walled collapsible entrance spout 30–40 mm long that normally closes automatically or is pulled closed. Below this spout is a hollow or depression in the wall (that looks like the nest entrance known as the 'false' entrance) and ridge on which the bird lands to open the closed spout entrance to nest. (**Gryskapokvoël**)

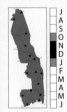

SOUTHERN BLACK TIT *Parus niger* **(554) 16 cm; 21 g**
SEXES *Similar, ♀ slightly duller and with greyer underparts. Noisy and easily detected species.* **JUV** *Like ♀ but buffier grey below and tail with narrower white tip.* **STATUS** Common resident, sedentary and usually in pairs or small family groups. Only tit occurring in KNP, obviating confusion with other species. Roosts singly in natural tree hole or behind loose bark on tree trunks. **HABITAT** Almost any woodland dominated by broad-leaved trees; less common in *Acacia*-dominated woodland. Fairly common in restcamp gardens. **FOOD** Mainly insects and their larvae. Opens *Combretum* and *Acacia* pods and thorns in search of larvae. Plucks fig fruit open to check for wasp larvae and also takes termite alates aerially. Recorded drinking nectar from aloes. **CALL** Main call harsh *diddy-jee-jee-jee-jee*; also shrill, buzzing *zeu-zeu-zeu-twit* and ringing *teeu teeu teeu* and *pitu pitu pitu*. **BR** Monogamous and commonly a co-operative br. Nests in natural tree cavities, or old woodpecker or barbet holes. Incubation by alpha ♀ only. Occasionally parasitised by Greater Honeyguide. (**Gewone Swartmees**)

SAND MARTIN *Riparia riparia*   **(532) 13 cm; 13 g**
SEXES *Alike. Distinguished from Brown-throated Martin by white throat and brown breast band. Most likely to be confused with Banded Martin (rare vagrant to Park) but lacks white eyebrow and is considerably smaller; in flight, underwing coverts dark (not white).* **JUV** *Feathers on upperparts edged cinnamon, rufous or creamy, especially on forehead.* **STATUS** Very uncommon non-br Palearctic migrant, present in KNP late Sept–Apr (mainly Oct–Mar). Recorded mainly in the northern parts of the Park. Sometimes perches on low bushes and even on the ground where it sun-bathes. Bathes by dipping into water surface during fluttering flight, and also in dew on foliage. Roosts with other swallows, especially Barn Swallow. **HABITAT** Favours drainage lines, marshes and grassland adjacent to water bodies; also occurs in Mopane woodland. **FOOD** Aerial insects taken mostly close to vegetation. **CALL** Harsh twittering. (**Europese Oewerswael**)

BROWN-THROATED MARTIN *Riparia paludicola*   **(533) 12 cm; 12,5 g**
SEXES *Alike. Distinguished from Sand, Rock and Banded martins by brown throat and whitish underparts. Small percentage of population (ca 2–20%) have entirely brown underparts. They can, however, be distinguished from slightly larger Rock Martin by dark (not pale) underwing coverts and lack of white spots in tail.* **JUV** *Like adult but with buff-tipped feathers on upperparts and underparts washed rufous-buff.* **STATUS** Common resident, nomadic in the non-br season; KNP populations may be supplemented in winter by short-distance migrant birds from the south and west. Usually gregarious and roosts communally in reedbeds at night, huddled in small groups. **HABITAT** Most common along the larger permanent rivers which offer suitable sandbanks for nesting. Occasionally found over dry land particularly when not br. **FOOD** Aerial insects. Although diurnal, forages till after sunset particularly at termite alate emergences. **CALL** Song high-pitched series of twitterings; contact call harsh *svee-svee*. **BR** Monogamous and usually colonial, excavating burrows in sandy river banks. Sometimes uses old abandoned burrows of other bird species such as bee-eaters. Incubation mainly by ♀, but ♂ and ♀ often in the burrow together at night. (**Afrikaanse Oewerswael**)

BARN SWALLOW *Hirundo rustica*   **(518) 14 cm (streamers = 6 cm); 20 g**
SEXES *Alike, but ♂ has slightly longer outer tail streamers, which are often absent in the austral summer months (non-br season). Diagnostic features are chestnut-rufous throat and forehead, blue-black breast band, glossy steel blue upperparts and deeply forked tail. White windows clearly visible in tail when spread.* **JUV** *Buffy forehead, paler chin and throat, and duller and browner upperparts.* **STATUS** Common to abundant non-br Palearctic migrant. Migration is diurnal, birds traveling in excess of over 300 km per day; first arrivals in Sept, most in Oct or early Nov. Highly gregarious, feeding by day in loose flocks and roosting communally (mostly in *Phragmites* reedbeds) in groups of a few hundred to thousands of birds. Large flocks gather on trees and telephone lines or before departing end Feb to early May. Susceptible to mass mortality during extreme weather conditions such as unseasonally cold weather and hail storms. **HABITAT** Forages over all KNP habitats. **FOOD** Mainly aerial insects, usually low over vegetation and often together with other swallow species. Most prey taken aerially but some prey (particularly caterpillars) plucked from vegetation. Drinks in flight by skimming surface of water. **CALL** Song of ♂ short bubbling twitter. (**Europese Swael**) **ALT NAME** *European Swallow*

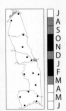

ROCK MARTIN *Hirundo fuligula*   **(529) 15 cm; 22 g**
SEXES *Alike. All-brown plumage distinctive. Likely only to be confused with very uncommon brown morph Brown-throated Martin. Tail, however, square (not slightly forked) and with white spots. Throat and breast pinkish-cinnamon; in flight has contrasting pale underwing coverts.* **JUV** *Feathers on upperparts buff-tipped.* **STATUS** Common but highly localised resident with probable altitudinal movements into KNP during winter months. Rarely drinks and sometimes roosts gregariously. **HABITAT** Rocky gorges, particularly the Luvuvhu Gorges but also the Olifants gorge. **FOOD** Aerial insects. Attracted to fires and attends termite alate emergences. **CALL** Song a soft melodious twittering. **BR** Monogamous and usually solitary. Nest a cup of mud pellets usually placed under rock overhang or under bridge. Incubation by both sexes. (**Kransswael**)

GREY-RUMPED SWALLOW *Pseudhirundo griseopyga* (531) 14 cm; 10 g
SEXES *Alike. Distinguished from Common House Martin by brown-grey (not blue) crown, greyish (not white) rump, and longer and more deeply forked tail. In flight, distinguished from Pearl-breasted Swallow by dark (not white) underwing coverts.* **JUV** *Chin and throat buffy, lacks blue gloss to upperparts, which appear scalloped with pale fringes to feathers.* **STATUS** Records suggest it is a scarce resident in KNP; numbers may be augmented by migrants during winter months, May–Sept when it breeds. Usually in pairs or small loose flocks, and roosts gregariously in reedbeds in non-br season. **HABITAT** Typically over alluvial floodplains, large woodland clearings or short grassland, often near water. **FOOD** Aerial insects, regularly taking termite alates. Attracted to fires. **CALL** Weak nasal *wha*, almost hissed. **BR** Monogamous. Br recorded in July in KNP; one of the few insectivorous bird species able to breed in mid-winter in the region. Nests underground in flat, short grassland, usually in deserted rodent burrows on alluvial soils along the major rivers, but sometimes in old kingfisher or Little Bee-eater burrow. (**Gryskruisswael**)

COMMON HOUSE-MARTIN *Delichon urbicum* (530) 14 cm; 18 g
SEXES *Alike. All-white underparts and white rump diagnostic. Confusion possible with Grey-rumped Swallow in flight but tail much less deeply forked, rump white (not greyish) and crown blue (not brown-grey).* **JUV** *Upperparts greyish-brown; rump off-white and narrower.* **STATUS** Scarce to locally common non-br Palearctic migrant; numbers fluctuate considerably from year to year. Present late Sept–Apr or May and occasionally overwinters. Usually forages at high altitudes and mixes with swifts and other swallows. Sometimes rests on trees, often with Barn Swallows. Probably sleeps on the wing, but occasionally reported roosting in reedbeds. Concern for declining populations in many parts of the w Palearctic br grounds. **HABITAT** Wide variety of habitats. **FOOD** Aerial insects. Attracted to bush fires. **CALL** High-pitched *prt* or *prt-prt* notes. (**Huisswael**) **ALT NAME** House Martin

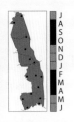

WIRE-TAILED SWALLOW *Hirundo smithii* (522) 14,5 cm (streamers = 4 cm); 13 g
SEXES *Similar. The only swallow in the Park with bright reddish-chestnut crown and all-white underparts. Black vent band conspicuous in flight and sometimes noticeable when perched; outer tail feathers long, slender and wire-like (shorter in ♀).* **JUV** *Crown ashy brown, underparts buffy, outer tail feathers short.* **STATUS** Locally common resident. Widespread and present at most of the Park's many bridges where it often perches on the railings or concrete structures. Usually in pairs; flight swift and graceful; roosts on nest edge at night, even when not br. **HABITAT** Associated with water bodies; forages over adjacent woodland and grassland. **FOOD** Aerial insects; forages singly or together with other swallows. **CALL** Twittering *chirrikweet, chirrik-weet*; also soft single notes. **BR** Monogamous. Nest a small open cup of mud pellets, lined with grasses and feathers and placed under an overhang or under bridge structures; incubation by both sexes. One of the few insectivorous birds in the Park able to breed during the winter months. (**Draadstertswael**)

PEARL-BREASTED SWALLOW *Hirundo dimidiata* (523) 13,5 cm; 12 g
SEXES *Alike. Only swallow in region with entirely blue-black upperparts and all-white underparts. Most similar to migratory Common House-Martin, but lacks white rump. In flight similar to Wire-tailed Swallow but lacks rufous crown and dark vent stripe.* **JUV** *Duller than adult.* **STATUS** Rare resident, with possible winter influxes of passage migrants in Apr, May and Oct. Aggregates with other swallow species during migration. Usually in pairs or small groups; flight fast and agile. **HABITAT** Usually associated with human structures and water bodies, mainly in broad-leaved woodland. **FOOD** Aerial insects; also recorded feeding on grass seeds. **CALL** Series of harsh sharp notes, *chip-chip-chip-cher-cher-chip-chip-cher* ... **BR** Monogamous. Single nesting record from Skukuza in December. Nest a shallow cup of mud pellets, strengthened with hair or dry grass and lined with feathers. Most nests in southern S Africa placed on buildings; in north of region such as in the KNP, nest sites vary from rock overhangs, bridges and buildings to mammal burrows and mine shafts. (**Pêrelborsswael**)

PLATE 62 Swallows, Saw-wing p220

LESSER STRIPED SWALLOW *Hirundo abyssinica*   **(527) 17 cm; 18 g**
SEXES *Similar. The only swallow in the Park with heavily streaked underparts. In flight from above, dark back and wings and rufous head and rump distinctive.* **JUV** *Duller, with breast and flanks washed buff.* **STATUS** Common br intra-African migrant. Most birds absent from KNP from Mar/Apr–July/Aug, with few birds remaining year-round. **HABITAT** Occurs throughout KNP in open savanna and riparian woodland regions. Habit of breeding on man-made structures has considerably improved the breeding potential, population size and distribution in KNP. Flocks during the non-br season, sometimes with other swallows and swifts. Roosts inside nest as soon as it is half-complete. **FOOD** Mainly aerial insects, but also recorded feeding on the small fruit of Pigeonwood *Trema orientalis*. **CALL** Sharp *chip* contact notes, and slow drawn-out *trip, trip, trreep, treep, treep* ... notes. **BR** Monogamous. Mud nest with entrance tunnel built by both sexes, usually placed under eaves or bridges, also under rock overhangs. Incubation by ♀ only. (**Kleinstreepswael**)

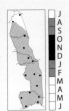

RED-BREASTED SWALLOW *Hirundo semirufa*   **(524) 22 cm; 31 g**
SEXES *Similar. Large size and rich rufous underparts very distinctive. Distinguished from Mosque Swallow by rufous (not whitish) cheeks and throat, and buff (not whitish) underwing coverts.* **JUV** *Brownish above, much paler below.* **STATUS** Locally common br intra-African migrant. Arrives Aug most departing by Mar for Equatorial Africa. The habit of using smaller road culverts for breeding has benefited and expanded the population and distribution within the KNP. Less common in the north where it may be replaced by the Mosque Swallow. Usually singly or in pairs; br adults roost in nest at night, even before eggs are laid. **HABITAT** Open savanna and sweet grassland. **FOOD** Aerial insects, attracted to bush or grass fires and feeds with other swallows with termite alate emergences. **CALL** Slow gurgled *chip-cheedle-urr* and a *peeeeurrr* whistle. **BR** Mostly monogamous. Mud nest with long entrance tunnel built by both sexes usually placed in narrow road culvert; also in natural sites such as Aardvark holes. Incubation by ♀ only. (**Rooiborsswael**)

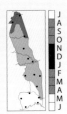

MOSQUE SWALLOW *Hirundo senegalensis*   **(525) 21 cm; 45 g**
SEXES *Alike, ♂>♀. Distinguished from Red-breasted Swallow by whitish (not rufous) cheeks and throat, and in flight by whitish (not buff) underwing coverts.* **JUV** *Upperparts dark brown; duller than adult and secondaries and tertials tipped buff.* **STATUS** Generally scarce, but locally common in the Baobab regions of the KNP. Largely resident, but may be a partial migrant as is the case in n Namibia, n Botswana and parts of Zimbabwe. Usually in pairs or small groups and often near large Baobabs *Adansonia digitata*. **HABITAT** Tall broad-leaved woodland with scattered Baobabs. **FOOD** Aerial insects. Forages just above vegetation, sometimes in association with other aerial feeders, and attends termite alate emergences. **CALL** Notes include nasal *naaaah* and piping *pyuuuu*. **BR** Monogamous. Mud nest with entrance tunnel built by both sexes and placed under overhang or in hollow of large tree (especially Baobabs). (**Moskeeswael**)

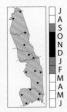

BLACK SAW-WING *Psalidoprocne holomelas*   **(536) 15 cm; 12 g**
SEXES *Similar, ♀ with shorter outer tail feathers. Only all-black member of swallow family in the KNP.* **JUV** *Brown sheen to plumage and shorter outer tail feathers. Underwing coverts sooty grey to blackish.* **STATUS** Very rare vagrant or passage migrant, recorded only at a few widespread localities. Present in s Africa as a breeding migrant, present Aug/Sept–Apr/May, with small numbers remaining year-round. Found singly or in pairs. **HABITAT** Fringes of woodland and forest, especially riparian. **FOOD** Aerial insects. Forages with graceful slow flight just above the canopy and between tall riverine trees. **CALL** Weak *zzer-eet* or *skreea*. **BR** Monogamous. Nest tunnel excavated by both sexes into low earth-bank, but no br records for KNP. (**Swartsaagvlerkswael**) **ALT NAME** *Black Saw-wing Swallow*

DARK-CAPPED BULBUL *Pycnonotus tricolor* (568) 20 cm; 39 g
SEXES *Alike, ♂>♀. An abundant species easy to identify with dark blackish head that has a slight crest, pale brownish body and bright yellow undertail.* **JUV** *Upperparts have rusty wash; otherwise similar to adult.* **STATUS** One of the most common and conspicuous birds in KNP; sedentary and usually in pairs or small family parties. **HABITAT** All KNP habitats and restcamps. **FOOD** Mainly fruit, also insects, nectar and flower petals. **CALL** Loud series of 3–7 disjointed notes such as *quick-chop-toquick* or *'wash-your-feet-Gregory'*. Occasionally mimics other species. **BR** Monogamous. Cup nest built by ♀ and well concealed in foliage, usually 2–4 m up. Parasitised by Jacobin Cuckoo. (**Swartoogtiptol**) **ALT NAME** *Black-eyed Bulbul*

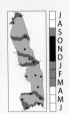

SOMBRE GREENBUL *Andropadus importunus* (572) 21 cm; 31 g
SEXES *Alike, ♂>♀. Dull olive-green upperparts and olive-grey underparts without any conspicuous features other than pale whitish eyes.* **JUV** *Brownish eyes, paler bill and orange gape for first 3 months.* **STATUS** Common resident, generally sedentary and found singly or in pairs. Usually found in riparian thickets and dense cover. **HABITAT** All dense riparian woodland and forest habitats; also common in wooded gulleys in hilly south-west and in restcamps. **FOOD** Mostly small fruit, also insects, snails, leaf buds and aloe nectar. **CALL** Contact call ringing *wi-llie*. Song rapid jumble of loud notes ending in nasal slurr, *'willie, come-and-have-a-fight, sca-a-a-ared'*. **BR** Monogamous. Thin cup nest well concealed in foliage; placed 1–3 m up. Host of Jacobin Cuckoo. (**Gewone Willie**) **ALT NAME** *Sombre Bulbul*

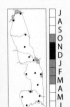

YELLOW-BELLIED GREENBUL *Chlorocichla flaviventris* (574) 22 cm; 40 g
SEXES *Alike, ♂>♀. Distinguished from smaller Sombre Greenbul by combination of yellower underparts, reddish eyes and nasal call. Other differences include white eye-ring, shaggy crest and, in flight, bright yellow underwing.* **JUV** *Duller and paler than adult; eyes initially grey.* **STATUS** Common but localised resident. Sedentary and in pairs or small groups during non-br season. Joins mixed-species foraging flocks. **HABITAT** Mostly riparian forests of the Luvuvhu/Limpopo floodplain in Nwambiya sand forest and scattered localities along the Lebombo hills. Can usually be seen in Olifants restcamp, Olifants Trails Camp and around Nwanetsi. A few records exist for Skukuza. **FOOD** Predominantly fruit; also insects, seeds and flowers. **CALL** Short series of very nasal yapping notes, rather like the call of the Black-backed Jackal. **BR** Monogamous. Nest a thin flimsy cup usually concealed 2–3 m above ground on leafy branch. Incubation period 14 days by ♀ only. (**Geelborswillie**) **ALT NAME** *Yellow-bellied Bulbul*

EASTERN NICATOR *Nicator gularis* (575) 23 cm; 47 g
SEXES *Alike, ♂ considerably larger than ♀. Slightly longer-tailed than bulbuls and with a heavier bill that is more hooked at the tip. Pale yellow wing spots conspicuous but with its shy and skulking behaviour may not easily be seen. Best located by its loud piercing call from upper canopy of a prominent tree.* **JUV** *Similar to adult.* **STATUS** Fairly common resident with a patchy distribution; sedentary. Not uncommon in the Malelane and Berg-en-dal areas. **HABITAT** Lowland riverine forest and thickets especially in Nwambiya sandveld and *Androstachys* woodlands and mixed thornveld. **FOOD** Insectivorous. Follows mammals such as Nyala and Warthog, taking flushed insects; also occasionally gleans ticks from mammals. **CALL** Calls frequently from tree-top song post. Loud series of repetitive bulbul-like bubbling notes, *wip chop chop rrup chopchopchop krrip krrr*; sometimes includes mimicry. **BR** Monogamous. Nest a dove-like platform concealed in thicket 1–2 m above ground. (**Geelvleknikator**) **ALT NAME** *Yellow-spotted Nicator*

TERRESTRIAL BROWNBUL *Phyllastrephus terrestris* (569) 21 cm; 36 g
SEXES *Alike, ♂>♀. Upperparts dull darkish brown, chin white and belly pale off-white. In small groups of 3–6 birds in lower 2 m of thick undergrowth. Usually heard before seen, when it gives itself away by scolding, ratchet-like alarm calls.* **JUV** *Paler with rufous-edged feathers; gape yellow.* **STATUS** Common resident throughout KNP wherever suitable habitat occurs, largely sedentary and in small groups in non-br season. **HABITAT** Understorey of evergreen and riparian forest, and dense woodland thickets. **FOOD** Mostly insects, also snails, small lizards and some fruit and seeds. **CALL** Song a series of low husky chattering notes; when disturbed, group gives series of harsh chattering churring notes. **BR** Monogamous. Cup nest suspended in thin horizontal fork and bound by roots, moss and *Marasmius* fungus, giving a blackish appearance. Uncommon host of Jacobin Cuckoo. (**Boskrapper**) **ALT NAME** *Terrestrial Bulbul*

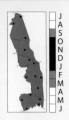

LONG-BILLED CROMBEC *Sylvietta rufescens*   **(651) 11 cm; 11 g**
SEXES *Alike. Conspicuous species with very short tail (appears almost tailless), longish bill, pale face and eyebrow.* **JUV** *Similar to adult.* **STATUS** Common resident throughout KNP, sedentary and usually in pairs or family groups. Regularly joins mixed-species foraging parties. **HABITAT** Dry woodland with well-developed shrub and bush growth. Favours *Acacia* thickets, but also occurs in broad-leaved woodland and restcamp gardens. **FOOD** Mainly insects; also seeds, fruit and aloe nectar. Most prey gleaned from twigs and foliage but hawks insects such as termite alates aerially, returning to perch to eat them. **CALL** A very vocal species that calls year-round. Tripping *chit-chirrit-chirrit-chirrit-chirrit* ... repeated several times. **BR** Monogamous. Hanging purse nest built by both sexes usually suspended from branch of bush or low tree, 1–2 m above ground. Parasitised occasionally by Klaas's Cuckoo. (**Bosveldstompstert**)

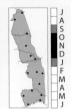

YELLOW-BELLIED EREMOMELA *Eremomela icteropygialis*   **(653) 11 cm; 8 g**
SEXES *Alike. Similar to Grey Penduline-Tit, but has longer tail, yellow (not buff) underparts and greyer throat and chest. Differs from Green-capped and Burnt-necked eremomelas by dark (not pale) eye and grey throat.* **JUV** *As adult; belly duller yellow.* **STATUS** Uncommon resident, mostly sedentary and usually in pairs or small groups up to 8 that move restlessly through the woodland. **HABITAT** Shrublands and broad-leaved woodland throughout KNP. **FOOD** Mainly insects; also fruit, seeds and aloe nectar. Gleans insects from twigs and foliage, working its way up from the base of the shrub to the top; during the non-br season joins mixed-species feeding flocks. **CALL** Short quick phrases, *chicku-chicku-chee*, and similar notes. Known to mimic other species especially during the non-br season when giving quiet sub-song. **BR** Monogamous. Nest a neat thin-walled cup placed about 1–1,5 m above ground between horizontal twigs in the outer branches of shrub or sapling, Incubation by both sexes; parasitised by Klaas's Cuckoo. (**Geelpensbossanger**)

GREEN-CAPPED EREMOMELA *Eremomela scotops*   **(655) 12 cm; 9 g**
SEXES *Alike, ♀>♂. Key features are whitish eyes, pale green to yellowish-grey crown and face, pale yellow breast and grey back. Lacks white eye-ring of white-eyes and has almost white (not yellow) belly.* **JUV** *Paler than adult.* **STATUS** Uncommon to locally fairly common. Sedentary and usually in pairs or small family parties in the non-br season. Interactions between neighbouring groups involve continual chases around and above canopy. Group members roost together in row on branch, often returning to same site on successive nights. Baths in dew and droplets in canopy leaves. **HABITAT** Favours broad-leaved and riparian woodland with tall trees. **FOOD** Insects gleaned from tree canopy. Forages mainly in canopy leaves and twigs. **CALL** Loud and monotonous *twip-twip-twip-twip-twip* ... usually from the top of a tall tree and often from dawn till after sunrise. **BR** Probably monogamous and br co-operatively. Small cup nest built by all members of the group is usually suspended at tip of branch, usually 6–8 m up. Incubation also by all group members. (**Donkerwangbossanger**)

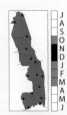

BURNT-NECKED EREMOMELA *Eremomela usticollis*   **(656) 11 cm; 9 g**
SEXES *Alike. Overall pale appearance with grey upperparts and whitish underparts. Longer bill and whitish (not buff) underparts distinguish it from Grey Penduline-Tit. Extent of narrow chestnut lower throat band variable, and usually absent in winter months.* **JUV** *Lacks chestnut cheeks and collar.* **STATUS** Fairly common resident in suitable habitat. Sedentary and in pairs or groups of 4–6. Highly active, moving restlessly through canopy. **HABITAT** A habitat-specific species that favours fine-leaved *Acacia* (particularly *Acacia tortilis*) woodland. Uncommon in broad-leaved woodland. **FOOD** Insects. Recorded feeding on aloe nectar. Joins mixed-species foraging flocks with other warblers, penduline-tits and white-eyes. **CALL** Song a very high-pitched and rapid *twee-twip-ti-ti-ti-ti-ti-ti* ... Call a sibilant *di-di-di-di*. **BR** Monogamous. Thin-walled cup nest placed in upper foliage of *Acacia* tree. Incubation by both sexes. (**Bruinkeelbossanger**)

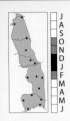

LITTLE RUSH-WARBLER *Bradypterus baboecala* **(638) 14 cm; 14 g**
SEXES *Alike. Dark brown upperparts, conspicuous dark broadish tail and dull rufous (not greyish-brown) undertail coverts distinguish it from Broad-tailed Warbler (very uncommon vagrant to Park).* **JUV** *Underparts tinged yellow.* **STATUS** Rather uncommon resident that is secretive and heard more often than seen. Sometimes perches prominently in the early mornings, especially in the winter months. Largely sedentary and found singly or in pairs. **HABITAT** Sedges and both *Typha* and *Phragmites* reedbeds, spending much time in lower strata of wetland vegetation. **FOOD** Insects. **CALL** Song a series of accelerating notes, *krak krak krak krak-krak-krak-krak-krak ...*, likened to a stick in the spokes of a bicycle wheel that is turning faster and faster. **BR** Monogamous. Bulky cup nest placed in low vegetation, usually just above water. (**Kaapse Vleisanger**) **ALT NAME** *African Sedge Warbler*

SEDGE WARBLER *Acrocephalus schoenobaenus* **(634) 13 cm; 12 g**
SEXES *Alike. Only warbler in region with boldly streaked crown; broad creamy eyebrow bordered by black above and below. Streaking on back inconspicuous, lower back and rump plain tawny-brown (good field characteristic). Water's edge species, seldom far from aquatic margins.* **JUV** *Overall warmer tawny or buff-coloured, and yellower below.* **STATUS** Rare non-br Palearctic migrant; arrives from late Oct–Jan, departs Mar/Apr. Usually found singly. **HABITAT** Mainly emergent aquatic vegetation at both perennial and ephemeral wetlands, also in adjacent weedy growth. **FOOD** Insects. Prey gleaned from aquatic vegetation, also takes insects aerially. **CALL** Contact call sharp *tuc*. Song fast series of churring and warbling notes (including mimicry). (**Europese Vleisanger**) **ALT NAME** *European Sedge Warbler*

AFRICAN REED-WARBLER *Acrocephalus baeticatus* **(631) 13 cm; 10 g**
SEXES *Alike. A nondescript species with no distinct characteristics that distinguish it from migratory Marsh Warbler in field, but usually occupies a different habitat to the latter. Warm brown upperparts paler below and with indistinct pale eyebrow.* **JUV** *Slightly more rufous.* **STATUS** Uncommon in KNP. Mostly resident north of 26° S; br intra-African migrant further south, present Aug–May. Probably more common during the winter months when birds from the escarpment to the west of the Park undertake altitudinal migration to lower levels. **HABITAT** Waterside margins, especially reeds and bulrushes; in drier habitats in non-br season. **FOOD** Insects gleaned from vegetation and hawked aerially. **CALL** Sustained series of guttural, churring notes, frequently changing pitch. Alarm note a scolding *churr*. **BR** Monogamous. Thick-walled cup nest placed in reedbeds over or close to water. Incubation by both sexes. (**Kleinrietsanger**) **ALT NAME** *African Marsh Warbler*

LESSER SWAMP-WARBLER *Acrocephalus gracilirostris* **(635) 17 cm; 15 g**
SEXES *Alike, ♂>♀. Plain brown upperparts, whitish underparts and eye-stripe, pleasant liquid call and likely only to be seen in reedbeds. Possible confusion only with larger and far less common Great Reed-Warbler but is weaker-billed and with less buffy flanks.* **JUV** *Similar to adult.* **STATUS** Common resident where suitable habitat occurs, largely sedentary. **HABITAT** Found almost exclusively in bulrushes on the margins of man-made dams, and *Phragmites* reedbeds. **FOOD** Insects and small frogs. **CALL** Beautiful liquid warbling notes; very vocal. **BR** Monogamous. Cup nest placed over or close to water. Incubation by both sexes. (**Kaapse Rietsanger**) **ALT NAME** *Cape Reed Warbler*

GREAT REED-WARBLER *Acrocephalus arundinaceus* **(628) 20 cm; 30 g**
SEXES *Alike. Largest member of warbler family in region, similar-looking to smaller Lesser Swamp-Warbler that is whiter below and has a weaker bill. Shy and secretive and very difficult to see, gives presence away by loud grating call.* **JUV** *Overall darker than adult.* **STATUS** Uncommon but widespread non-br Palearctic migrant; most arrive Dec/Jan and depart late Feb–Apr. **HABITAT** Usually in reedbeds and thickets near water; but often also in adjacent dense woodland. **FOOD** Mostly insects, but also small frogs. **CALL** Loud, harsh grating, crackling and creaking notes, with frequent use of *kar-kar-kar* and *kee-kee-kee*. Most vocal prior to departure during the months of Jan and Feb. (**Grootrietsanger**)

PLATE 66 Warblers p221

MARSH WARBLER *Acrocephalus palustris* (633) 13 cm; 11 g
SEXES *Alike. Field differentiation from African Reed-Warbler extremely difficult, but has different habitat preference and, unlike them, seldom found in emergent aquatic vegetation and reedbeds.* **JUV** *Upperparts warmer brown.* **STATUS** Uncommon but secretive non-br Palearctic migrant recorded throughout KNP; arrives late Nov, departs late Mar/Apr. **HABITAT** Favours dense thickets and rank tangled vegetation in lush woodland. **FOOD** Mostly insects. **CALL** Song quieter and less harsh than Great Reed-Warbler; rich and lively blend of warbling notes mixed with imitations of both Palearctic and African species. Usually sings from low down in thickets and rank vegetation within 2 m of the ground. Contact call *tuk*; alarm call longer *tcchhh*. (**Europese Rietsanger**) **ALT NAME** *European Marsh Warbler*

GARDEN WARBLER *Sylvia borin* (619) 14 cm; 19 g
SEXES *Alike. Only brownish-grey warbler in region without a noticeable eyebrow or any obvious features. Brown morph Icterine Warbler has longer bill and short but obvious pale eyebrow; Marsh Warbler warmer brown, and lacks greyish tinge to plumage. Presence usually given away by loud singing in woodland or restcamps (Jan–Mar).* **JUV** *Like adult, underparts more rusty brown.* **STATUS** Uncommon to locally common br Palearctic migrant, most arrive Oct/Nov and depart Mar/Apr. Many return annually to same sites in region. **HABITAT** Dense vegetation such as riparian forest edges, woodland thickets, dense secondary bush and well-wooded restcamp gardens. **FOOD** Insects, also berries and fruit. **CALL** Sustained mellow but forceful warbling song; alarm call hard *vik* or *vik-vik*. (**Tuinsanger**)

OLIVE-TREE WARBLER *Hippolais olivetorum* (626) 17 cm; 18 g
SEXES *Alike. Largest 'grey' warbler in region; slightly smaller and considerably greyer than Great Reed-Warbler but vocally similar. Blackish wings conspicuous against whitish underparts and with pale wing panel. Gives presence away by loud grating call in dry Acacia habitats.* **JUV** *Similar to adult.* **STATUS** Uncommon non-br Palearctic migrant, present Nov–Apr; solitary in non-br grounds. Many sites occupied year after year, suggesting non-br site fidelity. **HABITAT** Thickets in dry *Acacia* woodland, particularly Umbrella Thorn *A. tortilis* and Sickle-bush *Dichrostachys cinerea* thickets. Occasionally found in broad-leaved woodlands. **FOOD** Insectivorous. **CALL** Series of deep, harsh and grating *krek-krek, krak, krak, krok ...* notes. (**Olyfboomsanger**)

ICTERINE WARBLER *Hippolais icterina* (625) 13 cm; 13 g
SEXES *Alike. Larger than Willow Warbler, heavier-billed and longer-winged. Lacks long pale eyebrow that extends behind eye as in Willow Warbler, and has distinctive pale wing panel; blue-grey legs and feet diagnostic. Occurs in 2 colour phases: yellow phase birds lemon-yellow below and with yellow eye-ring; pale phase birds almost white below, greyish-brown above.* **JUV** *Pale wing panel more buffy-yellow, and not as prominent.* **STATUS** Uncommon non-br Palearctic migrant. Peak arrivals Nov, departure quick and almost simultaneous in second half of Mar and early Apr. **HABITAT** Prefers tall well-developed mixed woodland, particularly with scrub layer. **FOOD** Mostly insects. **CALL** Sings for prolonged periods throughout non-br season. Song a sustained warble, slower and more drawn out than Marsh Warbler (that is more likely to sing low down in rank vegetation) and comprising both harsh and melodious notes. Frequently mimics. Contact call a sharp *tec*. (**Spotsanger**)

WILLOW WARBLER *Phylloscopus trochilus* (643) 12 cm; 9 g
SEXES *Alike. Smaller and thinner-billed than Icterine Warbler, with eyebrow extending well behind eye, and pinkish-brown (not blue-grey) legs. Garden Warbler lacks conspicuous pale eye-stripe and has blue-grey (not pinkish-brown) legs.* **JUV** *Like adult but yellower below.* **STATUS** Common non-br Palearctic migrant; most arrive Oct/Nov, and depart Mar/Apr. Usually found singly but sometimes joins mixed-species foraging parties. **HABITAT** Wide range of woodland, including mixed broad-leaved woodland; also open *Acacia* savanna, lowland forest and restcamp gardens. **FOOD** Gleans insects from foliage and small twigs in canopy. **CALL** Presence often given away by frequently given contact call, a soft ascending *huitt*. Song a musical warbling, *si-si-sisi-swee-swee-su-su-sweet-sweet-sweetu*; most vocal between Feb and Apr, prior to departure. (**Hofsanger**)

PLATE 67 Hyliota, Babblers, White-eyes p221

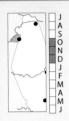

SOUTHERN HYLIOTA *Hyliota australis* **(624) 11,5 cm; 11 g**
SEXES *Similar.* **M** *Velvety-black upperparts, white wingbar, pale yellow wash to throat and breast and very distinct habitat preference makes this species unmistakable.* **F** *Similar but duller; back warm dark brown. White wingbar and yellow-washed throat and breast similar to ♂, outer tail feathers white.* **JUV** *Like ♀ but feathers buff-tipped above.* **STATUS** Very rare and localised br resident. Usually in pairs or small family groups; joins mixed-species foraging flocks. **HABITAT** In KNP strongly associated with the Wild Syringa *Burkea africana*, whose distribution is confined to sandveld areas around Punda Maria. Usually seen on the Mahonie Loop near Matukwala Dam. **FOOD** Insects gleaned from canopy of trees. **CALL** Series of tuneless squeaky whistles. **BR** Monogamous. Well-camouflaged cup-shaped nest placed on branch in upper canopy of tree. (**Mashonahyliota**) **ALT NAME** *Mashona Hyliota*

ARROW-MARKED BABBLER *Turdoides jardineii* **(560) 24 cm; 72 g**
SEXES *Alike. Streaked or 'arrowed' head and breast very distinctive. Two-tone eyes (yellowish centre and reddish rim) give an orange appearance.* **JUV** *Similar to adult but lacks white streaks; spotted dusky below; brown eyes.* **STATUS** Common and conspicuous resident, largely sedentary, usually in flocks of 4–8 birds. **HABITAT** Dry woodland, especially broad-leaved woodland; often close to riparian thickets and common in most restcamp gardens. **FOOD** Insectivorous; also small reptiles, seeds and small fruit. Recorded feeding on nectar. **CALL** Harsh bubbling chatter, often in chorus *ra, ra, ra, ra, ra ...* , rising to a crescendo. **BR** Monogamous and co-operative br. Bulky open cup nest placed 2–4 m above ground and built by all group members. Clutch 2–5 eggs; incubation and feeding of young shared by all group members. Main host of Levaillant's Cuckoo. (**Pylvlekkatlagter**)

CHESTNUT-VENTED TIT-BABBLER *Parisoma subcaeruleum* **(621) 14 cm; 16 g**
SEXES *Alike. Chestnut undertail coverts diagnostic at all ages. A skulking species often unwilling to expose itself in the open. Eyes white, upperparts slightly darker.* **JUV** *Undertail coverts paler chestnut; streaking on throat indistinct.* **STATUS** Very rare vagrant from the drier west with isolated records throughout the Park; usually recorded only in drier years. **HABITAT** Recorded mainly in thornveld. **FOOD** Mainly insects; also fruit, seeds and nectar. **CALL** Sharp *cherrri-tik-tik*; song melodious and bubbling. **BR** Monogamous; br peak varies regionally. Small thin-walled cup nest placed in thin outer branches of low tree or bush. Recorded (not in KNP) as a rare host of Jacobin Cuckoo. (**Bosveld-tjeriktik**) **ALT NAME** *Tit-Babbler*

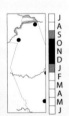

AFRICAN YELLOW WHITE-EYE *Zosterops senegalensis* **(797) 12 cm; 10 g**
SEXES *Alike. Underparts bright yellow, not washed green or grey as in Cape White-eye. Forehead and crown yellow (crown olive-green in Cape White-eye), upperparts paler and more yellow-green (not green-yellow).* **JUV** *Greener above than adult.* **STATUS** Uncommon resident, largely sedentary with local movements in response to food availability. Joins mixed-species feeding parties in non-br season. In KNP recorded only in the Luvuvhu and Limpopo riparian woodlands. **HABITAT** Favours riverine forest **FOOD** Insects, fruit and nectar. **CALL** Melodious, whistled warble that may last several minutes: *tsee-tseer-tsi-tsi-tsi ...* **BR** Monogamous. Thin cup nest placed in leafy foliage of tree. Incubation shared by both sexes. Parasitised by Green-backed Honeybird. (**Geelglasogie**) **ALT NAME** *Yellow White-eye*

CAPE WHITE-EYE *Zosterops virens* **(796) 12 cm; 11 g**
SEXES *Alike. Greener crown and greenish-yellow (not pure yellow) wash to underparts distinguish it from African Yellow White-eye.* **JUV** *White eye-ring develops after about 5 weeks.* **STATUS** Very common endemic, mostly sedentary, and found in small parties in non-br season. Common in most restcamps. Bathe gregariously on edge of small ponds and pools. Flock readily mobs owls disturbed during the day, and joins other birds in 'mobbing' snakes and other potential predators. Invariably roosts in twos, huddled together in the foliage of an outer tree branch. **HABITAT** Favours riparian zones but found in all woodland. **FOOD** Insects, fruit and nectar from both trees and aloes. Most insects gleaned from foliage, but also hawks prey in flight; joins mixed-species feeding parties. **CALL** Series of long warbling phrases, varying in pitch, volume and tempo; often includes mimicry. **BR** Monogamous. Small deep cup nest placed in leafy foliage of tree or bush. Incubation by both sexes. (**Kaapse Glasogie**)

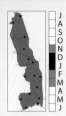

RED-FACED CISTICOLA *Cisticola erythrops* **(674) 14 cm; 15 g**
SEXES *Alike, ♂>♀.* **BR** *Only cisticola in region with uniform unmarked upperparts that are olive-grey from crown to tail. Sides to face pale rufous, flanks buff to light grey, and chin, breast and belly cream-white.* **NON-BR** *Crown and nape browner and more olive on upperparts.* **JUV** *Browner than adult.* **STATUS** Locally common and widespread resident that is highly vocal and usually gives its presence away by loud and ringing call; sedentary and usually in pairs. **HABITAT** Rank growth along edges of rivers, streams, and at the base of moist hillside slopes. Usually singly, in pairs or in small family groups after breeding. **FOOD** Insectivorous. **CALL** Distinctive loud ringing song ending in crescendo *weet-weet-weet-WEET-WEET-WEET-WEET* ... **BR** Monogamous. Nest unlike any other cisticola in the Park. Ball of dry grass leaves sewn with cobwebs into broad leaves of shrub or small tree, usually within 1 m of the ground. Incubation mostly by female. Parasitised in KNP by Klaas's Cuckoo. (**Rooiwangtinktinkie**)

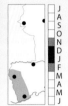

LAZY CISTICOLA *Cisticola aberrans* **(679) 14 cm; 14 g**
SEXES *Alike, ♂>♀. Best identification feature is habit of cocking its longish tail, prinia-like. Tail carried higher than other cisticolas and usually held vertical when calling. Confusion with prinias unlikely, however, as none of the prinias have rufous crowns.* **BR** *Crown dull russet, back olive-grey and upper tail coverts russet; underparts dull white.* **NON-BR** *Crown richer rufous-brown, underparts darker buff.* **JUV** *Duller than non-br adult.* **STATUS** Locally common resident, usually found in pairs or small groups after br season. Recorded only in the hilly south-western corner of KNP. Hops or runs mouse-like across rocks and boulders. **HABITAT** Preference for rank grassy vegetation amongst rocky outcrops. **FOOD** Insectivorous diet includes caterpillars, grasshoppers, small beetles and butterflies. **CALL** Distinctive whining or wailing notes; also crescendo *tu-whee-tu-whee* ... **BR** Monogamous. Oval-shaped nest built of dry grass and cobweb placed low down in grass or low bush. Parasitised by Brown-backed Honeybird. (**Luitinktinkie**)

RATTLING CISTICOLA *Cisticola chiniana* **(672) 15 cm; 16 g**
SEXES *Alike, ♂>♀. Robust, with few distinguishing features other than mottled greyish-brown back; call and habitat, however, distinctive.* **BR** *Crown russet, underparts greyish-brown, and flanks with greyish wash.* **NON-BR** *Crown more reddish, and underparts darker buff.* **JUV** *Similar to adult, variable and paler.* **STATUS** Very common and widespread resident. **HABITAT** Dry *Acacia* savanna, and broad-leaved woodlands. **FOOD** Insectivorous diet includes beetles, termites, grasshoppers, crickets, ants and butterfly larvae. Also recorded taking nectar from aloes. **CALL** One of the most characteristic sounds of the bushveld: *chi chi chi C H I R R R R R*. Calls from prominent perch high on top of shrub or small tree. Very vocal species that calls most of the year. When alarmed gives harsh *cheee, cheee, cheee* ... **BR** Monogamous. Ball nest, with side entrance, built of dry grass and concealed low down in thick grass or shrub cover. Incubation by ♀ only. Parasitised by Brown-backed Honeybird and Diderick Cuckoo. (**Bosveldtinktinkie**)

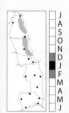

RUFOUS-WINGED CISTICOLA *Cisticola galactotes* **(675) 13 cm; 13 g**
SEXES *Alike. Rufous wingbar distinctive.* **BR** *With heavily streaked black back and rufous wingbar, unlikely to be confused with any other cisticola in the Park. Call diagnostic. Rattling Cisticola has greyer back that is lightly streaked black and lacks the prominent rufous wingbar.* **NON-BR** *Upperparts more rufous.* **JUV** *Underparts light lemon-yellow.* **STATUS** Rare and localised resident recorded mainly at a few localities in the northern parts of KNP. One record at Leeupan suggests it may have a wider distribution; near-endemic. **HABITAT** In KNP usually associated with the robust Vleigrass *Sporobolus consimilis* in vleis on basaltic soils. **FOOD** Insectivorous; prey includes grasshoppers, crickets, small beetles and caterpillars; some seeds also recorded. **CALL** Loud penetrating *chwik, chwik, chwik* ... (sometimes single notes given), followed by *chit chit chit* ... **BR** Monogamous. Ball nest made of dry grass placed just above damp ground in grass tuft, sometimes over water. (**Swartrugtinktinkie**) **ALT NAME** *Black-backed Cisticola*

PLATE 69 Cisticolas, Neddicky p222

CROAKING CISTICOLA *Cisticola natalensis* **(678) 15 cm; 20 g**
SEXES *Similar, ♂>♀.* **M (BR)** *Large size, whitish underparts, mottled brown back and black bill diagnostic.* **F (BR)** *Slightly smaller than ♂, mottled brown back, buff underparts and brown bill.* **NON-BR** *Underparts brighter buff.* **JUV** *Duller than adult, underparts washed yellow.* **STATUS** Generally uncommon but recorded at widely scattered localities, probably sedentary but perhaps with some winter altitudinal movements. Unobtrusive, usually giving its presence away during the summer months by loud croaking call given in flight. Male perches prominently on small bush in grassland between bouts of aerial cruises and display flights. **HABITAT** Open moist grassland and edges of seasonally flooded grassland, sometimes with scattered shrubs and trees. **FOOD** Mainly insects; seeds also recorded. **CALL** Diagnostic. In flight loud croaking *qrrrr, qrrrr, qrrrr* ... **BR** Monogamous. Oval nest with side entrance placed low down in tuft of grass. Incubation by ♀ only; parasitised by Cuckoo Finch. (**Groottinktinkie**)

NEDDICKY *Cisticola fulvicapilla* **(681) 11 cm; 9 g**
SEXES *Alike, ♂>♀. Small plain-backed cisticola with greyish underparts that gives itself away by distinctive call. Southern and eastern races grey below with dull rufous crown. Northern (KNP) race slightly more buffy below, longer-tailed and with more rufous crown.* **STATUS** Common resident, usually found singly or in pairs and in family groups at the end of breeding season. **HABITAT** Broad-leaved and mixed woodland, and shrubby grassland. Favours lightly wooded grassland. **FOOD** Insectivorous; also takes nectar. Forages low down, often on the ground; also takes insects aerially. **CALL** Monotonous *gedick, gedick, gedick* .... onomatopoeic from which it probably derives its name. Usually calls from a prominent perch at the top of a tree or shrub for long periods during the br season. **BR** Monogamous. Oval-shaped nest placed low in a grass tuft or shrub. Parasitised by Brown-backed Honeybird, Klaas's Cuckoo and Cuckoo Finch. (**Neddikkie**)

ZITTING CISTICOLA *Cisticola juncidis* **(664) 11 cm; 9 g**
SEXES *Alike. Common, very small short-tailed cisticola, usually heard giving diagnostic 'klink' calls over grassland.* **BR** *White-tipped tail has black subterminal tail band (distinctive in all plumages); back is boldly streaked blackish-brown.* **NON-BR** *Breast yellowish-buff.* **JUV** *Lacks yellowish-buff breast.* **STATUS** Very common resident, more often heard than seen. Less common in dry years when long grass cover absent. **HABITAT** Grassland, and well-developed grassy patches. **FOOD** Insectivorous. Forages low down in grass or forbs. **CALL** Monotonous *klink, klink, klink* ... or *zit zit zit* in circular display flight by ♂ only, and usually 30–40 m above the ground. The aerial cruise is undulating with steep dips between each *zit* that is given on the rise of each loop. **BR** Monogamous. Unique vertical pear-shaped soda-water 'bottle' nest with top entrance. The meaning of the genus name *Cisticola* is derived from the inhabitant of a water tank or reservoir, referring to the bottle-shaped nest of this species! (**Landeryklopkloppie**) ALT NAME *Fan-tailed Cisticola*

DESERT CISTICOLA *Cisticola aridulus* **(665) 11 cm; 9 g**
SEXES *Alike. Similar to Zitting Cisticola but is overall paler, lacks conspicuous black subterminal tail band, and occupies drier habitats. Easily distinguished from Zitting Cisticola by call.* **STATUS** Locally common resident. Usually singly or in pairs, sometimes in small groups in non-br season. **HABITAT** Well-developed grassland, particularly on eastern basaltic soils. **FOOD** Insectivorous. Forages low in grass layer or on ground. **CALL** Fast *ting-ting-ting-ting-ting* ... (in display flight) and staccato clicks *Bchick-Bchick* ... Sings from top of small bush or in bouncing display flight with wing-snap at each bounce. **BR** Monogamous. Pear-shaped nest with large side-top entrance placed low down in grass tuft. (**Woestynklopkloppie**)

PLATE 70 Prinia, Camaropteras, Apalis, Wren-warbler p222

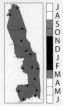

TAWNY-FLANKED PRINIA *Prinia subflava* **(683) 13 cm; 9,5 g**
SEXES *Alike. Long tail (often held erect), buffy underparts and flanks, and rufous wing panel distinctive. Non-br birds have horn-coloured bills.* **JUV** *Tinged lemon-yellow below; bill yellowish.* **STATUS** Common and widespread resident. Usually in pairs or small family parties. **HABITAT** Rank grass and shrubs, particularly along watercourses; also in restcamps. **FOOD** Insectivorous; also recorded taking nectar from aloes. **CALL** Monotonous *przzt-przzt-przzt* and loud piping *teep-teep-teep*. **BR** Monogamous. Thin-walled pear-shaped nest woven or knitted and usually about 1 m above ground. Parasitised by Cuckoo Finch. (**Bruinsylangstertjie**)

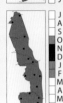

GREEN-BACKED CAMAROPTERA *Camaroptera brachyura* **(657) 13 cm; 11 g**
SEXES *Alike, ♂>♀. Greenish back easily distinguishes it from Grey-backed Camaroptera. Dark reddish eyes distinguish it from eremomela species with greenish upperparts. Tail often cocked upwards.* **JUV** *Yellow wash on underparts; tail brown.* **STATUS** Widespread and common resident, sedentary and found singly or in pairs. **HABITAT** Forest, riverine bush, moist savanna woodland and wooded thickets in restcamps. **FOOD** Insectivorous. **CALL** Loud penetrating *kwit-kwit-kwit* ... , like 2 stones knocking together. **BR** Monogamous. Ball nest concealed in leaves sewn together with fibres and cobweb. In KNP probably the host of Klaas's and Diderick cuckoos, and perhaps also Brown-backed Honeybird. (**Groenrugkwêkwêvoël**) ALT NAME *Bleating Warbler*

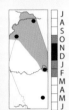

GREY-BACKED CAMAROPTERA *Camaroptera brevicaudata* **(-) 13 cm; 10 g**
SEXES *Alike, ♂>♀.* **BR** *Easily distinguished from Green-backed Camaroptera during br season by grey back and crown.* **NON-BR** *Upperparts ashy brown, flight feathers duller yellow.* **JUV** *Upperparts olive-brown.* **STATUS** Records exist for the KNP in the western parts of the Luvuvhu and Limpopo rivers. This represents the easternmost limit of its distribution. May hybridise with the Green-backed Camaroptera in this part of its range. **HABITAT** Similar wooded habitats and thickets to Green-backed Camaroptera. **FOOD** Insectivorous. **CALL** Like Green-backed Camaroptera, a loud and penetrating *kwit-kwit-kwit* ... **BR** Monogamous. Parasitised by Klaas's and Diderick cuckoos, and Brown-backed Honeybird. (**Grysrugkwêkwêvoël**) ALT NAME *Bleating Warbler*

YELLOW-BREASTED APALIS *Apalis flavida* **(648) 12,5 cm; 8 g**
SEXES *Similar, ♂>♀.* **M** *Combination of yellow breast patch and red eyes diagnostic.* **F** *Similar to ♂ but lacks central black breast band.* **JUV** *Duller green above with paler yellow breast.* **STATUS** Common resident. Largely sedentary. Usually in pairs or small family groups and may maintain territory year-round. **HABITAT** Woodland, especially along riparian fringes in thornveld, and in Mopane and Tambotie *Spirostachys africana* thickets. **FOOD** Mainly insects; also recorded taking fruit and nectar. **CALL** ♂ gives repetitive *krunk-krunk-krunk* ... , replied by ♀, *krik-krik-krik* ... Calls vary individually, and structure of duet probably pair-specific. **BR** Monogamous. Oval nest with side-top entrance usually placed 1–3 m above ground. (**Geelborskleinjantjie**)

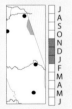

RUDD'S APALIS *Apalis ruddi* **(649) 13 cm; 10 g**
SEXES *Similar, ♀ with narrower breast band. Lacks yellow breast patch of Yellow-breasted Apalis and with dark (not red) eye; call distinctive.* **JUV** *Paler than adult; breast band indistinct.* **STATUS** Near-endemic and listed as Near-threatened. Locally common resident; sedentary and usually in pairs or small family groups. **HABITAT** In KNP occurs only in the Sand Camwood (Nyandu) *Baphia massaiensis* community of Nwambiya sandveld in the far north-eastern part of the Park. **FOOD** Insects; also small fruit and flower buds. **CALL** Rapid *tok-tok-tok-tok-tok* ..., usually replied by ♀, *kli-kli-kli-kli* ... **BR** Monogamous. Nest an upright oval structure with side-top entrance, concealed in outer branches of thorn tree, 1–3 m from ground. (**Ruddse Kleinjantjie**)

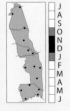

STIERLING'S WREN-WARBLER *Calamonastes stierlingi* **(659) 14 cm; 13 g**
SEXES *Alike. Finely barred brown and white underparts and warm brown upperparts unmistakable.* **JUV** *Breast has yellowish wash, upperparts more reddish-brown.* **STATUS** Fairly common resident throughout KNP, usually in pairs or small groups. Inconspicuous when not calling. **HABITAT** Thickets in broad-leaved and mixed woodland. **FOOD** Insects. **CALL** Cricket-like *whe-ee-ee, whe-ee-ee, whe-ee-ee* ... **BR** Monogamous. Thick-walled oval nest stitched into living leaves and usually placed about 3 m above ground. (**Stierlingse Sanger**) ALT NAME *Stierling's Barred Warbler*

PLATE 71 Larks p222

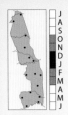

MONOTONOUS LARK *Mirafra passerina* **(493) 14 cm; 24 g**
SEXES Alike. *Most likely to be confused with Rufous-naped Lark, but much smaller, not resident in the Park and only found some years in summer months. Difficult to identify unless singing (song very distinctive) when white throat shows prominently.* **JUV** *Darker above and with buff-fringed feathers.* **STATUS** A summer visitor, nomadic and irrupts erratically, sometimes in considerable numbers into KNP grassland areas in wetter summers when grass plentiful. May be absent in drier years. Near-endemic to s Africa. **HABITAT** Open savanna and shrub with well-developed grass layer. **FOOD** Insects and grass seeds. **CALL** Monotonous and repetitive *'for syrup is sweet'* that may continue into the night in br season. Sings from the top of small bush or tree, occasionally in short display flight. **BR** Monogamous. Nest has partly or completely domed roof and is placed between grass tufts. (**Bosveldlewerik**)

RUFOUS-NAPED LARK *Mirafra africana* **(494) 17 cm; 44 g**
SEXES Alike. *The largest of the 'brown' larks in the Park. Fairly stocky build with short but distinctive erectile rufous crest, and rufous wing panel that is usually more visible in flight than on ground. Sings conspicuously from top of fence post, termite mound or shrub.* **JUV** *Darker above with buff edging to feathers.* **STATUS** Rather uncommon in KNP but a widespread resident, sedentary and usually solitary or in pairs. **HABITAT** Open grassland or lightly wooded savanna, often near overgrazed areas around waterholes. Avoids very dense grassland unless scattered with bare patches and termitaria. **FOOD** Insects and seeds. May stand on termite mounds, eating termites as they emerge; also hawks termite alates in flight. **CALL** Song a clear, loud and ringing *tsuwee-tsweeoo*, repeated every 2–4 secs, characteristically vibrating its wings after each phrase. Flight song a series of rambling whistles sometimes with imitations of other species. **BR** Monogamous. Partly or completely domed cup nest concealed under grass tuft. (**Rooineklewerik**)

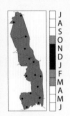

FLAPPET LARK *Mirafra rufocinnamomea* **(496) 15 cm; 26 g**
SEXES Alike. *Smallish dark lark with rufous-edged wings and distinctive wing-clapping. Attracts attention with aerial wing-flapping display flight high in the air. In flight appears dumpy with narrow, shortish tail. Loud fluttering sound produced by clapping wings together under body during flight.* **JUV** *Darker above with pale-tipped feathers.* **STATUS** Fairly common resident, sedentary and usually solitary or in pairs. Inconspicuous in non-breeding season when not displaying; rarely perches on bushes and trees. **HABITAT** Grassy clearings in woodland. **FOOD** Invertebrate diet includes grasshoppers, mantids, butterfly larvae, beetles and termites. Also grass seeds. **CALL** Bursts of wing-flapping in aerial display without loud whistle diagnostic. Whistles, when heard, are faint. **BR** Monogamous. Cup nest completely or partly domed and concealed under grass tuft. (**Laeveldklappertjie**)

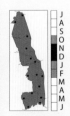

SABOTA LARK *Calendulauda sabota* **(498) 14,5 cm; 23 g**
SEXES Alike. *Medium-sized lark with heavily streaked breast and white eyebrow that extends well behind eye. Prominently marked head features include black eye-stripe, white crescent below eye, and moustachial and malar stripes. Subspecies in KNP has narrower bill than races that occur in the north-western regions of s Africa.* **JUV** *Like adult but with more spotted appearance.* **STATUS** Very common near-endemic, usually in pairs or small family parties. Generally sedentary, locally nomadic when environmental conditions are altered, eg fires and droughts. Often perches on trees and bushes, and when disturbed flies to next tree or bush. **HABITAT** Arid savanna to open woodland. Occurs in most KNP habitats including broad-leaved, *Acacia* and Mopane; avoids riparian zones. **FOOD** Seeds (60%) and insects. Not known to drink water. **CALL** Melodious song rambling and variable; excellent mimic, able to imitate the call and noises of a wide range of other species. **BR** Monogamous. Cup nest sometimes domed, and usually concealed under grass tuft, under aloe, or next to a stone. (**Sabotalewerik**)

PLATE 72 Larks, Sparrowlarks p222

DUSKY LARK *Pinarocorys nigricans* (505) 19 cm; 40 g
SEXES *Alike. Overall dark appearance (especially above), heavily streaked underparts, short stout beak and shorter legs distinguish it from similarly patterned Groundscraper Thrush that also has heavily streaked underparts. Flight undulating and underwings all-dark.* **JUV** *Duller than adult, and breast buffier.* **STATUS** Sporadic and fairly uncommon non-br intra-African migrant. Found singly, in pairs or in small flocks during migration. May be absent in drier years. Breeds south of the Equator in c Angola, s DRC, n Zambia and w Tanzania. Present Dec–May. **HABITAT** Favours short sparse grass in mixed woodland. Often on gravel roads and overgrazed areas around waterholes. **FOOD** Mainly insects, also seeds. **CALL** Largely silent in non-br areas; gives soft *wek-wek ...* call. (**Donkerlewerik**)

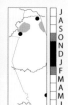

FAWN-COLOURED LARK *Calendulauda africanoides* (497) 15 cm; 23 g
SEXES *Alike. Face less boldly patterned and breast more lightly streaked than Sabota Lark. Underparts white.* **JUV** *Buff-spotted above.* **STATUS** Fairly common but very localised endemic. Sedentary or nomadic in response to environmental conditions. Solitary or in pairs, usually seen in open bare patches on the ground but often perches on bushes or trees. **HABITAT** Almost exclusively on sandy soils in shrublands and savanna. In KNP restricted to sandveld habitats in the Nwambiya and to the north of Punda Maria. **FOOD** Insects and seeds. **CALL** Melodious whistles and canary-like trills; includes mimicry. Sings from perch on bush or in aerial display. **BR** Monogamous. Cup nest usually domed and concealed under grass tuft or shrub. (**Vaalbruinlewerik**)

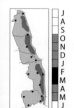

CHESTNUT-BACKED SPARROWLARK *Eremopterix leucotis* (515) 13 cm; 22 g
SEXES M *White ear-patch and rich chestnut back and wings very distinctive.* **F** *Paler chestnut-brown upperparts and light collar that extends round to nape.* **JUV** *Like ♀ but white tips to upperpart feathers give more mottled appearance.* **STATUS** Locally fairly common. Resident but subject to nomadic fluctuations; in small flocks most of the year, particularly in non-br season. **HABITAT** Semi-arid short grassland and savanna, particularly on the eastern basaltic soils. **FOOD** Mainly grass seeds, also insects. Drinks regularly. **CALL** Soft *kree kree kree hu-hu*, first note high-pitched. **BR** Monogamous. Breeds mostly in open or overgrazed areas, nest placed on south-eastern side of grass tuft or stone for shade. (**Rooiruglewerik**) ALT NAME *Chestnut-backed Finchlark*

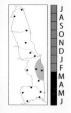

GREY-BACKED SPARROWLARK *Eremopterix verticalis* (516) 13 cm; 17 g
SEXES M *White ear-patch and sandy-grey back and wings distinctive.* **F** *Lightly streaked breast and blackish belly patch; pale collar extends around ear coverts.* **JUV** *Like ♀; more reddish-brown above and with darker chestnut spots.* **STATUS** Locally abundant near-endemic. Nomadic in response to rainfall and drought conditions elsewhere in the country. Mainly a bird of the drier western parts of South Africa but enters KNP in large numbers when extreme drought conditions prevail in their usual habitats. Gregarious and usually in loose flocks, even in br season. Widespread when present, but most records in the region of the S90 road (n-e of Satara). **HABITAT** Usually occurs in arid and semi-arid grassland and shrublands, when present in KNP may be found in any habitat. **FOOD** Mainly seeds, also insects. **CALL** Song a soft twittering; flight song repetitive *twip-twip-chik*. **BR** Monogamous. Br not recorded in Park. Cup nest, with distinctive ramp of small stones, usually concealed next to grass tuft or shrub. (**Grysruglewerik**) ALT NAME *Grey-backed Finchlark*

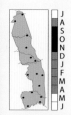

RED-CAPPED LARK *Calandrella cinerea* (507) 16 cm; 24 g
SEXES *Alike. Rufous crown and side breast patches very distinctive. Short erectile hind-crown crest normally held flat, but raised when alarmed or hot; underparts variable but usually white, sometimes with buffy flanks.* **JUV** *Appears spotted above; buffy below with brown spots.* **STATUS** Present in the Park mainly as an uncommon nomad or a partial migrant. May have declined in numbers owing to closure of waterholes and resultant recovery of associated overgrazed areas. **HABITAT** Short grassland and shrublands, particularly sparsely vegetated and overgrazed areas. Favours recently burnt grasslands and road verges. **FOOD** Seeds and insects, drinks regularly. Hawks moths and termite alates aerially. **CALL** Song a jumble of melodious phrases. Frequently sings in high dipping display flight, often includes mimicry. **BR** Monogamous. Cup nest placed next to grass tuft, shrub or stone. Incubation by ♀ only. (**Rooikoplewerik**)

PLATE 73 Thrushes, Flycatchers p223

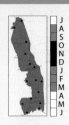

KURRICHANE THRUSH *Turdus libonyana* (576) 22 cm; 63 g
SEXES Alike. Dark malar stripes and bright orange bill distinctive field characteristics. **JUV** Underparts densely spotted blackish-brown; wing coverts tipped orange-buff. **STATUS** Locally common resident, generally sedentary and usually found in pairs. **HABITAT** Occurs throughout but favours denser, well-developed woodland regions, particularly thickets in riparian zones, also in restcamp gardens where it is most likely to be seen running beneath tall trees in typical thrush manner, stopping and listening. **FOOD** Insectivorous, with some fruit. **CALL** Repetitive loud tuneful whistled notes: *tyeoo-weet-weeit*. **BR** Monogamous. Cup nest often mud-lined and placed in main fork of large tree. Incubation by ♀ only. (**Rooibeklyster**)

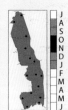

GROUNDSCRAPER THRUSH *Psophocichla litsitsirupa* (580) 21 cm; 76 g
SEXES Alike. Unlikely to be confused with any other species except Dusky Lark. Grey (not dark brown) upperparts, heavily blotched underparts and upright stance are unmistakable features. **JUV** Flecked with white above; buff-spotted below. **STATUS** Fairly common resident, usually in pairs. Shy in natural habitats, bolder and more confiding around human habitation. Often runs fast across open ground with head held low, stopping abruptly in upright stance to flick its wings. **HABITAT** Occurs throughout KNP but favours open broad-leaved woodland, often in restcamp gardens. **FOOD** Mainly insects and earthworms. **CALL** Mixture of harsh and whistled phrases. **BR** Monogamous. Cup nest placed in large fork of tree. Incubation by both sexes. (**Gevlekte Lyster**)

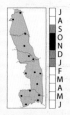

PALE FLYCATCHER *Bradornis pallidus* (696) 16 cm; 22 g
SEXES Alike. Medium-sized inconspicuous brown plumaged flycatcher with underparts only marginally paler than upperparts and not as contrasting as Marico Flycatcher that has whitish underparts. **JUV** Underparts buff, streaked brown; upperparts scaled off-white. **STATUS** Rather uncommon but a localised resident, sedentary, sometimes solitary but usually in pairs or small family groups. Flicks tail on settling and sometimes spreads and closes wings; often feeds on the ground. **HABITAT** Mainly in broad-leaved and mixed broad-leaved/*Acacia* woodland, uncommon in Mopane. **FOOD** Insects, especially beetles, caterpillars and termites; occasionally small fruit. **CALL** High-pitched warble of raspy notes. **BR** Monogamous. Nest a small flimsy cup usually placed 2–3 m high on lateral branch. Incubation by ♀ only. Uncommon host of Diderick Cuckoo. (**Muiskleurvlieëvanger**) **ALT NAME** *Pallid Flycatcher*

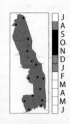

SOUTHERN BLACK FLYCATCHER *Melaenornis pammelaina* (694) 20 cm; 30 g
SEXES Similar, ♀ duller than ♂, with slight brownish tinge to black plumage and eye lighter brown. Lacks deep red eye of drongos, deeply forked tail of Fork-tailed Drongo (page 119), and differs in habitat. Plumage similar to larger ♂ Black Cuckooshrike (page 131) but has square (not rounded) tail and lacks orange gape at base of bill. **JUV** Dull blackish-brown, spotted and streaked above with rufous, blotched buff below. **STATUS** Common resident, mostly sedentary and usually in pairs or small family groups. **HABITAT** Wide range of open woodlands and restcamp gardens. **FOOD** Insects; occasionally nectar from aloes and small fruit. **CALL** High-pitched whistles, often a 3-phrase *tseep-tsoo-tsoo*. **BR** Monogamous. Nest usually placed in shallow tree cavity or behind bark of tree trunk, especially where tree has been fire-blackened. Not a regular host of any of the cuckoos, only one record of parasitism by Red-chested Cuckoo and another by Black Cuckoo. (**Swartvlieëvanger**) **ALT NAME** *Black Flycatcher*

MARICO FLYCATCHER *Bradornis mariquensis* (695) 18 cm; 24 g
SEXES Alike. Strong contrast between brown upperparts and almost-white underparts easily distinguishes it from Pale Flycatcher. **JUV** Upperparts buff-spotted; off-white underparts streaked brown especially on chin, throat and breast. **STATUS** Very uncommon near-endemic vagrant to the Park, present during the dry winter months. A species that favours the drier western regions of the country but expands its range eastwards into KNP in the drier years. Usually in pairs or small family parties. **HABITAT** Most common in *Acacia* savanna but also in mixed woodland and Mopane. **FOOD** Insects, especially termites, caterpillars and grasshoppers. **CALL** Series of monotonous harsh unmusical chirps, somewhat sparrow-like. **BR** Monogamous. Br peak later in Namibia. Builds small flimsy cup nest; incubation by ♀ only. Occasional host of Diderick Cuckoo. No breeding records for KNP, but juvenile recorded with adults along the Sweni River. (**Maricovlieëvanger**)

PLATE 74 Flycatchers, Tit-Flycatchers p223

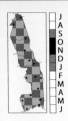

FISCAL FLYCATCHER *Sigelus silens* (698) 18 cm; 26 g
SEXES *Similar, ♂ upperparts black; ♀ upperparts dark grey-brown, chin and throat 'dirty' off-white. Confusion most likely with Common Fiscal (recorded only as a rare vagrant to the Park), but lacks heavier hooked bill and white 'V' on back and shoulder. Both sexes show white wingbar.* **JUV** *Upperparts dull as in ♀, spotted rufous-brown; underparts mottled grey-brown.* **STATUS** Uncommon endemic, with altitudinal movements undertaken by small numbers of individuals into KNP in winter, especially into the southern regions. **HABITAT** Open woodland. **FOOD** Insects; also small fruit and nectar from aloes. **CALL** High-pitched wheezing notes and whistles. Mimics other species. **BR** Monogamous. Bulky cup nest built by ♀ and usually placed in thorn tree. Br not recorded in Park. (**Fiskaalvlieëvanger**)

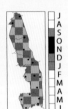

AFRICAN DUSKY FLYCATCHER *Muscicapa adusta* (690) 13 cm; 11 g
SEXES *Alike. Slightly smaller and darker than Spotted Flycatcher, with more indistinct and diffused streaking on breast. Gives presence away by diagnostic 'tsirrrt' call, and flicks wings at intervals when perched. Crown lacks light streaks characteristic of Spotted Flycatcher.* **JUV** *Tail and upperwing coverts tipped with buff.* **STATUS** Small resident population, but numbers increase substantially during winter months with influx of altitudinal migrants from escarpment to warmer low-lying areas. In KNP usually found singly. **HABITAT** Most woodlands and forest margins. Found in restcamp gardens. **FOOD** Predominantly small flying insects, occasionally fruit. **CALL** Repeated and fast *tsip-tsirrrt* … , also high-pitched *tseeeu*. **BR** Monogamous. Nest a bulky cup, built in natural tree cavity, in crevice or on ledge of building. Incubation by ♀ only. No br records for KNP; outside Park occasional host of Red-chested and Klaas's cuckoos. (**Donkervlieëvanger**) **ALT NAME** *Dusky Flycatcher*

SPOTTED FLYCATCHER *Muscicapa striata* (689) 14 cm; 15 g
SEXES *Alike. Slightly larger than African Dusky Flycatcher, with paler underparts, slightly longer-tailed and giving a less dumpy appearance. Streaking on breast fairly indistinct. Pale streaking on crown diagnostic (unstreaked in African Dusky Flycatcher, which is also overall more grey-brown). Characteristic wing-flicking when returning to perch.* **JUV** *Paler tips to upper tail and wing coverts.* **STATUS** Common and widespread non-br Palearctic migrant, present Oct–Mar/Apr (latest May). Usually found singly and perched low on an outer branch in search of aerial insects. Some birds return to same locality in successive years. **HABITAT** Most open woodlands, and restcamp gardens. **FOOD** Insects, occasionally small fruit. Most prey caught aerially, bees and wasps beaten against branch to dislodge sting. **CALL** Rarely heard in non-br range: thin sibilant *seep* or harsh *chirrt*. (**Europese Vlieëvanger**)

ASHY FLYCATCHER *Muscicapa caerulescens* (691) 15 cm; 18 g
SEXES *Alike. Pale blue-grey plumage and white eye-ring distinctive. Easily distinguishable from Grey Tit-Flycatcher by upright posture, grey tail (which it does not fan) and black loral stripe.* **JUV** *Spotted dark brown above and speckled blackish below.* **STATUS** Fairly common resident in KNP with perhaps some altitudinal movements; usually in pairs. Easily overlooked, but not shy and easily located by call. **HABITAT** Shows a preference for riparian forest; but also found in most denser woodland types. **FOOD** Insects, mostly taken in flight with audible snap of beak; occasionally small fruit. Joins mixed-species foraging flocks. **CALL** Sibilant descending 4-syllable *pit PIT pit-pit*, and variations thereof. **BR** Monogamous. Nest placed in tree cavity, on rock face or on building ledge. Infrequent host of Klaas's Cuckoo. (**Blougrysvlieëvanger**) **ALT NAME** *Blue-grey Flycatcher*

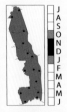

GREY TIT-FLYCATCHER *Myioparus plumbeus* (693) 14 cm; 21 g
SEXES *Alike. Black tail with white outer tail feathers frequently fanned and flicked from side to side, a distinctive feature that helps distinguish it from Ashy Flycatcher. Chest greyer than Ashy Flycatcher and lacks white eye-ring and black loral stripe.* **JUV** *Spotted brown and buff, both above and below.* **STATUS** Fairly common resident found singly or in pairs; easily located by call. **HABITAT** Both mixed *Acacia* and broad-leaved woodland, especially along riverine margins and thickets. **FOOD** Insectivorous. Gleans insects from foliage and twigs but also hawks prey within canopy. Sometimes joins mixed-species foraging flocks. **CALL** Plaintive, high-pitched, drawn-out *peely-peeerr*. **BR** Monogamous. Constructs flimsy cup nest in old woodpecker or barbet nest or natural tree hole. (**Waaierstertvlieëvanger**) **ALT NAME** *Fan-tailed Flycatcher*

PLATE 75 Robin-Chats, Scrub-Robins p223

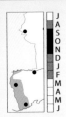

CAPE ROBIN-CHAT *Cossypha caffra* **(601) 17 cm; 28 g**
SEXES *Alike. Pale orange throat and grey sides to neck, breast and belly diagnostic. Unlikely to be confused with any other robin in the Park.* **JUV** *Upperparts heavily mottled buff; no white eyebrow; tail rufous.* **STATUS** Very rare, entering the south-western KNP through winter altitudinal movements down riparian corridors, but usually only in severe winters. **HABITAT** Typically a bird of thickets. **FOOD** Insectivorous. **CALL** Distinctive guttural *wur-de-dur* call. Song a series of short melodious phrases, including well-known *Jan-Frederik*; also a capable mimic. **BR** Monogamous. Deep cup nest usually placed on or close to ground; no br records for Park. Common host of Red-chested Cuckoo. (**Gewone Janfrederik**) **ALT NAME** *Cape Robin*

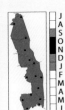

WHITE-THROATED ROBIN-CHAT *Cossypha humeralis* **(602) 15,5 cm; 23 g**
SEXES *Alike, ♂>♀. White throat and breast, white eyebrow and shoulder bar, and orange rump distinctive.* **JUV** *Upperparts mottled blackish-brown, with no white eyebrow and shoulder bar.* **STATUS** Locally common endemic, sedentary and usually in pairs. **HABITAT** Thickets in *Acacia* and broad-leaved woodland; usually near watercourses. **FOOD** Insectivorous, taking mainly beetles, ants, termites, moths and caterpillars, occasionally fruit. **CALL** Song a series of high-pitched melodious phrases, commonly incorporating mimicry. Contact call likened to squeak of bicycle tyre being pumped. **BR** Monogamous. Cup nest mostly placed on ground (94%). Parasitised by Red-chested Cuckoo. (**Witkeeljanfrederik**) **ALT NAME** *White-throated Robin*

WHITE-BROWED ROBIN-CHAT *Cossypha heuglini* **(599) 18,5 cm; 35 g**
SEXES *Alike, ♂>♀. White eyebrow, black crown and bright orange underparts unmistakable.* **JUV** *Mottled buff on brown above (tail orange). Moults into orange adult plumage after about 6 weeks.* **STATUS** Locally common and conspicuous resident, sedentary and usually in pairs. **HABITAT** Mostly in evergreen thickets adjacent to river courses (especially in dry regions), and restcamp gardens. **FOOD** Mainly ants, beetles, moths, caterpillars and termites, with some fruit. **CALL** Series of flute-like phrases increasing in volume and tempo, '*don't-you-do-it*' (or variations thereof), particularly at dawn and dusk. Unlike other robin-chats, seldom includes mimicry but does include alarm calls of other birds. **BR** Monogamous. Cup nest built by ♀, mostly 1–2 m above ground. Incubation by ♀ only. Parasitised by Red-chested Cuckoo. (**Heuglinse Janfrederik**) **ALT NAME** *Heuglin's Robin*

RED-CAPPED ROBIN-CHAT *Cossypha natalensis* **(600) 16,5 cm; 32 g**
SEXES *Alike. Only robin-chat with orange face and light grey-blue wings and back.* **JUV** *Upperparts mottled buff on dark brown, tail orange except for central pair of feathers (as in adult).* **STATUS** Fairly common localised resident. **HABITAT** Locally common in well-developed riverine forests and thickets. Also occurs in some well-wooded restcamp gardens (particularly Skukuza and Letaba). **FOOD** Insectivorous. **CALL** 2-syllabled contact call *see-saw, see-saw* ... Song loud, rich and with impressive repertoire that includes mimicry of many bird calls and local sounds, eg yapping dogs! **BR** Monogamous. Open cup nest usually within 1,5 m of ground, often in natural tree holes or behind roots on an earth bank. Infrequent host of Red-chested Cuckoo. (**Nataljanfrederik**) **ALT NAME** *Natal Robin*

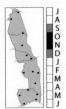

BEARDED SCRUB-ROBIN *Cercotrichas quadrivirgata* **(617) 17 cm; 26 g**
SEXES *Alike. Only robin in region with rufous upper breast and flanks, and bold head markings showing prominent white eye-stripes and conspicuous black malar stripes.* **JUV** *Similar to adult but with brown mottling to upper and undersides.* **STATUS** Fairly common localised resident, often difficult to locate. **HABITAT** Mostly riverine forest and thickets along the major rivers, especially the Sabie and Luvuvhu. **FOOD** Beetles, ants and termites make up bulk of invertebrate diet. **CALL** Fine songster; series of loud melodious whistled phrases. **BR** Monogamous. Cup nest usually in hollow stump. Occasionally parasitised by Red-chested Cuckoo. (**Baardwipstert**) **ALT NAME** *Bearded Robin*

WHITE-BROWED SCRUB-ROBIN *Cercotrichas leucophrys* **(613) 15 cm; 20 g**
SEXES *Alike. Streaked breast, white eyebrow and wingbars, and rufous rump diagnostic.* **JUV** *Mottled above eye-stripe, wingbars buffy, and chest mottled not streaked.* **STATUS** Very common and widespread resident, usually in pairs. **HABITAT** Woodland, especially *Acacia*. **FOOD** Mainly insects. **CALL** Almost monotonous series of loud melodious whistled phrases. **BR** Monogamous. Deep cup nest placed from ground level to 1,6 m high. Occasionally parasitised by Diderick and Red-chested cuckoos. (**Gestreepte Wipstert**) **ALT NAME** *White-browed Robin*

PLATE 76 Palm-Thrush, Nightingale, Stonechat, Chats, Cliff-Chat   p223

COLLARED PALM-THRUSH *Cichladusa arquata*   (603) 20 cm; 35 g
SEXES *Alike, ♂>♀. Buff throat and upper breast bordered by black necklace diagnostic. Pale eyes and mostly rufous upperparts also aid identification.* **JUV** *Necklace less distinct and speckled blackish below. Throat buffier.* **STATUS** Very rare and localised br resident, recorded regularly in the Shingwedzi restcamp over the past 10 years, less conspicuous (possibly even absent) in dry season. An older record exists for Letaba. **HABITAT** Palm and other dense thickets in Shingwedzi restcamp. **FOOD** Insects, including bugs, beetles, grasshoppers and termites. **CALL** Beautiful staccato and melodious ringing song, often interrupted with chat-like cries. **BR** Monogamous. Base of cup-shaped nest a mixture of mud and palm and grass fibres, well hidden in palm tree. Incubation by both adults. (**Palmmôrelyster**)

THRUSH NIGHTINGALE *Luscinia luscinia*   (609) 16,5 cm; 25 g
SEXES *Alike. Like a nondescript large warbler or small bulbul. Difficult to identify (easiest by song). Upperparts dull brown, below dull white with brown-flecked chest and dull russet tail. Smaller than Terrestrial Brownbul.* **JUV** *Like adult with pale spots to wing coverts.* **STATUS** Non-br Palearctic migrant; uncommon, skulking, shy and inconspicuous, recorded along the Sabie and Shingwedzi rivers and in Pafuri region. Best located by song. Present from late Dec and departs late March. Generally solitary. **HABITAT** Favours dense riparian thickets such as *Capparis* and *Lantana*. **FOOD** Mostly insects. **CALL** Rich and loud melodious song from undergrowth or thicket. (**Lysternagtegaal**)

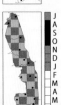

AFRICAN STONECHAT *Saxicola torquatus*   (596) 14 cm; 15 g
SEXES M *Black head and throat, white neck patch and rufous breast very distinctive, and unlikely to be confused with any other species in the region. Flight jerky and low, flashing conspicuous white rump and wing patches.* **F** *Streaked brown and buff upperparts with white wing patch and whitish rump.* **JUV** *Similar to ♀ but more mottled above, some mottling on breast, rump pale rufous.* **STATUS** Rather uncommon winter altitudinal migrant usually recorded May–Aug. Restless, frequently flicking wings and tail. **HABITAT** Mostly open grassland with few shrubs. **FOOD** Mainly insectivorous. **CALL** Alarm sharp *seep-chak-chak*, song canary-like with trilling notes. **BR** Monogamous. No br records for KNP. (**Gewone Bontrokkie**) **ALT NAME** Stonechat

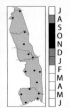

FAMILIAR CHAT *Cercomela familiaris*   (589) 15 cm; 22 g
SEXES *Alike, ♂>♀. Fairly nondescript brownish-grey bird with orange rump and outer tail feathers (especially noticeable in flight) and habit of flicking its wings.* **JUV** *Scaly appearance; spotted buff above.* **STATUS** Uncommon sedentary resident with patchy distribution but widely recorded. Usually in pairs or small family groups. Flicks wings constantly, and on landing usually raises tail once or twice. **HABITAT** Usually associated with human habitation, rocky slopes and outcrops, but also in sparse woodland, and drainage lines. **FOOD** Mainly insects. Formerly ate grease from wagon axles, giving rise to Afrikaans name 'Spekvreter' (fat-eater). **CALL** Series of unmelodious notes. **BR** Monogamous. Cup nests placed mainly in natural holes in trees and earth banks. (**Gewone Spekvreter**)

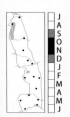

ARNOT'S CHAT *Myrmecocichla arnoti*   (594) 18 cm; 33 g
SEXES *♂>♀.* **M** *The only all-black bird in the Park with a white cap and shoulder patch.* **F** *White throat (with black flecks) distinguishes it from ♂.* **JUV** *Dull black; white shoulder patch.* **STATUS** Locally common resident; sedentary. Usually in pairs. **HABITAT** In KNP restricted to tall Mopane woodlands in the northern KNP. For the past few years birds have not been found at an isolated locality near the Timbavati where they were once regularly recorded, perhaps a result of elephant-induced changes in this habitat. **FOOD** Insectivorous, especially ants, beetles and spiders. **CALL** Rich jumble of whistles, squeaks and rasping notes, including mimicry. **BR** Monogamous (sometimes co-operative). Usually nests in natural hole, incubation by ♀ only. (**Bontpiek**)

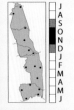

MOCKING CLIFF-CHAT *Thamnolaea cinnamomeiventris*   (593) 22 cm; 48 g
SEXES M *Distinctive bright plumage unlikely to be confused with any other species, especially with characteristic habit of tail-lifting.* **F** *Upperparts, throat and chest dull grey-brown, lower chest and belly dull rufous.* **JUV** *Duller than adult.* **STATUS** Locally common resident. Usually in pairs or small family groups of 4–5 birds. **HABITAT** Well-wooded granite inselbergs, cliffs and boulder-strewn slopes usually not far from water. **FOOD** Insects and fruit; also aloe nectar. **CALL** Rich and beautiful liquid notes, sometimes in duet. **BR** Monogamous. Nests in used swallow nest with tunnel removed, in cave or under bridge. Incubation by ♀ only. (**Dassievoël**) **ALT NAME** Mocking Chat

PLATE 77 Starlings p224

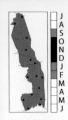

CAPE GLOSSY STARLING *Lamprotornis nitens* **(764) 23 cm; 90 g**
SEXES *Alike, ♂>♀. Sheen more blue than green, is less glossy and lacks dark ear coverts of the Greater Blue-eared Starling. Eyes orange-yellow.* **JUV** *Duller than adult; eyes yellow.* **STATUS** Common and widespread resident; sedentary and forms flocks in winter. Tame and familiar at picnic sites. **HABITAT** Wooded savanna, especially riverine bush in drier regions. **FOOD** Fruit, insects, nectar from aloes and scraps from picnickers. Regularly associates with ungulates, benefiting from disturbed insects. Hawks insects from backs of Impala and White Rhinoceroses, also recorded removing ectoparasites from game such as Sable Antelope. **CALL** Song a sustained warbling; flight call *turrrreeu*. **BR** Monogamous and sometimes a co-operative breeder. Nest constructed by both sexes and usually placed in natural tree cavity, old woodpecker or barbet hole, and also recorded in a hole in a river bank. Incubation by ♀ only. Host of Great Spotted Cuckoo and, rarely, Greater Honeyguide. (**Kleinglansspreeu**) **ALT NAME** *Glossy Starling*

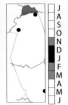

MEVES'S STARLING *Lamprotornis mevesii* **(763) 34 cm; 75 g**
SEXES *Alike, ♂>♀. More slender and smaller-bodied than Burchell's Starling; tail much longer but distinctly tapered. Eyes dark. Ranges of these two species do not overlap in KNP.* **JUV** *Underparts matt black and tail shorter than adult.* **STATUS** Localised but common and conspicuous resident recorded only in the far northern parts of the Park in the Pafuri area. Non-br birds roost communally in large *Acacia* trees. **HABITAT** Tall woodland, especially Mopane with scattered Baobab trees. **FOOD** Mainly insects; also fruit and flower petals of trees such as Ana-tree *Faidherbia albida* and Jackal-berry *Diospyros mespiliformis*. Most prey caught on the ground where it spends much of the time foraging. Catches insects disturbed by large mammals such as elephants. **CALL** Series of churring notes. **BR** Monogamous. Most nests in natural tree cavities. Incubation by ♀ only. Occasionally parasitised by Great Spotted Cuckoo. (**Langstertglansspreeu**) **ALT NAME** *Long-tailed Starling*

GREATER BLUE-EARED STARLING *Lamprotornis chalybaeus* **(765) 22 cm; 90 g**
SEXES *Alike, ♂>♀. Sheen on upperparts more green than blue. Key features royal blue flanks and belly, and broad dark (blue) ear coverts. Usually 2 rows of black spots on wing coverts, upper row less evident on some birds or even absent. More glossy than Cape Glossy Starling, and call distinctive.* **JUV** *Duller than adult.* **STATUS** Common and widespread resident, easily seen at picnic sites and in restcamps. Sedentary, forming large flocks in non-br season when it roosts communally in large *Acacia* trees or reedbeds. **HABITAT** Open savanna woodland, occurring in both *Acacia* and Mopane. **FOOD** Fruit, insects and aloe nectar during winter. Forages both on the ground and in trees, and hawks insects such as termite alates aerially. **CALL** Typical call drawn-out nasal *wraanh*, but song a series of rambling notes. **BR** Monogamous. Nests in tree cavities; also uses old woodpecker and barbet holes. Incubation by ♀ only. Parasitised by Great Spotted Cuckoo and Greater Honeyguide. (**Grootblouoorglansspreeu**)

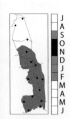

BURCHELL'S STARLING *Lamprotornis australis* **(762) 32 cm; 105 g**
SEXES *Alike, ♂>♀. Largest and heaviest glossy starling in region. Comparable only to Meves's Starling but tail heavier, rounded and not graduated. Ranges in KNP do not overlap. Eyes dark.* **JUV** *Lacks iridescence of adult.* **STATUS** Locally common near-endemic; found singly or in small groups (often associates with other starlings), absent from the northern parts of KNP where it is replaced by Meves's Starling. Roosts in reedbeds or *Acacia* trees. **HABITAT** Mostly in tall, open *Acacia* woodland, especially where there are large Knob Thorn *Acacia nigrescens* trees, avoids most pure broad-leaved woodland habitats. **FOOD** Mainly insects; also vertebrates such as small mice, fruit and flowers of *Acacia*. Feeds mainly on ground and commonly scavenges scraps at picnic sites. **CALL** Various harsh notes and whistled phrases. **BR** Monogamous. Nests in natural tree cavity or large woodpecker hole. Incubation by ♀ only Occasionally parasitised by Great Spotted Cuckoo and Greater Honeyguide. (**Grootglansspreeu**)

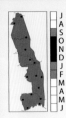

RED-WINGED STARLING *Onychognathus morio* **(769) 30 cm; 140 g**
SEXES M *Black body and red wings very distinctive, unlikely to be confused with any other species in KNP.* **F** *Grey head and rufous wingbar.* **JUV** *Like adult ♂ but duller.* ♀ *attains grey on head only after 6 months.* **STATUS** Common resident where suitable habitat occurs. Usually in pairs; in flocks in non-br season. Resident pairs roost at nest site year-round. Normally aggressive towards other starlings. **HABITAT** Prefers mountainous cliffs and rocky regions; has adapted well to some restcamps where it br on building structures. **FOOD** Mainly fruit and insects; also aloe nectar. **CALL** Musical, whistled *twee-twee* contact call very characteristic. Song a combination of whistles and warbles. **BR** Monogamous. Pairs remain together for years. Bulky bowl nest has mud incorporated into base; placed on cliff ledge or crevice, or ledge on building. Incubation mostly by ♀. Parasitised by Great Spotted Cuckoo. (**Rooivlerkspreeu**)

VIOLET-BACKED STARLING *Cinnyricinclus leucogaster* **(761) 18 cm; 45 g**
SEXES M *Iridescent amethyst or 'plum' colour on throat and upperparts very distinctive.* **F** *Heavily streaked below; eyes yellow. Thrush or chat appearance.* **JUV** *As ♀, eyes brown.* **STATUS** Common br intra-African migrant, present throughout KNP from Oct–Apr. Usually in pairs but after br form small flocks. **HABITAT** Riverine forest and savanna woodland. **FOOD** Mainly fruit; also insects. Insects gleaned from foliage and branches; also hawks prey aerially. **CALL** Single long high-pitched whistle by ♂, song a series of nasal whistles. **BR** Monogamous. Nests in natural tree holes, also fence posts. Incubation by ♀ only. Occasionally parasitised by both Greater and Lesser honeyguides. (**Witborsspreeu**) **ALT NAME** *Plum-coloured Starling*

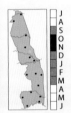

YELLOW-BILLED OXPECKER *Buphagus africanus* **(771) 20 cm; 60 g**
SEXES *Alike. Very similar to Red-billed Oxpecker. Best identification features pale rump, and broader bill with red tip and yellow base.* **JUV** *Bill of immature birds brown; plumage darker than adult.* **STATUS** Historically occurred in the Park but it is suspected that it became extinct as a breeding species as a result of the decimation of its host species by rinderpest in 1896. Birds returned in 1979 (presumably from Zimbabwe when less regular cattle dipping allowed birds to move south via Gona-re-zhou). The first recorded breeding in the Park in 1985, now a fairly common resident in the northern KNP, and also present further south. More strongly associated with larger, coarse-haired ungulates such as Buffalo and Giraffe; listed as Vulnerable in S Africa. **HABITAT** Open savanna woodland. **FOOD** Mostly ticks, also other ectoparasites, flying insects and blood from open wounds. Leaves host briefly to drink water. **CALL** Hissing, crackling *krus, krus ...* **BR** Monogamous. Nests in tree cavities, occasionally with nest helpers. (**Geelbekrenostervoël**)

RED-BILLED OXPECKER *Buphagus erythrorhynchus* **(772) 20 cm; 50 g**
SEXES *Alike. Red bill, large yellow wattle around eye and dark rump key field characteristics.* **JUV** *Similar to juv Yellow-billed Oxpecker; dark rump diagnostic.* **STATUS** Locally common resident. Range drastically reduced elsewhere in SA, but many populations re-established from capture and translocation operations out of KNP. Regarded as Near-threatened in S Africa. Spends most of the day perched on hosts. May be found on most of the larger herbivores in the KNP, most of which tolerate their activities. Elephants, however (unless in poor physical condition), reject oxpeckers. Roost if flocks in trees away from host animals; uses Northern Lala-palm *Hyphaene petersiana* in Letaba and Shingwedzi. **HABITAT** Open woodland where ungulate hosts present. **FOOD** Ticks and other ectoparasites, also other invertebrates and blood from open wounds. Captive birds have been recorded eating estimated 12 500 tick larvae, or 100 engorged adult ticks per day. Also flying insects. **CALL** Sharp hissing *ksss* and staccato *tsik tsik*, also given in flight. **BR** Monogamous and co-operative br, nests in tree cavities. Incubation by both sexes. (**Rooibekrenostervoël**)

WATTLED STARLING *Creatophora cinerea* **(760) 21 cm; 70 g**
SEXES M (BR) *Head featherless with large black wattle on forehead and throat, and bright yellow hind crown.* **F & M (NON-BR)** *Pale whitish-grey plumage (incl head of ♂) with diagnostic white rump. Wing and tail feathers black in ♂, brownish in ♀.* **JUV** *Like ♀; bill initially yellow.* **STATUS** An irruptive nomad in KNP, some years present in very large numbers, but may be absent in others. Gregarious and always in flocks. Often roosts in association with other starling species. **HABITAT** Open dry woodland, usually with short grass cover. **FOOD** Insects; also fruit, nectar from trees and aloes and seeds. **CALL** Series of jumbled wheezy phrases. **BR** Irregular br in KNP, monogamous and colonial, sometimes thousands of nests covering many hectares. Untidy ball nest made of sticks; usually placed in thorn trees, particularly shrub Knob Thorn *Acacia nigrescens*. Incubation by both sexes. (**Lelspreeu**)

PLATE 79 Sunbirds p224

SCARLET-CHESTED SUNBIRD *Chalcomitra senegalensis* (791) 14 cm; 13,6 g
SEXES M *Bright scarlet breast, iridescent green throat and forecrown, and velvety-blackish body unmistakable. Imm ♂ similar to ♀, but has mottled scarlet breast patch. Lacks pectoral tufts.* F *Overall darker plumage, whitish alula (shoulder edge) and lack of pale eyebrow distinguish it from ♀ Amethyst Sunbird.* **JUV** *Like ♀, underparts yellower and barred rather than streaked. Gape whitish or pale yellow, turning orange then finally black.* **STATUS** Common and widespread resident. Nomadic in response to food resources during the non-br season and highly irruptive. **HABITAT** Open woodland, especially thornveld; common in restcamps. **FOOD** Nectar, insects and spiders. **CALL** Loud and penetrating *tip, teeu* notes, also fast *syip-syip-syip* ... **BR** Monogamous. Oval or pear-shaped nest usually suspended about 5 m up in tree, often within mid-canopy. Nest construction and incubation by ♀ only. Common host to Klaas's Cuckoo, and possibly irregular host of Diderick Cuckoo. (**Rooiborssuikerbekkie**)

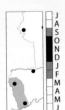

AMETHYST SUNBIRD *Chalcomitra amethystina* (792) 14,5 cm; 15 g
SEXES M *All-black body, metallic green forecrown, iridescent reddish-purple throat and shoulder patches very distinctive. Lacks pectoral tufts. Immature ♂ similar to adult ♀ but with reddish-purple throat.* F *Pale eyebrow and lack of white alula distinguish it from ♀ Scarlet-chested Sunbird.* **JUV** *Throat blackish; underparts more barred than streaked. Gape pale yellowish.* **STATUS** Uncommon resident; confined to the higher-lying ground in the south-western corner of KNP (vagrant elsewhere). Sedentary, with some possible local altitudinal movements into KNP in winter. **HABITAT** Open woodland and restcamps (particularly Pretoriuskop). **FOOD** Nectar, insects and spiders. Forages at nectar sources alone, and joins mixed-species foraging flocks when hunting for insects such as spiders. Hawks soft-bodied flying insects in flight. **CALL** Song a sustained series of high-pitched twittering notes. Call note a penetrating *chak* ... **BR** Monogamous. Nest a thick-walled oval with hooded entrance. Nest construction and incubation by ♀ only. Parasitised by Klaas's Cuckoo. (**Swartsuikerbekkie**) **ALT NAME** Black Sunbird

COLLARED SUNBIRD *Hedydipna collaris* (793) 10,5 cm; 8 g
SEXES *Smallest and only short-billed sunbird in the Park.* M *Bright metallic green throat and upperparts, yellow belly and bluish-purple breast band distinctive. Pectoral tufts yellow.* F *Metallic green upperparts and yellow underparts make it unmistakable.* **JUV** *Like ♀; gape initially bright yellow, fading gradually.* **STATUS** Common resident in suitable habitat, mostly sedentary and usually found in pairs. **HABITAT** Mainly riparian forest and thickets, and restcamp gardens. **FOOD** Insects and nectar, sometimes small fruit. Often joins mixed-species foraging flocks; more insectivorous than other sunbirds in KNP. 'Steals' nectar from flowers with large corolla tubes by slitting tube down to base, or piercing from the side. **CALL** Song high-pitched *chirri, chirri, chirri* ... , contact call thin *tsip*. **BR** Monogamous. Oval nest often placed near wasp nest for protection from monkeys. Nest construction and incubation by ♀ only. Parasitised by Klaas's Cuckoo. (**Kortbeksuikerbekkie**)

WHITE-BELLIED SUNBIRD *Cinnyris talatala* (787) 11 cm; 7,5 g
SEXES ♂>♀. M **(BR)** *Only sunbird in region with white belly and iridescent green throat and back.* M **(NON-BR)** *Variable number of brown feathers in green plumage.* F *Grey-brown above; whitish below.* **JUV** *Like ♀ with yellow wash to belly; gape orange-yellow.* **STATUS** Common and widespread resident. Highly nomadic throughout its range. **HABITAT** Dry *Acacia* and other woodlands, and restcamp gardens. **FOOD** Nectar, insects, spiders. Joins mixed-species foraging flocks. At nectar sources, usually dominated by Marico Sunbird. Seldom drinks water, usually only when very hot. **CALL** Characteristic *chewy-chewy-chewy* phrase and other strident notes. **BR** Monogamous. Builds pendant nest, construction and incubation by ♀ only. Parasitised by Klaas's Cuckoo. (**Witpenssuikerbekkie**)

MARICO SUNBIRD *Cinnyris mariquensis* (779) 13,5 cm; 11 g
SEXES ♂>♀. M *Overall dark appearance and broad purple-maroon breast band (which may appear reddish at some angles) distinctive. Lacks pectoral tufts. Territorial males call regularly from same prominent perch in a tall tree.* F *Drab grey-brown and fairly heavily streaked below; belly straw yellow.* **JUV** *Similar to ♀; juv ♂ with black throat.* **STATUS** Fairly common resident. **HABITAT** Dry *Acacia* and riparian woodland; gardens. **FOOD** Nectar and invertebrates. Regularly drinks water. **CALL** Rapid series of canary-like notes; contact call a hard *chip-chip, jik-jik-jik*. **BR** Monogamous. Builds pear-shaped nest, construction and incubation by ♀ only. Usually placed in canopy of thorn tree; occasionally parasitised by Klaas's Cuckoo. (**Maricosuikerbekkie**)

PLATE 80 Masked-Weavers, Weavers p224

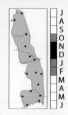

LESSER MASKED-WEAVER *Ploceus intermedius* (815) 14 cm; 20,5 g
SEXES ♂>♀. **M (BR)** *Small weaver with whitish-yellow eyes and grey legs. Black mask extends to centre of crown. Lower edge of black bib on throat rounded (not pointed as in Southern Masked-Weaver).* **M (NON-BR)** *Lacks black face mask and has duller plumage; bill dark horn-coloured, and legs blue-grey (not brownish-pink as in Southern Masked-Weaver and Village Weaver).* **F** *Less yellow below than non-br ♂; eyes whitish-yellow. Belly becomes whitish in non-br season.* **JUV** *Resembles non-br ♀.* **STATUS** Locally common resident, gregarious and nomadic in non-br season. **HABITAT** Savanna woodland, especially thornveld; usually not far from water. **FOOD** Insects, seeds and nectar. **CALL** Typical rasping weaver swizzles. **BR** Polygynous and colonial. Nesting colonies usually in trees (often over water), sometimes in reedbeds. To avoid predation, nests sometimes constructed close to human activity on busy restcamp buildings such as thatched verandahs. Nest very small with short, very narrow entrance tunnel (that may help limit access by Diderick Cuckoo). Nest construction by ♂, incubation by ♀ only. Occasional host to Diderick Cuckoo. (**Kleingeelvink**)

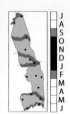

SPECTACLED WEAVER *Ploceus ocularis* (810) 15,5 cm; 29 g
SEXES ♂>♀. **M** *Sharply pointed black beak, pale yellow eye, with black eye-stripe from beak through and behind eye, distinctive. Dark brownish wash on throat forms narrow black bib.* **F** *Lacks black bib; throat and upper breast sometimes with chestnut wash. Bill remains black throughout year.* **JUV** *Similar to ♀, but without black line through eye. Bill pale brown, eyes brown.* **STATUS** Fairly common resident, sedentary and usually in pairs year-round. Does not join flocks or roost with other weavers. **HABITAT** Mainly riparian forest and associated woodland. **FOOD** Largely insectivorous; also feeds on spiders, fruit, seeds and nectar. **CALL** Liquid descending *tee-tee-tee-tee ...* **BR** Monogamous and solitary. Nest, with long entrance tunnel, usually built by ♂ alone, is strategically suspended at end of branch or vine. Parasitised by Diderick Cuckoo. (**Brilwewer**)

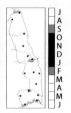

GOLDEN WEAVER *Ploceus xanthops* (816) 17 cm; 41 g
SEXES ♂>♀. **M** *Combination of heavy black bill, pale yellow eyes and lack of black facial markings diagnostic. Chin and throat may have pale chestnut-brown wash; back olive-green.* **F** *Similar but duller than ♂.* **JUV** *Duller than ♀, with buffy-yellow underparts, and brown bill and eyes.* **STATUS** Very uncommon, recorded along the Crocodile River and at a few widely scattered localities. **HABITAT** Savanna woodland and riparian bush, normally singly or in pairs. **FOOD** Insects, fruit, seeds and nectar. **CALL** Swizzling weaver notes. **BR** Monogamous (occasionally polygynous). In KNP nesting recorded only in reedbeds, mostly solitary; nest lacks entrance tunnel. Host of Diderick Cuckoo. (**Goudwewer**)

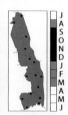

SOUTHERN MASKED-WEAVER *Ploceus velatus* (814) 15 cm; 34 g
SEXES ♂>♀. **M (BR)** *Distinguished from slightly larger Village Weaver by plain (not blotched) back. Distinguished from smaller Lesser Masked-Weaver by red (not whitish-yellow) eyes, heavier bill, yellow (not black) crown and brownish-pink (not grey) legs.* **M (NON-BR)** *Lacks black face mask; similar to ♀ with red-brown eyes and considerable variation in shades of plumage.* **F** *Yellow-buff above, throat and breast yellowish. Non-br birds paler below with whitish belly.* **JUV** *Resembles non-br ♀.* **STATUS** Fairly common resident. Gregarious, in small flocks and fairly sedentary. **HABITAT** Open savanna, especially *Acacia* and Mopane. **FOOD** Insects, seeds, fruit, plant material and nectar. Also table scraps from picnic sites. **CALL** Series of buzzy swizzling and churring phrases, also sharp *zik* note. **BR** Polygynous. Often a solitary breeder in trees away from water, nest constructed by ♂, incubation by ♀ only; regularly parasitised by Diderick Cuckoo. (**Swartkeelgeelvink**) **ALT NAME** *Masked Weaver*

VILLAGE WEAVER *Ploceus cucullatus* (811) 16 cm; 37 g
SEXES ♂>♀. **M (BR)** *Black mask extends only to forehead in southern race, and to crown in northern race* nigriceps *(that occurs just north of KNP boundary). Distinguished from other masked weavers by heavily blotched back.* **M (NON-BR)** *Resembles ♀, but larger.* **F** *Throat washed yellow, belly whitish, back mottled dull olive and eyes red or brown.* **JUV** *Resembles ♀.* **STATUS** Very common resident, gregarious and generally sedentary. **HABITAT** Wide range of woodlands, and restcamp gardens; usually near water. **FOOD** Insects, seeds and nectar. Feeds at tables for scraps at picnic sites. **CALL** Rolling swizzling song, *cheee cheee shrrrrr, zzzzzrrr, cheee, ch-ch-ch-ch*. **BR** Polygynous and colonial. Coarsely woven nest usually suspended from tree or in reeds, often over water. Nest construction by ♂, incubation by ♀ only. Parasitised by Diderick Cuckoo. (**Bontrugwewer**) **ALT NAME** *Spotted-backed Weaver*

PLATE 81 Buffalo-Weaver, Quelea, Weavers p225

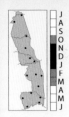

RED-BILLED BUFFALO-WEAVER *Bubalornis niger* **(798) 23 cm; ca 78 g**
SEXES ♂>♀. **M** *Red bill and black plumage with lightly white-flecked shoulder patches and primary edges unmistakable. White wing patches conspicuous in flight.* **F** *Duller than ♂, more brownish-black, especially on underparts, which have white mottling.* **JUV** *Resembles ♀ but underparts white, streaked dark brown.* **STATUS** Locally common summer visitor to KNP, though some birds overwinter. Gregarious; often in association with starlings and other weaver species. Usually seen in vicinity of large stick nests in which they roost communally at night. **HABITAT** Woodland, especially in *Acacia* savanna with scattered Baobab trees. **FOOD** Insects, seeds, fruit and food scraps. **CALL** ♂ gives loud *lookatit-lookatit-lookatit* or *widdla-widdla-widdla-widdla*; ♀ gives musical *chwee*. **BR** Colonial br with both polygamous and co-operative br systems. Nest a bulky multi-chambered structure of thorny twigs that may be occupied by different females. Large raptors such as White-backed Vulture and Bateleur sometimes build their nests on buffalo-weaver structures which probably provides some measure of protection from predators for the buffalo-weavers, and a degree of camouflage for the raptors. Incubation by the ♀ only. (**Buffelwewer**)

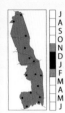

RED-BILLED QUELEA *Quelea quelea* **(821) 12 cm; 19 g**
SEXES M (BR) *Face mask variable; most commonly black but may be white, surrounded by yellowish or pink band.* **M (NON-BR)** *Bright head features turn brown; throat white, bill remains red.* **F** *Similar to non-br ♂, but bill and eye-ring dull yellow in br season.* **JUV** *Like adult ♀, bill horn-brown.* **STATUS** One of the most abundant bird species on earth, and the most abundant species in the Park when at their peak with estimated numbers reaching 33 million birds. Nomadic and highly gregarious; particularly abundant in some wetter years, but can be sparse during drought years. Usually drinks daily, often huge flocks visiting waterholes. Roost communally, sometimes with weavers. **HABITAT** Most abundant in semi-arid savanna. **FOOD** Mainly seeds of grasses, also insects. When feeding on the ground in flocks, they move in spectacular 'roller feeding' movements with birds from behind flying in waves to the front of the flock. Termite alates taken in flight. **CALL** Chattering warble. **BR** Monogamous and colonial. Usually nests in huge colonies in *Acacia* trees (particularly shrub Knob Thorn *Acacia nigrescens*) in aggregations up to millions. Single trees may have from a few to hundreds of nests, and chicks fledge within 25 days of egg-laying. Breeding colonies are a major attraction to a wide range of predators; avian predators include raptors, herons and storks that feed mainly on nestlings. Incubation mostly by ♀. (**Rooibekkwelea**)

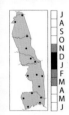

THICK-BILLED WEAVER *Amblyospiza albifrons* **(807) 18 cm; 56 g**
SEXES ♂ *considerably larger than ♀.* **M** *Overall dark brown plumage, enormous bill and white wing and forehead patches very distinctive.* **F** *Heavy horn-coloured (not black) bill distinguishes it from more slender-billed ♀ Violet-backed Starling.* **JUV** *Like ♀; bill initially yellow, becoming darker.* **STATUS** Rather rare in KNP; but has expanded its range in the past 50 years owing to construction of man-made dams. Forms small flocks during the non-br season: drop in numbers during winter suggest movements out of KNP in search of better food sources elsewhere. Roosts in reedbeds in small groups, often with other weavers, bishops and waxbills. **HABITAT** Rank grass and reeds in and adjacent to wetlands and rivers during br season; forests, forest margins and woodlands in non-br season. **FOOD** Seeds of fruiting trees, fruit, grass seeds and insects incl termite alates. **CALL** Various clicking notes and trills. **BR** Polygynous and usually colonial. Very neat finely woven oval nest usually suspended over water and between two reeds or bulrushes by ♂; nest lining and incubation by ♀. (**Dikbekwewer**)

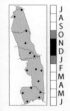

RED-HEADED WEAVER *Anaplectes melanotis* **(819) 14,5 cm; 22 g**
SEXES ♂>♀. **M (BR)** *Combination of bright scarlet head and breast, orange or reddish bill and white underparts are features unlike any other species in region.* **F & M (NON-BR)** *Bill orange to reddish, head and breast lemon-yellow; yellow-edged wing feathers.* **JUV** *Like adult ♀.* **STATUS** Uncommon to fairly common resident; joins mixed-species foraging flocks in woodland during winter months. Usually solitary or in pairs. **HABITAT** Broad-leaved woodland and *Acacia* savanna. **FOOD** Mainly insects and spiders; also seeds and fruit. Most food obtained from within the canopy of trees, but sometimes forages on the ground and hawks termite alates aerially. **CALL** Series of high-pitched, squeaky, swizzling notes. **BR** Usually monogamous, sometimes polygynous. Nest typically weaver-shaped but constructed of thin pliable twigs, vines and leaf petioles, giving shaggy appearance; placed on outer branches of trees, particularly Baobabs and often near Red-billed Buffalo-Weaver colonies, on telephone wires and around human habitation. Occasionally parasitised by Diederick Cuckoo. Incubation mostly by ♀. (**Rooikopwewer**)

PLATE 82 Bishops, Widowbirds p225

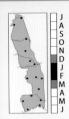

YELLOW-CROWNED BISHOP *Euplectes afer* **(826) 11 cm; 15,5 g**
SEXES ♂>♀. **M (BR)** *Yellow crown and small size distinctive.* **F & M (NON-BR)** *Drab brown plumage, similar to other non-br* Euplectes *species but with whitish (not buff) underparts, broad eyebrow and dark ear coverts contrasting with paler face.* **JUV** *Buffier than* ♀. **STATUS** Rare but recorded at widespread localities throughout KNP, but usually only in very wet years. Male very conspicuous in br plumage, especially when performing aerial 'bumble-bee flights' with fluffed plumage displaying prominent yellow back and rump. **HABITAT** Closely associated with marshes and seasonally flooded wetlands. **FOOD** Seeds and insects. **CALL** Harsh *zeep*, and *zzzeeet* trill notes. **BR** Polygynous. Nest an oval ball concealed in grass or sedges, incubation by ♀ only. **(Goudgeelvink) ALT NAME** *Golden Bishop*

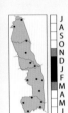

SOUTHERN RED BISHOP *Euplectes orix* **(824) 13 cm; 23 g**
SEXES ♂>♀. **M (BR)** *Bright red and black plumage unmistakable.* **F & M (NON-BR)** *Bill brown; best distinguished from non-br widows by small size, compact shape and shorter tail. Breast and flanks fairly heavily streaked. Underparts buffier and eyebrow narrower than smaller* ♀ *and non-br* ♂ *Yellow-crowned Bishop.* **JUV** *Upperparts buffier than* ♀. **STATUS** Fairly common resident, gregarious and sedentary; construction of man-made dams has favoured distribution and increased numbers. Typically roosts in reedbeds, often with other ploceids. **HABITAT** Marshy grassland and wetlands. **FOOD** Seeds, insects and nectar. Hawks insects aerially. **CALL** ♂ gives sizzling song. **BR** Polygynous. Oval woven nest constructed by male and lined by female; usually built over water. Incubation by ♀ only. Commonly parasitised by Diderick Cuckoo. **(Rooivink) ALT NAME** *Red Bishop*

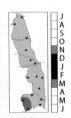

RED-COLLARED WIDOWBIRD *Euplectes ardens* **(831)** ♂ **35 cm,** ♀ **12 cm; 19 g**
SEXES ♂>♀. **M (BR)** *Distinctive black plumage, long-tailed widowbird with red collar across upper breast. Rarely, this may be orange, yellow, or even absent. Tail long and narrow, bill black.* **M (NON-BR)** *Similar to* ♀*, but upperparts streaked black. Retains black flight feathers; bill brown.* **F** *Like non-br* ♂*, plumage drab brown. Underparts buffy and unstreaked, upperparts streaked brown (not black).* **JUV** *Like* ♀*, upperpart feathers with broad buffy margins.* **STATUS** Fairly common only in the s-w, uncommon elsewhere; recorded widely, especially in wetter years. Gregarious and often with other widowbirds and bishops. **HABITAT** Tall moist grasslands and savanna. **FOOD** Mainly seeds, also insects, small berries and nectar. **CALL** In display flight, ♂ gives soft husky *hizz zizz zizz*, also rapid *screep-screep* and a drawn-out *sscherz*. **BR** Polygynous. Oval nest often woven into small bush or shrub; also uses tall grass tufts. Incubation by ♀ only. Occasionally parasitised by Diderick Cuckoo. **(Rooikeelflap) ALT NAME** *Red-collared Widow*

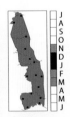

WHITE-WINGED WIDOWBIRD *Euplectes albonotatus* **(829) 15 cm; 20 g**
SEXES ♂>♀. **M (BR)** *Combination of yellow epaulet and white in wings distinctive. Bill silver-grey.* **M (NON-BR)** *Like* ♀*, but retains black wings with white margins. Bill blue-black.* **F** *Flight feathers brown and without white bases. Underparts whiter than other* Euplectes *species, and with largely unstreaked breast.* **JUV** *Like* ♀*, but upperparts sometimes with yellow wash.* **STATUS** Rare to common resident, sedentary, with some local movements which are dependent on rainfall; can be very common in years of good rainfall. Gregarious, forming large flocks in non-br season, often with other widowbirds and bishops. **HABITAT** Usually in moist or rank grassland. **FOOD** Mainly grass seeds, also insects and aloe nectar. **CALL** Song *squee-squi-squeege*, flight call *squi-squi-squeege-squeege*. **BR** Polygynous. Oval nest woven into tall thick grass. Incubation by ♀ only. Occasionally parasitised by Diderick Cuckoo. **(Witvlerkflap) ALT NAME** *White-winged Widow*

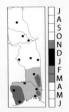

FAN-TAILED WIDOWBIRD *Euplectes axillaris* **(828) 15 cm; 26 g**
SEXES ♂>♀. **M (BR)** *Black plumage with red epaulets and, in flight, longish fan-shaped tail distinctive.* **M (NON-BR)** *Similar to* ♀ *but wings black and retains red epaulets. Bill blue and tail feathers brown, rounded.* **F** *Wings brown, with orange-brown epaulets less conspicuous than in* ♂*, bill brown.* **JUV** *Like* ♀*, with buffy upperpart feathers.* **STATUS** Fairly common localised resident, most prominent in wetter years, recorded only on the basaltic grasslands in the southern half of the Park, and in the higher-lying ground in the south-west. **HABITAT** Tall moist grassland, marshes and bushed grassland. **FOOD** Mainly seeds, also insects and occasionally nectar. **CALL** Song a series of high-pitched twittering and chirping notes, also a sizzling call. **BR** Polygynous. Woven oval nest built into dense grass, often over marshy ground; incubation by ♀ only. Occasionally parasitised by Diderick Cuckoo. **(Kortstertflap) ALT NAME** *Red-shouldered Widow*

PLATE 83 Waxbill, Quailfinch, Finch, Twinspot p225

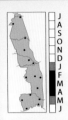

ORANGE-BREASTED WAXBILL *Sporaeginthus subflavus* **(854) 9,5 cm; 7,5 g**
SEXES *In flight, call and bright orange-red rump distinguish it from African Quailfinch.* **M** *Red eyebrow and bright yellow-orange underparts unmistakable.* **F** *Duller below than ♂, and lacks red eyebrow.* **JUV** *Duller than ♀, with black bill.* **STATUS** Very rare in KNP, but recorded widely and mostly only in wetter years. Sometimes in pairs but most often in family groups and small flocks. Roosts communally in reedbeds; drinks and bathes regularly. Flock movements in flight well synchronised. **HABITAT** Moist grassland and wetland margins flanking reedbeds. **FOOD** Mostly small grass seeds, also insects and recorded taking soft shoots. **CALL** Song a series of *chip, chit* and *chink* notes; flight call *tink, tink* ... **BR** Monogamous and often solitary. Usually occupies old nest of Southern Red Bishop (less often those of widowbirds and weaver species; rarely cisticolas and prinias) for br after ♀ has re-lined the nest with dry grass and feathers. Breeds late summer to synchronise with availability of vacant bishop nests. Rarely builds own nest. Occasional host of Pin-tailed Whydah. (**Rooiassie**)

AFRICAN QUAILFINCH *Ortygospiza atricollis* **(852) 10 cm; 11,5 g**
SEXES *Very small, short-tailed, terrestrial finch, seen and heard in flight more than on the ground. Difficult to locate in grassland; usually gives its presence away by distinctive double-note flight call.* **M** *Dark facial markings, black throat and barred breast and flanks. Lacks orange or red on rump as in Orange-breasted Waxbill.* **F** *Paler than ♂, and lacks black throat.* **JUV** *Paler than ♀, with black bill.* **STATUS** Fairly common but inconspicuous resident, particularly on the eastern grasslands, gregarious and nomadic in non-br season. Usually only flushes at last minute, flying up (often in pairs) in short, jerky movements giving characteristic call. **HABITAT** Favours short open grassland near water but also in well-grassed mesic regions. **FOOD** Mostly small grass seeds, also insects. **CALL** Flight call most often heard: *drink-drink* or *djink-djink* notes given in quick succession. **BR** Monogamous and solitary. Ball-shaped nest built by both sexes placed below grass tuft with a clearing in front of entrance. Incubation by both sexes, and pair roosts in nest at night. (**Gewone Kwartelvinkie**) **ALT NAME** *Quail Finch*

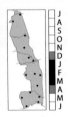

CUT-THROAT FINCH *Amadina fasciata* **(855) 12 cm; 18 g**
SEXES M *Red throat band and barred crown diagnostic.* **F** *Similar to ♂ but lacks red throat crescent and rich brown belly. Head and upperparts barred.* **JUV** *Similar to ♀, but juv ♂ has pale red throat stripe.* **STATUS** Uncommon to fairly common resident, most easily seen in the north of the Park (eg Shingwedzi) where Red-headed Weaver nests are readily available for breeding during the late summer months; nomadic and in flocks during non-br season. Associates with weavers and queleas in mixed-species flocks. Roosts in old weaver and woodpecker nests, also natural holes in trees. **HABITAT** Semi-arid savanna woodland. **FOOD** Seeds and termites; drinks regularly. **CALL** Sparrow-like chirping, also buzzing notes. **BR** Monogamous and usually solitary. Almost invariably br in old (used) weaver nests (especially Red-headed Weavers) but also in other nests such as Red-billed Buffalo-Weaver and even woodpecker holes. Incubation by both sexes and recorded roosting together during incubation. (**Bandkeelvink**)

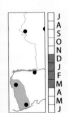

GREEN TWINSPOT *Mandingoa nitidula* **(835) 11 cm; 9,5 g**
SEXES *Small, shy species easily overlooked in shaded forest undercanopy, best located by high-pitched 'tsit' or 'tick' call. As with other twinspots, name derived from the double spot on each feather, one on either side of the feather shaft.* **M** *Red face and mostly green upperparts diagnostic.* **F** *Face yellowish.* **JUV** *Dull green above, greyish-olive below; white spotting develops at almost 1 year old, starting with just a few initially on flanks and upper belly region.* **STATUS** Very uncommon resident, recorded only in the south-western parts of KNP, usually in small parties, and in pairs during the br season. Usually found on or near the ground and flushes into canopy where it sits quietly till danger has passed. **HABITAT** Riparian forest and well-developed and shaded restcamp gardens. **FOOD** Mainly seeds of forest grasses but also recorded gleaning insects from foliage and taking insects aerially. **CALL** Contact call high-pitched *tsit, tsit*. **BR** Monogamous. Oval nest built in forest vines and foliage. May breed in first summer of 'adulthood', sometimes still in partial juv plumage (incomplete spotting of underparts). Incubation by both sexes and recorded roosting together in nest during incubation (**Groenkolpensie**)

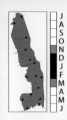

COMMON WAXBILL *Estrilda astrild* (846) 12 cm; 8 g
SEXES Similar. Brownish upperparts and bright red bill and eye-stripe distinctive. ♀ has paler red belly stripe and reddish-orange bill when br. **JUV** Paler than adult; pale red to orange eye-stripe and blackish bill. **STATUS** Common resident, sedentary and in pairs or small flocks of 30 or more birds. Roosts communally in reedbeds or bushes, sometimes in hundreds, perched singly or in rows. **HABITAT** Rank vegetation along watercourses and dams, and also in restcamp gardens. **FOOD** Mainly grass seeds, also insects, ripe figs and flowers of grasses. **CALL** Harsh *di-di-di-JEE* and *tcher-tcher-preeee* notes. **BR** Monogamous. Pear-shaped ball nest built by both sexes; usually placed on ground often well concealed under grass tuft. Many nests have false or 'cock's' nest built on roof of main structure. Animal fur and scats sometimes added to false nest, apparently to reduce risk of nest predation. Incubation by both sexes, and both roost in nest at night. Main host of Pin-tailed Whydah. (**Rooibeksysie**)

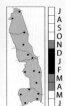

VIOLET-EARED WAXBILL *Granatina granatina* (845) 14 cm; 12 g
SEXES M One of the most colourful seed-eaters in region with violet cheeks, red bill, blue rump and forehead, and rich chestnut body. **F** Paler than ♂, with fawn underparts. **JUV** Paler than ♀, and lacks bright facial colours; bill black. **STATUS** Rare but widespread resident, sedentary or locally nomadic. Usually in pairs or small family groups; often associates with Blue Waxbill. **HABITAT** Dry shrublands, *Acacia* and broad-leaved woodland with thickets. **FOOD** Mainly seeds but also insects and recorded feeding on flesh of fruit and nectar from aloes. Takes termite alates aerially, and drinks water regularly at waterholes. **CALL** Soft twittering notes. **BR** Monogamous, with lifelong pair bond. Nest an oval ball of dry grass, usually placed 1–2 m high in shrub or tree. Incubation by both sexes. Parasitised by Shaft-tailed Whydah. (**Koningblousysie**)

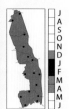

BLUE WAXBILL *Uraeginthus angolensis* (844) 12,5 cm; 10 g
SEXES Similar, ♀ with paler and less extensive blue on face and underparts. Only small waxbill in region with pale blue underparts, tail and rump. Extremely rarely, males have scarlet cheek patches. **JUV** Blue restricted to face, throat, breast, rump and tail. **STATUS** Common and widespread resident. In flocks of up to 40 or more in non-br season. Dependent on surface water, so nomadic during periods of low rainfall. **HABITAT** Favours open *Acacia* but also present in broad-leaved woodland such as Mopane. Typically roosts in canopy of dense *Acacia*. **FOOD** Mainly seeds, also insects (termite alates taken aerially) and small fruit; drinks regularly. **CALL** Loud contact call, *sweep-sweep* or *tseep-tseep*. Song, *chreu-chreu-chittywoo-weeoo-wee*. **BR** Monogamous. Ball nest with side entrance built of dry grass and usually placed in thorny tree, almost invariably close to wasp nest. Occasionally uses old weaver or sunbird nest in which to breed; incubation by both sexes. Occasionally parasitised by Shaft-tailed Whydah. (**Gewone Blousysie**)

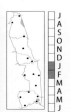

PINK-THROATED TWINSPOT *Hypargos margaritatus* (838) 13 cm; 13 g
SEXES M Pink face, throat and breast, brown upperparts and white spotted underparts unmistakable. **F** Duller than ♂ with greyish face, throat and breast. Central belly whitish. **JUV** Underparts pale buff, lacking spots. **STATUS** Generally shy and uncommon endemic, regarded as Near-threatened in s Africa. In KNP found only in the Nyandu (Sand Camwood *Baphia massaiensis*) community of the Nwambiya sandveld. Usually in pairs or small flocks, shy and wary, seldom venturing far from cover; darts into dense thickets when disturbed. Gives presence away by soft contact calls. **HABITAT** Dry scrub, woodland and sand forest. **FOOD** Grass seeds, also insects. **CALL** Song, 2–6 flute-like notes; call includes trills. **BR** Monogamous. Ball-shaped nest usually placed within 1 m of ground. Incubation by both sexes. (**Rooskeelkolpensie**)

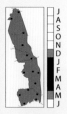

GREEN-WINGED PYTILIA *Pytilia melba* (834) 13 cm; 15 g
SEXES M Grey (not red) ear coverts and olive-green (not orange) wings distinguish it from Orange-winged Pytilia (that is recorded as a rare vagrant to the Park). **F** Key features olive-green wings, and dark grey and white barring below. **JUV** Head and throat olive; plain greyish below, bill black. **STATUS** Fairly common and widespread resident. Probably nomadic in the dry season with dependence on surface drinking water year-round. **HABITAT** Semi-arid *Acacia* savanna and dry mixed woodland. **FOOD** Seeds and insects, especially termites obtained by breaking open mud workings. **CALL** 2–3 water droplet sounds followed by trill *prrreeeeeoooo* ... **BR** Monogamous. Ball nest placed 1–2 m high, usually in thorny bush. Incubation by both sexes. Parasitised by Long-tailed Paradise-Whydah. (**Gewone Melba**) **ALT NAME** *Melba Finch*

PLATE 85 Firefinches, Mannikins p226

RED-BILLED FIREFINCH *Lagonosticta senegala*   **(842) 9,5 cm; 9 g**
SEXES *Only firefinch in the Park with combination of pinkish bill and red rump.* **M** *Crown, nape, face and breast pinkish-red. Lower belly and undertail coverts deep buff.* **F** *Lores, upper tail and rump pink, rest of plumage sandy-brown.* **JUV** *Similar to ♀ but bill black and lacks pink lores and white spotting.* **STATUS** Common resident, sedentary and usually in pairs or small groups, sometimes as many as 30 birds. Gives its presence away by contact *fweet* call. Pairs remain together during non-br season and pair bond remains strong, even in feeding flocks. Bathes and drinks regularly and also sips water droplets from leaves. **HABITAT** Rank grass and thickets in *Acacia* and broad-leaved woodland. **FOOD** Mainly grass seeds and also insects, including termite alates. **CALL** Soft fluty *dwee* or *fweet* notes. **BR** Monogamous. Same mate may be retained in successive br seasons. Ball-shaped nest built by ♂ and placed near ground. Incubation by both sexes. Parasitised by Village Indigobird. (**Rooibekvuurvinkie**)

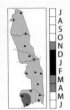

AFRICAN FIREFINCH *Lagonosticta rubricata*   **(840) 11,5 cm; 10 g**
SEXES *Distinguished from Jameson's Firefinch by greyish (not pinkish) crown and nape, and by darker brown back and wings.* **M** *Distinguished by dark bluish-black bill, greyish head and nape, and grey-brown back and wings.* **F** *Duller than ♂, with pinkish-brown wash below.* **JUV** *Like ♀ but duller.* **STATUS** Fairly common resident in the higher-lying south-west, rare elsewhere in KNP. Least common of the firefinches. Sedentary and usually in pairs or small groups. Flushes into cover with short 'bouncy' flight and trill call when disturbed. **HABITAT** Riparian forest margins and rank vegetation. **FOOD** Mainly grass seeds, also insects including termite alates. **CALL** Trills, tinkles and *chit* notes. **BR** Monogamous. Nest a ball of dry grass built by male, usually placed 0,5 m above ground in rank growth. Incubation by both sexes during the day and by the ♀ alone at night. Host of Dusky Indigobird. (**Kaapse Vuurvinkie**) **ALT NAME** *Blue-billed Firefinch*

JAMESON'S FIREFINCH *Lagonosticta rhodopareia*   **(841) 11 cm; 9 g**
SEXES M *Distinguished from African Firefinch by pinkish (not grey-brown) wash to upperparts, and from Red-billed Firefinch by bluish-black (not pinkish) bill.* **F** *Paler than ♂ and with pinkish-buff throat and breast, and less spotting below. Reddish loral patch distinguishes it from ♀ African Firefinch.* **JUV** *Duller and browner than ♀.* **STATUS** Fairly common resident, sedentary; usually in pairs or in small groups (sometimes with other seedeaters) during the non-br season. Drinks regularly, and dependent on surface water. Sips water droplets from foliage. **HABITAT** Favours drier woodland than African Firefinch. **FOOD** Mainly grass seeds, also insects, including termite alates taken aerially; opens mud tunnels to expose termite workers. **CALL** Main call trilling *we-we-we-we-we-we*. **BR** Monogamous. Oval nest mainly dry grass and usually built by both sexes, usually within 1 m of ground. Incubation by both sexes during the day and by the ♀ alone at night. Parasitised by Purple Indigobird. (**Jamesonse Vuurvinkie**)

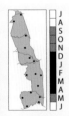

BRONZE MANNIKIN *Spermestes cucullata*   **(857) 9,5 cm; 10 g**
SEXES *Alike. Small size, black head and bib, barred flanks, white breast and bicoloured bill unmistakable. Bronzy-green shoulder patch most evident in sunlight. Back brown (not rich chestnut) as in Red-backed Mannikin.* **JUV** *Overall drab grey-brown, paler below.* **STATUS** Fairly common but localised resident, more common in the southern parts of the Park. Generally sedentary and usually in small flocks. Drinks regularly and roosts communally in specially constructed non-br roosting nests. Br nests sometimes relined and used for roosting. **HABITAT** Grassy patches in woodland and riverine margins; common in some restcamp gardens. **FOOD** Mainly seeds, also insects, plant material and nectar. **CALL** *Tchrie* or *chie* notes. **BR** Monogamous. Both sexes build untidy ball nest. Incubation mostly by the ♀ during the day and by both sexes at night. Occasionally parasitised by Pin-tailed Whydah. (**Gewone Fret**)

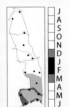

RED-BACKED MANNIKIN *Spermestes bicolor*   **(858) 9,5 cm; 9 g**
SEXES *Alike. Rich chestnut back and wing coverts very distinctive.* **JUV** *Warm brown upperparts distinguish it from other drab brownish mannikin juvs in region.* **STATUS** Generally uncommon and localised resident recorded mainly in south-western parts of KNP. In pairs when br otherwise in small flocks. Roosts communally in loosely constructed roosting nests, or in reedbeds. **HABITAT** Mainly along fringes of well-developed riparian woodland; also in restcamp gardens, especially Skukuza. **FOOD** Mainly grass seeds, also insects, flowers and nectar. Joins mixed-species foraging flocks with Bronze Mannikin and Common Waxbill. **CALL** Most common call *seeet-seeet*. **BR** Monogamous. Oval ball nest usually concealed in tree or low bush. Incubation by both sexes. (**Rooirugfret**)

PLATE 86 Paradise-Whydah, Whydahs, Indigobirds p226

LONG-TAILED PARADISE-WHYDAH *Vidua paradisaea* (862) ♀ 15 cm, ♂ 36 cm; 20 g
SEXES ♂>♀. **M (BR)** *Not likely to be confused with any other species in the region. Long tail feathers, black head, golden neck, rich chestnut breast and straw underparts unmistakable.* **F & M (NON-BR)** *Brownish-grey with distinct dark stripes on crown.* ♀ *similar, but less heavily streaked than non-br* ♂. *Male bill black; remains black in non-br dress.* ♀ *has dark 'C' mark around edge of ear coverts, and black bill in br season.* **JUV** *Similar to adult* ♀. **STATUS** Fairly common and conspicuous (when breeding) resident, gregarious. **HABITAT** Dry, open *Acacia* or mixed savanna woodland. **FOOD** Mainly seeds, also insects. **CALL** Song a jumble of sparrow-like chirps; mimics song of host. **BR** Polygynous. Host-specific, parasitising Green-winged Pytilia.
(Gewone Paradysvink) ALT NAME *Paradise Whydah*

SHAFT-TAILED WHYDAH *Vidua regia* (861) ♀ 11 cm, ♂ 28 cm (incl tail); 14 g
SEXES ♂>♀. **M (BR)** *Tawny-yellow underparts and long, narrow, flag-tipped tail distinctive.* **F & M (NON-BR)** *Dark brown crown, with rusty-buff streaks (not bold stripes), distinguishes it from similarly plumaged Pin-tailed Whydah. Pink bill and feet distinguish it from all other* ♀ *viduine species except Village Indigobird, which has smaller bill, longer tail, and lacks rufous on head.* **JUV** *Similar to* ♀ *but with black bill, brownish legs and uniform brown head.* **STATUS** Rather rare resident and near-endemic. **HABITAT** *Acacia* and Mopane savanna; recorded mainly in the northern half of KNP. **FOOD** Seeds. **CALL** Series of rapid warbling notes, and fluting whistles, such as *tiu-woo-wee*. Song includes mimicry of host. **BR** Polygynous. Principal host is Violet-eared Waxbill, but occasionally also Blue Waxbill. **(Pylstertrooibekkie)**

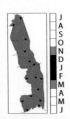

PIN-TAILED WHYDAH *Vidua macroura* (860) ♀ 12,5 cm, ♂ 30 cm (incl tail); 15 g
SEXES M (BR) *Black and white plumage, with long tail and red bill distinctive.* **F & M (NON-BR)** *More boldly striped on head than similarly plumaged Shaft-tailed Whydah and indigobirds. Black moustachial stripe and red bill fairly distinctive.* **JUV** *Like adult* ♀, *upperparts and tail mouse-brown, paler below; best distinguished from juv Long-tailed Paradise-Whydah by slightly bolder striping on crown.* **STATUS** Common resident, sedentary. **HABITAT** Open savanna woodland, grassland and restcamp gardens. **FOOD** Mainly seeds, also termite alates. **CALL** High-pitched jerky repetitive notes; does not mimic call of host species. **BR** Polygynous. Primary host is Common Waxbill, Orange-breasted Waxbill a secondary host. **(Koningrooibekkie)**

VILLAGE INDIGOBIRD *Vidua chalybeata* (867) 11,5 cm; 13 g
SEXES M (BR) *Combination of red bill and legs, and mimicry of Red-billed Firefinches, distinguish it from the two other indigobirds in the Park.* **F & M (NON-BR)** *Plumage brown, with dull red bill and legs.* **JUV** *Like* ♀, *with grey bill and pink legs.* **STATUS** Fairly common resident; most common indigobird in KNP. **HABITAT** Favours open *Acacia* savanna, but also in broad-leaved woodland such as Mopane. **FOOD** Grass seeds; also insects incl termite alates. **CALL** Song a jumble of fairly harsh notes that incorporates calls of 2 host species. **BR** Polygynous. Brood parasite; host-specific in KNP, parasitising Red-billed Firefinch. **(Staalblouvinkie) ALT NAME** *Steelblue Widowfinch*

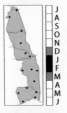

DUSKY INDIGOBIRD *Vidua funerea* (864) 11,5 cm; 15 g
SEXES M (BR) *Bill whitish, sometimes tinged pale pink, legs and feet orange-red; plumage not as glossy as Village Indigobird.* **F & M (NON-BR)** *Streaked brown above, paler below, bill white.* **JUV** *Lighter streaking above than* ♀. **STATUS** Rather uncommon resident, sedentary and locally nomadic in non-br season. Males perch conspicuously on high prominent perch, singing much of the day in br season. **HABITAT** Open woodland, particularly along riparian margins. **FOOD** Mainly grass seeds, also termite alates. **CALL** Scratchy song includes mimicry of host. **BR** Polygynous. Parasitises African Firefinch; egg(s) of host may be eaten by laying ♀. **(Gewone Blouvinkie) ALT NAME** *Black Widowfinch*

PURPLE INDIGOBIRD *Vidua purpurascens* (865) 10,5 cm; 13 g
SEXES M (BR) *Combination of white bill, whitish legs (pale pink or mauve from close), and mimicry of Jameson's Firefinch song, diagnostic. Plumage purplish or bluish-purple.* **F & M (NON-BR)** *Streaked brown above, paler below.* **JUV** *Less streaked above than* ♀. **STATUS** Uncommon but widespread resident; rarest of the indigobirds in KNP. Often in close proximity to host species which it follows to food and water sources. **HABITAT** Favours dry woodland with tall grasses. **FOOD** Grass seeds, also termite alates. **CALL** Scratchy song includes mimicry of host species. **BR** Polygynous. Parasitises Jameson's Firefinch. **(Witpootblouvinkie) ALT NAME** *Purple Widowfinch*

PLATE 87 Finch, Sparrows, Petronia p226

CUCKOO FINCH *Anomalospiza imberbis* **(820) 13 cm; 21 g**
SEXES M (BR) *Differs from canaries in short, deep, black (not grey or horn-coloured) bill, stocky build and rapid, direct (not undulating) weaver-like flight.* **M (NON-BR)** *Duller and with greener upperparts, and dusky-brown bill.* **F** *Resembles ♀ bishop or quelea; best distinguished by very short, heavy brownish bill.* **JUV** *Similar to ♀. Bill black above, yellowish below.* **STATUS** Uncommon and possibly resident, but appears to be an irregular visitor in KNP where it is usually solitary. Roosts communally in reedbeds elsewhere in s Africa but this habit not recorded to date in KNP. **HABITAT** Open or lightly wooded grassland. **FOOD** Mainly grass seeds, rarely insects. **CALL** Song a jumble of repeated chirps *tissiway, tassaway* decending in pitch; also other short rasping song phrases that do not include mimicry of hosts. Three-note call incl *dzi-bee-chew* or *choop-ee-chew*. **BR** Probably polygynous. Parasitises wide range of cisticola species, also prinias. Although evidence of breeding not recorded to date in KNP, probable hosts are Rattling and Zitting cisticolas, Neddicky and Tawny-flanked Prinia. (**Koekoekvink**)

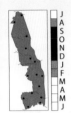

HOUSE SPARROW *Passer domesticus* **(801) 14,5 cm; 26 g**
SEXES M *The only sparrow in the Park with a black bib, so unlikely to be confused with any other species. A tame and confiding species that has upperparts rufous and rump grey. Non-br birds have horn-coloured bills and reduced bib size.* **F** *Dull grey-brown with pale, inconspicuous eyebrow, unlike Yellow-throated Petronia that has a bold, broad white eyebrow.* **JUV** *Similar to ♀.* **STATUS** Common resident, unlikely to be found far from human habitation; generally sedentary. In pairs during the br season, otherwise in small flocks. **HABITAT** Widespread in KNP, found at almost all restcamps and many picnic spots. **FOOD** Seeds, invertebrates, aloe nectar, soft buds, and fruit. Termite alates taken aerially. **CALL** Sharp penetrating *chi-chip, chichiririp* and *cheep* call notes; advertisement call of ♂ *chirrup*. **BR** Monogamous. Untidy ball nest built by both sexes usually placed under eaves; also uses swallow nests, usually breaking away the long entrance tunnel. Sometimes displaces swallows from newly built nests. Incubation mostly by ♀. (**Huismossie**)

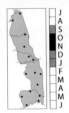

YELLOW-THROATED PETRONIA *Petronia superciliaris* **(805) 16 cm; 24 g**
SEXES *Alike. Broad, almost straight (not narrow and curved) pale buff or whitish eyebrow, plain brown (not grizzled) crown, and 2 pale wingbars differentiate it from Streaky-headed Seedeater. Yellow spot on lower throat difficult to see in field. Most evident when singing in display or when bird lifts head while drinking water. No seasonal change to bill colour like sparrows.* **JUV** *Like adult but lacks yellow throat spot.* **STATUS** Fairly common and widespread resident, usually in pairs. Unobtrusive, usually giving itself away with distinctive 2–4-note call. Usually in pairs and in small groups during non-br season when it joins other similar-sized seedeaters. Restless, often flicks wings and tail. Roosts in holes in trees year-round. **HABITAT** Open woodlands (including Mopane) with sparse grass. **FOOD** Seeds, insects and aloe nectar. **CALL** 2–4 loud, evenly pitched *tri-tri-tri* notes. **BR** Monogamous and apparently with permanent pair bonds. Nests in natural holes, or in old barbet and woodpecker holes in trees that are lined with feathers and hair. Incubation by ♀ only. Parasitised by Greater Honeyguide and Brown-backed Honeybird. (**Geelvlekmossie**) **ALT NAME** *Yellow-throated Sparrow*

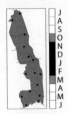

SOUTHERN GREY-HEADED SPARROW *Passer diffusus* **(804) 15 cm; 24 g**
SEXES *Alike. Pale grey head, chestnut shoulders and rump, and white wingbar distinctive. Bill black, changing to yellow-horn during the non-br season.* **JUV** *Duller than non-br adult. Bill like non-br adult and with yellow gape.* **STATUS** Common resident, sedentary with limited post-br dispersal. Gregarious in non-br season. **HABITAT** Favours dry *Acacia* but also in broad-leaved woodland; also favours human settlements and often present at picnic sites. **FOOD** Mainly seeds, also insects, fruit and aloe nectar. Hawks termite alates aerially. **CALL** Repetitive *khrip, kroep, krip* and similar notes. **BR** Monogamous. Nests in holes in trees and buildings, also in old swallow, swift and buffalo-weaver nests. Incubation by both sexes. Host of Diderick Cuckoo and Greater Honeyguide. (**Gryskopmossie**) **ALT NAME** *Grey-headed Sparrow*

PLATE 88 Wagtails, Longclaw p226

AFRICAN PIED WAGTAIL *Motacilla aguimp* (711) 20 cm; 27 g
SEXES Similar. Boldly pied plumage very distinctive and unlikely to be confused with any other species. In br plumage, ♀'s back duller than pitch black of ♂. In non-br plumage, ♂'s back duller, contrasting with black head, and ♀'s upperparts dark olive-grey. **JUV** Paler grey-brown version of adult, with whiter wings than Cape Wagtail. **STATUS** Common and widespread resident, sedentary, with some local movements according to water availability. Usually in pairs or small family parties after br. Regularly baths in shallow pools. Roosts communally during non-br season in trees, in reedbeds and on buildings. **HABITAT** Mainly aquatic margins but has adapted well to human-modified habitats such as dams and gardens. **FOOD** Mainly invertebrates. Sometimes seeds, worms, small crabs, tadpoles and even small fish. Hawks insects aerially. **CALL** Loud tuneful whistles often given from elevated perch; sometimes mimics other species. **BR** Monogamous. Cup nest built by both sexes often placed over water, also on variety of man-made structures. Incubation by both sexes or by ♀ only. Parasitised by Red-chested and Diderick cuckoos. (**Bontkwikkie**)

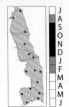

CAPE WAGTAIL *Motacilla capensis* (713) 20 cm; 21 g
SEXES Alike. Drab grey-brown plumage makes confusion unlikely with any other wagtail. Folded wing shows much less white than juv African Pied Wagtail. **JUV** Browner upperparts than adult, with pale yellow wash below. **STATUS** Very uncommon vagrant, or perhaps altitudinal migrant in winter. Usually found singly. Wags tail continuously with exaggerated movements after landing. Baths regularly. **HABITAT** Usually found close to water. Recorded in restcamp gardens and along river and dam margins. **FOOD** Mainly insects, also snails, small tadpoles, and fish. **CALL** Distinctive loud *tseep* and *tseep-eep* call notes. **BR** Monogamous. Deep cup nest placed either close to water or on buildings (especially verandas). Incubation by both sexes. Parasitised by Red-chested and Diderick cuckoos. (**Gewone Kwikkie**)

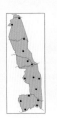

YELLOW WAGTAIL *Motacilla flava* (714) 17 cm; 17 g
SEXES Plumage variable with a number of subspecies visiting region. **NON-BR** All races have pale to deep yellow underparts. ♀ duller and browner above than ♂. **JUV** Browner above and washed buff below. **STATUS** Very uncommon non-br migrant, arriving Oct–Dec and departing Mar/Apr. Found singly or in loose groups; some return to same locality in successive years. Roosts in reedbeds or in trees standing in or next to water. **HABITAT** In KNP associated with aquatic margins particularly on overgrazed areas around man-made dams. After rains disperses more widely and away from aquatic margins. **FOOD** Mainly insects, also some plant matter. **CALL** Single *psie* or *pseeu* notes, but seldom calls in region. (**Geelkwikkie**)

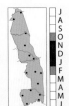

YELLOW-THROATED LONGCLAW *Macronyx croceus* (728) 21 cm; 48 g
SEXES Similar. The only longclaw in the Park. Bright yellow underparts and broad black necklace very distinctive. In flight displays prominent white tips to outer tail feathers. Distinguished from smaller Golden Pipit (very rare vagrant to Park and surrounding region) by black necklace (not breast band) and brown (not yellow) flight feathers. **JUV** Buff or buff-yellow below; necklace initially broken by spots and streaks. **STATUS** Fairly common resident, sedentary and usually in pairs or small family groups after breeding. Habitually turns its back to any threat, making it difficult to identify, and perches with difficulty on thin twigs, fluttering wings to keep balance. Roosts on ground in grass. **HABITAT** In KNP favours well-developed shortish moist grassland with scattered trees. Often along wetland margins. **FOOD** Mainly insects, also millipedes, molluscs and some plant material. **CALL** Loud, ringing *chu-weee-chuweee* ... given from prominent low perch on anthill, or in flight. **BR** Monogamous. Bulky cup nest built by ♀ and accompanied by ♂. Usually well concealed under grass tufts. All or most incubation by ♀. (**Geelkeelkalkoentjie**)

PLATE 89 Pipits p227

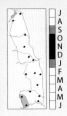

STRIPED PIPIT *Anthus lineiventris* (720) 18 cm; 34 g
SEXES Alike. Heavily streaked underparts (that is whiter than any of the other pipits in the Park) and, from close, yellow edges to folded wing are key features. Bill fairly long and eyebrow prominent. **JUV** Upperparts paler and more spotted. **STATUS** Fairly common resident in the hilly and rocky Pretoriuskop and Berg-en-dal areas, also in the rocky gorges of Pafuri with perhaps some seasonal movements. Usually singly or in pairs; sometimes in small family groups after br season. When disturbed may fly to trees, where capable of walking along branches and often perching parallel to branch, with bill pointing upwards. **HABITAT** Usually on steep, rocky, wooded hillsides. Seldom found far from rocky outcrops or boulder-scattered vegetation. **FOOD** Insects. **CALL** Song rich, varied series of loud, melodious notes. **BR** Monogamous. Cup nest placed on ground next to rock or grass tuft. (**Gestreepte Koester**)

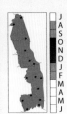

AFRICAN PIPIT *Anthus cinnamomeus* (716) 16,5 cm; 24,5 g
SEXES Alike. Most abundant pipit in KNP, being standard against which other pipits are compared. Key features conspicuous white outer tail feathers (best seen in flight), boldly streaked breast, distinct black streaking on back, yellowish base to lower mandible, and slender upright appearance. Dark malar stripe and bold facial markings also help identification. **JUV** Darker above, upper breast with diffuse spotting and base of lower mandible often pinkish. **STATUS** Common resident with little evidence of migratory movements recorded elsewhere in s Africa; possible local nomadic movements in response to rainfall and drought. Usually found singly or in pairs. Roosts on the ground, flying to habitual roost site daily. **HABITAT** Favours short grassland, dry floodplains, overgrazed areas, road verges and airstrips. **FOOD** Mostly insects. Occasionally drinks in dry weather. **CALL** Male territorial song series of *pli-pli-pli-pli ...* or *tree-tree-tree* notes; take-off call *chizz-ick*. **BR** Monogamous. Cup nest concealed at base of grass tuft, incubation mostly by ♀. (**Gewone Koester**) **ALT NAME** *Grassveld Pipit*

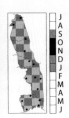

PLAIN-BACKED PIPIT *Anthus leucophrys* (718) 17 cm; 24 g
SEXES Alike. Key features are plain, unstreaked back and buffy outer tail feathers in flight. Distinguished from Buffy Pipit by yellowish (not pinkish) base to lower mandible, darker back, and greater contrast between pale supercilium and dark brown ear coverts. Shorter legs and less intense tail-wagging; chest not as protruding and with less upright stance. **JUV** Upperparts browner. **STATUS** Uncommon seasonal visitor to the Park. Probably an altitudinal migrant from the high eastern continental plateau. Usually solitary. **HABITAT** Short grassland, especially where burnt or overgrazed. **FOOD** Mainly insects and their larvae. **CALL** Monotonous series of double notes *chree-cheup, chree-cheup ...* **BR** Monogamous. Cup nest concealed under grass tuft, but no records for KNP. (**Donkerkoester**)

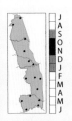

BUFFY PIPIT *Anthus vaalensis* (719) 18,5 cm; 29 g
SEXES Alike. Plain sandy-brown (not darkish brown-grey) unstreaked upperparts, pinkish (not yellowish) base to lower mandible, and more exaggerated and slower tail-wagging distinguish it from slightly smaller Plain-backed Pipit. Stance often upright with breast protruding when standing still. Outer tail feathers buffy, and with only faint breast streaking; rump slightly paler and more rufous. **JUV** Darker above and with bold spotting. **STATUS** Uncommon, but probably resident in KNP; some birds may be partially migrant. Solitary, in pairs or small groups. **HABITAT** Open grassy plains, favouring bare ground and overgrazed areas; also recently burnt grassland. **FOOD** Insects, also seeds. **CALL** Sparrow-like *chrep, chiri, chree, chreu*. **BR** Monogamous. Cup nest concealed under grass tuft or rock. (**Vaalkoester**)

BUSHVELD PIPIT *Anthus caffer* (723) 13,5 cm; 17 g
SEXES Alike. Considerably smaller than any of the other pipits in KNP; regarded as one of the diminutive pipits of Africa. Has narrow white eye-ring, shortish tail with fairly heavy streaking on breast, eyebrow fairly indistinct and buffish underparts. Call distinctive and with golden-brown upperparts and streaking on flanks less conspicuous than on breast. Lacks yellow in primaries of larger Striped Pipit, but also of semi-arboreal habits. **JUV** Upperparts paler and spotted. **STATUS** Locally common but localised resident occurring mainly in the southern two-thirds of the Park. Usually flies up to perch prominently on shrub or tree when disturbed. **HABITAT** Mainly in broad-leaved woodland on granitic soils but occasionally also in open mixed *Acacia* woodland. **FOOD** Insects. **CALL** Song *werrp-cheer, werrp-cheer ...* ; nasal *tszwee-aa* or *bzeeat* when flushed. **BR** Monogamous. Cup nest concealed below grass tuft. Incubation mostly by ♀. (**Bosveldkoester**)

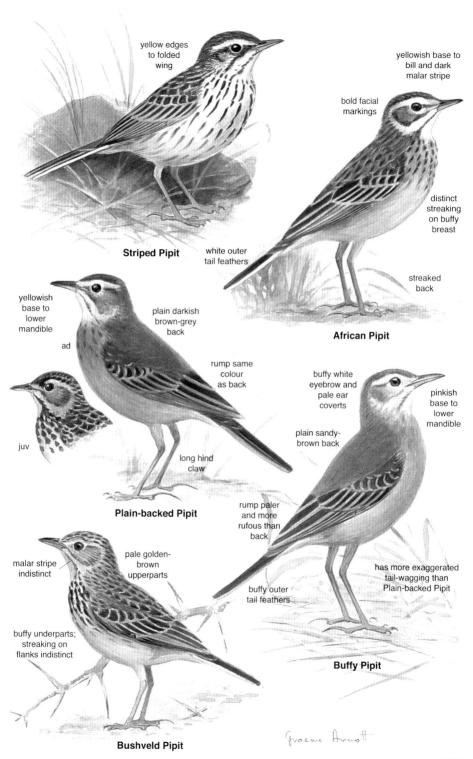

YELLOW-FRONTED CANARY *Crithagra mozambica* (869) 12 cm; 12,5 g
SEXES Similar, ♀ with paler plumage, especially during br season. Bold facial markings and grey crown distinctive. Eyes dark brown (not yellow as indicated in misleading former name), and in flight yellow rump and white-tipped tail (sometimes difficult to see) diagnostic. Distinguished from larger Brimstone Canary by grey (not green) crown, lighter bill and bolder facial markings. **JUV** Duller than ♀ with light streaking on breast. **STATUS** Common and widespread resident, mainly sedentary. Usually in pairs or small family groups during the br season, and in small flocks the rest of the year. Roosts in the foliage of trees; drinks and bathes regularly. **HABITAT** Open woodland; common on lawns in restcamp gardens. **FOOD** Grass and shrub seeds; also partial to tree seeds such as Pigeonwood *Trema orientalis*. Eats flowers of trees and forbs incl Flame Climbing Bushwillow *Combretum microphyllum* and takes nectar from aloes. Diet also includes termites taken both on the ground and aerially, as well as other invertebrates. **CALL** Typical canary-like series of trills and whistles. Song may include mimicry of other species. **BR** Monogamous. Thin cup nest placed on outer branches of tree or shrub. Incubation by ♀ only. (**Geeloogkanarie**) **ALT NAME** *Yellow-eyed Canary*

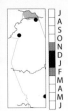

LEMON-BREASTED CANARY *Crithagra citrinipectus* (871) 12 cm; 11 g
SEXES M Combination of grey-brown upperparts, lemon-yellow throat patch and rump, and whitish belly diagnostic. Malar stripe conspicuous. Distinguished from ♂ Yellow-fronted Canary by whitish (not yellow) lower breast and belly. **F** Fairly nondescript plumage with drab grey-brown upperparts, conspicuous malar stripe and pale fawn-brown to whitish underparts. Pale yellow patch sometimes visible on throat. **JUV** Like adult ♀. **STATUS** Generally uncommon and localised near-endemic, recorded only in the Northern Lala-palm *Hyphaene petersiana* groves in the Pafuri area where it is probably sedentary, listed as Near-threatened in s Africa. Entire s African distribution linked to that of Lala-palm on which it breeds and from which it obtains nesting material. In pairs during the br season, and in small flocks the rest of the year, sometimes with Yellow-fronted Canary. **HABITAT** Found only where there are Northern Lala-palms in mixed lowland woodland at Pafuri. **FOOD** Mainly seeds, also insects. **CALL** Canary-like twitter. **BR** Monogamous. Breeds in the bottom of the 'V' of partly opened new palm fronds. All nesting material apart from cobweb obtained from old flower inflorescences and fronds of Lala-palm. Incubation by ♀ only. (**Geelborskanarie**)

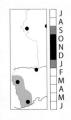

BRIMSTONE CANARY *Crithagra sulphurata* (877) 15,5 cm; 25 g
SEXES Similar, ♀ slightly duller. Large thickset canary, heavier-billed than Yellow-fronted Canary and with greenish (not grey) crown and less conspicuous facial markings. Green on crown extends down to bill. **JUV** Paler below with faint streaks on breast. **STATUS** Rather uncommon resident found only in the south-western parts of KNP. Quiet and unobtrusive when not calling. Usually in pairs or small family groups and in the non-br season associates loosely with other canaries in mixed-species foraging flocks. **HABITAT** Hilly shrublands. **FOOD** Seeds and kernels of fruiting trees, also flesh of fruit (the most frugivorous of all canaries in the region), also grass seeds and nectar. **CALL** Sustained trills and whistles interspersed with harsh *chirrups* and *chirrs*. Contact call *chee-u-wee*, or dry *tap-tap* like pebbles knocking together. **BR** Monogamous. Cup nest built by ♀ (accompanied by ♂) and placed in foliage of tree. Incubation by ♀ only. (**Dikbekkanarie**) **ALT NAME** *Bully Canary*

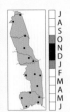

STREAKY-HEADED SEEDEATER *Crithagra gularis* (881) 15 cm; 20 g
SEXES Alike. Least colourful of the canaries in the Park with long, narrow, curved white eyebrow, brown (not yellow) rump and no white wingbars. Distinguished from Yellow-throated Petronia by much narrower and more curved white eyebrow. **JUV** Blotched or streaked below. **STATUS** Uncommon but widely recorded in KNP. Fairly quiet and unobtrusive; in pairs or small family groups during the br season, and in small flocks during the winter months. **HABITAT** Wide range of wooded habitats, especially rocky areas. **FOOD** Seeds, flower buds, fruit, flowers, insects and nectar. Associates loosely with Yellow-fronted and Brimstone canaries in mixed-species foraging flocks. Drinks regularly. **CALL** Loud vibrant song of twittering, warbling and whistling phrases. **BR** Monogamous. ♀ builds cup nest, accompanied by ♂. Nest placed in outer branches of tree or in leaves of aloe. Incubation by ♀ only. (**Streepkopkanarie**) **ALT NAME** *Streaky-headed Canary*

PLATE 91 Buntings p227

LARK-LIKE BUNTING *Emberiza impetuani* **(887) 15 cm; 15 g**
SEXES *Alike. Nondescript brownish bird, no bold diagnostic features. Pale eyebrow, cinnamon wash to underparts and pale rufous edge to wing feathers. Unlike similar-looking larks, hops rather than walks.* **JUV** *Similar to adult but slightly paler.* **STATUS** Usually absent but irrupts into KNP from the drier western regions of s Africa in very dry years; highly nomadic. Usually gregarious; in loose flocks from a few birds to hundreds. Shy and restless, associating with sparrowlarks and other canaries while feeding. **HABITAT** Dry open (usually stony) shrublands, sparse grassland and watercourses. **FOOD** Mainly seeds; also some insects. Drinks regularly. **CALL** Short monotonous canary-like song. **BR** Monogamous. Shallow cup nest built by ♀ and usually placed behind rock or tuft of grass. No evidence of breeding in the Park to date, and unlikely ever to breed regularly. (**Vaalstreepkoppie**)

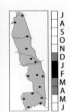

CINNAMON-BREASTED BUNTING *Emberiza tahapisi* **(886) 15 cm; 15 g**
SEXES M *Unlikely to be confused with any other species in the Park, with distinctive cinnamon underparts, blackish throat, yellowish-horn-coloured bill, and black-and-white-striped head.* **F** *Overall paler, head brown with white stripes.* **JUV** *Similar to ♀ but flecked with brown below.* **STATUS** Fairly common and widespread, probably with both resident and br summer migrant components to the KNP populations. Usually in pairs or small family groups during br season, but at other times of the year may aggregate in small flocks. **HABITAT** Woodland with rocky substratum and boulders, rocky ridges, mountainsides, dry watercourses. **FOOD** Seeds and some insects. **CALL** Fast, grating jumble of notes, repeated incessantly. **BR** Monogamous. Shallow cup nest built by ♀, accompanied by ♂. Nest placed on the ground, usually next to rock or grass tuft. Incubation mostly by ♀. (**Klipstreepkoppie**) **ALT NAME** *Rock Bunting*

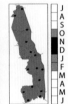

GOLDEN-BREASTED BUNTING *Emberiza flaviventris* **(884) 16 cm; 19 g**
SEXES *Similar.* **M** *Black head, with white stripes above and below eye diagnostic.* **F** *Duller than ♂, with brown head and white eye-stripes.* **JUV** *Similar to ♀, but duller and with brown breast streaks.* **STATUS** Common and widespread resident; usually in pairs but may form small flocks in non-br season. When disturbed, has the habit of moving to a nearby tree with dipping flight showing characteristic white outer tail feathers. Unobtrusive, but not particularly shy. Some birds develop habit of 'attacking' own image reflected in car mirrors, windows and hubcaps. These birds return daily to confront and defend territory. **HABITAT** Dry woodland; found in most KNP habitats. **FOOD** Insects, seeds and flower buds. Drinks regularly. **CALL** Repetitive *toodletee, chipchipchip*. **BR** Monogamous. Deep cup nest built by ♀, accompanied by ♂, and usually placed 1–3 m high in tree or shrub (especially *Acacia*). Incubation by ♀ only. Occasionally parasitised by Jacobin Cuckoo. (**Rooirugstreepkoppie**)

GLOSSARY

Acacia: deciduous trees of the Genus *Acacia*. In Africa these are thorny, with bipinnately compound leaves (each leaf is again divided into small leaflets) and small powderpuff-like or elongated flowers.

accipiter: sparrowhawks and goshawks. Long-tailed, short-winged raptors with long, unfeathered legs and long toes. They specialise in catching small birds (or small mammals in the larger species).

Afromontane: term used for the high mountainous ranges across the Afrotropical Region; mostly inland at high altitude and under temperate conditions.

alate: winged and able to fly; alates in this guide refer to winged (flying) termites.

albinism: white plumage and pink soft parts resulting from a complete lack of pigment melanin.

alien: introduced from another part of the world.

allopatric: the geographical range of one species does not overlap with that of another, similar species.

allopreening: preening of one bird by another; if both preen each other simultaneously, it is called mutual preening.

altitudinal migrant: seasonal movements from one altitude to another, usually from high-altitude breeding grounds in summer to lower altitudes in winter.

altricial: type of young bird that is more or less helpless at hatching and has to be fed in the nest by the parents until able to fly (synonymous with nidicolous: see **precocial**).

anisodactyl: toes 2, 3 and 4 pointing forwards, toe 1 backwards; found in most birds.

alula: four small feathers found on a bird's 'thumb'. They control airflow over the leading edge of the wing – the 'bastard wing'.

anting: bird lies on the ground with wings and tail spread, allowing ants to crawl over plumage, or picks up ants and rubs them over feathers; this is done perhaps to remove stale preening oil (with formic acid produced by ants), or glandular secretions of ants may act as antibiotics, protecting feathers.

antiphonal duetting: duetting by producing syllables alternately with such good synchronisation that the duet sounds like a single call, eg Black-collared Barbet.

aquatic: living in or on water.

arboreal: living in trees.

arid: dry, parched.

arthropod: invertebrate with segmented body, jointed limbs and hard exoskeleton; subdivided into insects (Hexapoda), millipedes (Diplopoda), centipedes (Chilopoda), crustaceans (Crustacea), spiders, scorpions, and related animals (Arachnida), sunspiders or solifugids (Solpugae), ticks and mites (Acari), and some other minor groups.

auricular patch: a distinct colour-patch of feathers over or about the ear.

austral: of the southern hemisphere.

axillaries: the axilla is the area where the underwing joins the body – the 'armpit'; the feathers in this area are known as axillaries.

Baikiaea **woodland:** a broad-leaved woodland type (usually tall), restricted to Kalahari sands and dominated by *Baikiana plurijuga*.

belly-wetting (belly-soaking): a habit of ground-nesting birds such as plovers, pratincoles and skimmers that cool their eggs with wet belly feathers, or carry water (sandgrouse) by soaking their belly feathers in water before flying back so that the chicks can drink.

bib: a rounded breast patch including the chin (= a gorgette).

Brachystegia **woodland:** a broad-leaved woodland type of Leguminous trees (belonging to the Pea Family); see **miombo**.

bracken: a robust fern that forms dense thickets in montane grassland and forest margins.

breeding endemic: species that only breed within southern Africa, but migrate (at least partially) outside the region in the non-breeding season.

breeding season: months in which egg laying has been recorded.

brood parasite: a bird species that lays its eggs in the nests of other (foster or host) species; the host rears the brood parasite's young. Examples of the brood parasites are cuckoos, honeyguides, whydahs, widowfinches and the cuckoofinch.

brood patch: feathers are shed at the onset of breeding forming a bare area of skin on the belly, which is well supplied with blood vessels and used to cover the eggs during incubation.

bush: refers to any terrain with trees of moderate height as opposed to thornveld where trees of the genus *Acacia* are dominant.

bushveld: a terrain with mixed trees of moderate height (5-10 m), where the trees frequently touch each other below canopy height; sometimes in dense thickets and usually with a grassy groundcover.

cainism: process in which older (first-hatched) chick

kills younger sibling, commonly found in eagles, ground-hornbills and birds such as boobies (see **siblicide**).

call: any vocalisation, but usually restricted to those not definable as **song.**

carpal joint: (flexure) the joint found between the 'arm' and the 'hand' of the wing.

casque: horny ridge on top of the bill of hornbills.

cere: bare wax-like or fleshy structure on the skull or bill.

cline: gradual geographic change in size or colour.

cock's nest: refers to the structure or false nest that is placed above the occupied nest; constructed by Common Waxbills as a possible decoy for predators; used by males for roosting and not for breeding purposes.

colonial: usually describes gregarious breeding habits in which nests are built together in more or less dense groups or colonies.

colour morph: different colouring within a single interbreeding population, unrelated to season, sex or age (and formerly known as colour phase), eg Western Reef Heron.

congener: species of the same Genus.

conspecific: belonging to the same species.

co-operative breeding: refers to non-breeding birds attending the nests of breeding members of their own species and assisting with some or all functions of breeding behaviour.

coverts: small feathers hiding/protecting the bases of larger ones.

covey: a group of francolin or spurfowl.

crepuscular: mainly active in the dim light of dusk and dawn.

cryptic: having protective colouring or camouflage.

culmen: ridge along top of bird's bill from tip to base of feathers at forehead.

dambo: seasonal wet patches of grassland that are found along drainage lines in woodland such as *Brachystegia*.

decurved: curved downward (eg bill of sunbird or ibis).

dentate: toothed or serrated.

diagnostic: having value in description for the purpose of classification or for positive identification.

diatoms: microscopic unicellular alga with silicified cell-walls; found as plankton.

diffuse: dispersed or spread widely; referred to here in the context of feather streaking.

dimorphism: (sexual) the occurrence of two distinct types of plumage colour and/or patterns between the sexes of the same species; also shape and size.

discontinuous: (distribution) usually applied to a species with gaps between populations where it does not occur.

distjunct: (in reference to ranges) geographically separate.

dispersal: movement of young bird away from birthplace.

diurnal: active during daylight hours.

donga: an erosion gully, usually with vertical sides, often used by burrow-nesting birds, such as starlings, chats and the Ground Woodpecker.

dorsal: on the back or upperside.

double-brooded: the laying of a second clutch in one breeding season after rearing the first brood successfully.

eclipse: (plumage) non-breeding plumage, usually duller-coloured than breeding plumage.

ectoparasite: an external parasite such as a tick.

egg-dumping: laying of egg(s) in a nest of a conspecific or other species.

emarginated: pertaining to the primary feather that is notched or abruptly narrowed, usually near the tip.

endemic: living in, and usually originating in, one geographical area only, and found nowhere else.

ephemeral: refers here to wetlands that remain dry for long periods, only filling during periods of high rainfall.

escarpment: the steep face of a tilted plateau. In southern Africa usually refers to the eastern escarpment, which forms the edge of the inland plateau or highveld.

estrildid finches: general term for members of the Family Estrildidae, which includes waxbills, twinspots, firefinches and manikins.

etymology: the study of the origins and history of words or names; how they evolved.

eutrophic: rich in nutrients.

exotic: introduced from another part of the world (ie not indigenous).

extralimital: beyond the borders of the geographical area under review.

faecal sac: a white gelatinous 'envelope' which contains the faeces of the young of many species (especially altricial chicks).

feral: having returned to the wild after domestication. An introduced animal, foreign to an environment.

fledgling: leaving the nest; partly or wholly feathered, flightless or partly flighted, but before full flight capacity.

floodplain: the base of a valley (usually grassland) which is under water when the river is in flood. These become progressively wider the closer the river gets to the sea.

forked: divided into two prongs or points (eg tail of many swallows, some swifts and terns).

frons: the forehead or feathered front of the crown, immediately above the base of the upper bill.

frontal shield: distinctive, unfeathered, horny or fleshy forehead that extends down to base of the upper mandible.

frugivorous: fruit-eating.

fynbos: a natural habitat in the winter-rainfall region from the Western Cape to about Port Elizabeth, Eastern Cape; comprising a vegetation type characterised by evergreen shrublets with hard, needle-shaped leaves: a Mediterranean-type scrub comprised of proteas, ericas and legumes, among other plants.

gape: the fleshy base of the beak, which is often cream, yellow or orange in young birds.

Genus (pl. **Genera**): a taxonomic category between Family and species, ie a group of closely related species.

gorget: throat band or broad necklace.

graduated: decreasing stepwise from long to short.

gregarious: living together in groups or flocks.

gular: of the throat. A gular pouch is distensible skin in the central area of the throat.

gular-fluttering: rapid fluttering of the skin of the mouth and upper throat; used to reduce heat load by evaporative cooling.

hallux: toe 1, usually directed backwards, sometimes reduced (eg coucals) or absent.

heronry: colonial breeding site of herons or egrets; sometimes also applied to other colonially-breeding wading birds such as ibises and spoonbills.

holarctic: refers to the Holarctic region; the combined northern hemisphere Nearctic and Palaearctic biogeographical regions.

host: the species that incubates the eggs and raises the young of avian **brood parasites**.

immature: (= sub-adult) all plumages that follow first moult until full breeding capacity and/or plumage is reached. Birds are usually independent of adults.

indigenous: native to a geographical area (ie not exotic or introduced).

invertebrate: an animal lacking a spinal column or 'backbone', eg insects, molluscs and worms.

inselberg: isolated, usually steep-sided hill rising from a plain in arid regions.

intra-African migrant: bird that migrates entirely within Africa.

iridescence: play of colours (in feathers) by light on feather structure; not a pigment colour.

irruption: a temporary migration into a new area, usually brought about by more favourable conditions in that region.

juvenile: a young bird under sub-adult age.

karoo: a semi-arid habitat of central and western South Africa consisting of low woody shrubs and little grass, on a largely stony substratum.

kleptoparasitise: to steal food from another individual.

kloof: a gully or ravine (often densely wooded), usually on a mountainside.

koppie: a small hill, often with a rocky summit.

krill: marine plankton consisting of minute crustaceans of the Order Euphausiacea, occurring in vast numbers in colder oceans.

lanceolate: pointed like the head of a spear.

LBJs: or 'little brown jobs', a collective term for difficult to identify species such as warblers and cisticolas.

leaf-gleaner: small insectivorous birds that work their way through leaves and *Acacia* or *Albizia* pods looking for grubs, aphids, beetles, scale insects, weevil larvae or other invertebrate fauna hiding in or under the leaves or pods. Other 'gleaners' are adapted to different niches, these include bark gleaners (Spotted Creeper), gravel gleaners (Long-billed Lark, turnstones) and carcass gleaners (vultures, crows etc).

lek: a display area used by males of polygynous birds, such as bustards and korhaans, to which females are attracted for mating.

leucism: reduction in intensity of feather-colouring pigments.

liana: rope-like stems of vines and plants, usually found in under-canopy of forests.

littoral: the region of lowland lying alongside the sea shore.

lores: area between the bill and the eye; may be bare or feathered.

Lowveld: the eastern part of southern Africa, which lies between an altitude of about 100 m and 900 m and comprises bushveld.

malar: (moustache) line from base of bill down sides of throat, often forming distinctive stripe in birds.

mandible: the upper or the lower half of a bird's bill.

mangrove: trees or shrubs of the Genera *Rhizophora* and *Avicennia* that grow mainly in tropical coastal swamps and have tangled roots that grow above ground forming dense thickets.

mantle: feathers forming a covering of the upper back and the base of the wings.

***Marasmius* fungus:** fungi that are often black, thread-like and tend to have soft stems that are sometimes covered in hairs giving a velvety appearance. Usually found on the ground in leaf litter and on rotting branches and used extensively for nest building by species such as Olive Sunbird and Terrestrial Brownbul.

mask: black or dark area that encloses the eyes and part of the face.

melanistic: tending to be black or blackish, resulting from an excess of the dark feather pigment melanin.

mesic: opposite of 'arid'. Areas of reasonably high rainfall, creating mesic grassland or mesic savanna.

migratory: of regular geographical movement.

miombo: broad-leaved woodland in which trees of the Genus *Brachystegia* dominate; common in Zimbabwe.

monogamous: a mating system in which a single male forms a pair bond with a single female to breed.

montane: mountainous country.

Mopane: a broad-leaved, deciduous tree, *Colophospermum mopane*. Forms a dense woodland in some regions; stunted on poorly drained soils but reaches a canopy height of 15-20 m in suitable areas. Leaves are rounded, heart-shaped and reddish when young.

morph: an alternative but permanent plumage colour.

moult: the process of shedding old feathers and replacing them with new ones.

mouth spot: spot (usually several) that forms part of a characteristic pattern inside mouths of nestlings of some species (eg waxbills).

Msasa: a common tree in miombo woodland: *Brachystegia spiciformis*.

nail: the hooked top of the upper mandibles of albatrosses and petrels.

nape: the back of a bird's neck.

neartic region: the biogeographical region comprising north of North America.

near endemic: species whose range extends only marginally outside southern Africa.

nestling: (= hatchling = downy) in or about the nest. Naked or downy, ie *before* feathers develop.

New World: the Americas; the western hemisphere.

nictitating membrane: a third 'eyelid' that can be drawn across the eye from the nasal side for protection, lubrication and cleaning the eye. Some are translucent; some have a clear central window so vision is not seriously impaired.

nidicolous: refers to a bird species in which the young remain in the nest for some time after hatching.

nidifugous: refers to a bird species in which the young leave the nest soon after hatching.

nocturnal: active at night.

nomad: a species with no fixed territory when not breeding.

nomadic: of variable, often erratic movement with regard to time and area.

non-passerine: not belonging to the Passeriformes.

notched: having central tail feathers slightly shorter than outer ones, forming a notch or shallow fork.

nocturnal: active at night.

nuchal collar: a collar across the nape (hind neck).

nuchal crest: crest positioned on the nape.

nuptial: of or pertaining to breeding, eg nuptial behaviour or nuptial breeding.

occipital plumes: the breeding or ornamental feathers (plumes) originating from the crown/nape of egrets.

Old World: the part of the world that was discovered before the Americas: Europe, Asia and Africa.

omnivorous: eating both plant and animal foods of many kinds.

onomatopoeic: the forming of names or words from sounds (created in this case from birds).

Palaearctic: pertaining to the zoogeographical region which includes Europe, North Africa and northern Asia east to eastern Siberia.

pamprodactyl: foot arrangement with all four toes directed forwards, eg swifts; toe 1 sometimes reversible, eg mousebirds.

parapatric: (applied to two or more species) having ranges that abut, but do not overlap.

parasite (parasitises): referring here to one bird species laying its eggs in the nest of another species, and relying on that species (the host) to rear its offspring.

partial migrant: a term applied to a species in which part only of the population migrates annually.

passage: (migration) concerning birds passing through while migrating from one point to another, but not stopping over.

passerines: the largest Order of birds; an Order that includes all the so-called 'songbirds' that are characterised by a complex syrinx (voice-box); or

'perching birds' that have feet adapted for perching, with 3 toes facing forward and one facing backward.
pectinate: comb-like; bearing numerous tooth-like projections as in the middle claw of nightjars.
pectoral patch: a well-defined dark area of plumage on either side of the breast.
pelagic: oceanic, living far from land except when nesting.
plumage: the feather covering of a bird.
plume: a long, showy, display feather, eg in egrets.
polyandry (adj. **polyandrous**): mating of one female with two or more males.
polychaetes: worms living in the mud and sand of estuaries and rivers; having bristle-like hairs on their foot-stumps, eg Annalids and leeches.
polygamy (adj. **polygamous**): a mating system in which an individual will have more than one sexual partner; **polyandry** and **polygyny** are specific variants of polygamy.
polygyny (adj. **polygynous**): having more than one female to each male.
polymorphic: having two (then called dimorphic) or more distinct, colour morphs within a species, independent of age, sexual, seasonal or subspecific variation.
precocial: type of chick able to leave the nest within minutes or hour of hatching; it is covered with down and able to thermoregulate and often able to feed by itself.
primaries: (primary **remex**) outer flight feathers of the wing.
race: see **subspecies**.
rachis: the shaft of a feather.
raptor: a bird with strong claws and sharp talons for tearing prey. Usually used with reference to the diurnal Falconiforms (hawks and relatives), but applies also to owls.
rectrix: (pl. **rectrices**) tail feathers.
recurved: curved upwards, like the bill of an Avocet.
remex: (pl. **remiges**): flight feathers of the wing.
renosterveld: fynbos shrubland, usually dominated by renosterbos *Elytropappus rhinocerotis*.
restio: grasslike plants in the Families Cyperacea and Restionaceae, which are components of fynbos.
rictal bristles: stiff, whisker-like protrusions about the base of the bill.
riparian: of or on river banks.
roost: a resting or sleeping place; perch for birds; term also applied to a group of resting waders.
rufous: reddish-brown.

rump: the squarish area between the lower back and base of the tail.
Sahel: semi-desert zone south of the Sahara Desert.
savanna: grassy plain in tropical-subtropical regions; mainly a vegetation type that comprises open woodland with an understorey of grass.
scapular: feathers that lie along the dorsal shoulder (base of the wing) of a bird.
secondaries: (secondary remex) flight feathers of the wing attached to the forearm (specially the ulna).
sedentary: refers in this guide to birds that live in the same place year-round; not travelling far.
serrate: finely toothed.
sexual dimorphism: differences in appearance between the male and the female of the same species. This may be in size, shape, plumage or colour, or sometimes a combination of these.
shaft: the main stem (rachis) of any feather.
shoulder: general term for upperwing coverts.
shrubland: vegetation dominated by short woody plants less than 2 m tall.
siblicide: the dominance and subsequent death of smaller and weaker nestlings in a brood (also known as **cainism**). The uncommon phenomenon where the older chick invariably kills the younger chick is known as obligate siblicide (eg Vereauxs' Eagle). Facultative siblicide is far more common and usually linked to the dominance of the eldest chick that out competes the younger chick(s) for food, often resulting in their starvation.
site fidelity: faithfulness to a particular site.
soft parts: unfeathered areas of the body: bill, eyes, legs, feet and any bare skin, wattles etc.
song: vocalisation used to advertise territory, as well as availability as a mate; usually employed by male birds, and not necessarily melodious to the human ear.
species: the division of classification into which a Genus is divided, the members of which can interbreed among themselves, ie the taxonomic rank below Genus.
speculum: iridescent, reflective dorsal patch on a duck's wing; contrasts with the rest of the wing.
spish: to attract birds by making a hissing noise between the teeth.
spur: sharp bony projection on the wing or leg.
strandveld: fynbos and Karoo shrubland restricted to the coastline along the southern and Western Cape.
striated: (striations) streaked; usually dark marks aligned on bird's long axis.

subspecies: a geographic race of a species; a population morphologically and geographically defined. The subspecies of a species may interbreed where (and if) their ranges overlap.

sub-terminal: not right at the end or tip of a structure (usually refers to tail).

supercilium: 'eyebrow' or part of the head immediately above the eye; in many birds marked by a **superciliary stripe**.

superspecies: one of a group of very closely related species with non-overlapping ranges.

supra: a prefix meaning 'above'. Supraloral lines are those above the lores.

sympatric: (distribution) occurring together in the same area.

syndactyl: toes 3 and 4 fused at the base, eg kingfishers and hornbills (3 toes forward, 1 backwards). In the case of trogons, toes 3 and 4 are forward and fused at the base, and toes 1 and 2 face backwards. See **zygodactyl**.

talon: sharply hooked claw used for holding and killing prey, eg all birds of prey.

tarsus (pl. **tarsi**) the **tarsometatarsus**; the lower part of the leg of a bird, usually bare of feathers.

taxonomy (adj. **taxonomic**) the science of classification of plants and animals according to their natural relationships.

terminal: at the end or tip of a structure.

terrestrial: living on land (not aquatic).

territorial: behaviour whereby an area is defended, usually for courtship, breeding purposes and protection of food resources.

tertiary: (feathers) row of inner flight feathers on a bird's 'upper arm'.

thicket: a number if shrubs or trees growing very close together.

tibia (pl. **tibiae**) the **tibiotarsus**, the 'drumstick' of the leg; see **tarsus**.

tomium: cutting edge of bill.

tongue spot: spot (usually several) that forms part of a characteristic pattern on tongues of nestlings of certain species (eg warblers and waxbills).

trailing edge: the back or hind edge of a wing or flipper.

Trichoptiles: the wiry hair-like natal feathers on coucal and malkoha chicks.

ubiquitous: present everywhere; universal.

underparts: the chin, throat, breast, belly, underwing, flank, vent and undertail; the ventral surface of a bird.

upperparts: frons, lores, face, crown, nape, mantle, back, upperwing, rump, base of tail, uppertail; the dorsal surface of a bird.

Uapaca (or ***Uapaca* woodland**)*:* this is a group of trees in the *Euphorbia* family of plants. In parts of Mozambique, Zimbabwe, Zambia and Malawi, *U. kirkiana* forms the dominant species within the woodland.

***Usnea*:** a group of fungi growing on trees in forests and woodland that are often called 'old man's beard' due to the grey colour of the fine stems that hang down off the tree branches. Usually found in high rainfall areas, especially in the cooler mist belt areas of escarpments.

vagrant: a bird found in an area that is not its usual habitat, having strayed there by mistake, eg through disorientation or by adverse winds.

veld: a term used loosely in reference to various types of terrain, thus grassveld, bushveld, thornveld etc.

vent: the cloaca; includes anus, oviduct and sperm duct openings. Also refers to patch of feathers around this.

ventral: pertaining to the undersurface of the body (see **underparts**).

vestigial: describes the structure or behaviour pattern that is so reduced through long disuse as to be almost absent.

vermiculations: densely patterned with fine winding or wavy lines.

viduine: of the Family Viduidae, represented in southern Africa by paradise-whydahs, whydahs and indigobirds.

vlei: seasonally flooded moist or marshy depression in grassland or savanna; an Afrikaans word for wetland or swamp.

wattle (wattled): bare, fleshy structure around eye, base of bill, throat or elsewhere on head of bird.

wing formula: mathematical representation of the relative lengths of primaries of a bird's wing; used to identify some species in the hand (eg *Acrocephalus* warblers).

wing-fripping: audible rapid flapping of the wings in flight during display, eg Black-backed Puffback.

wingspan: the shortest distance between the wingtips; the greatest extent of the spread wings.

xanthochroism (adj. **xanthochroic**) abnormal and excessive yellow pigmentation in feathers.

zygodactyl: toes 2 and 3 pointing forwards, and toes 1 and 4 pointing backwards, eg woodpeckers, cuckoos, coucals and barbets.

ILLUSTRATED GLOSSARY

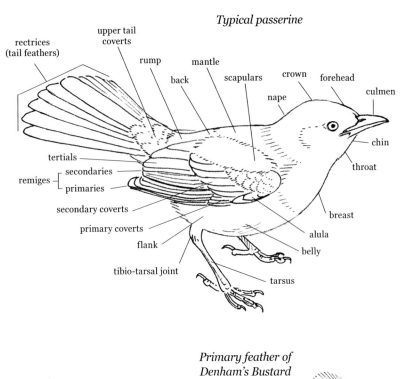

Typical passerine

Primary feather of Denham's Bustard

Yellow-billed Duck

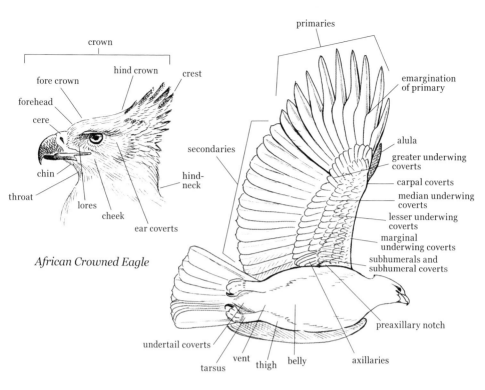
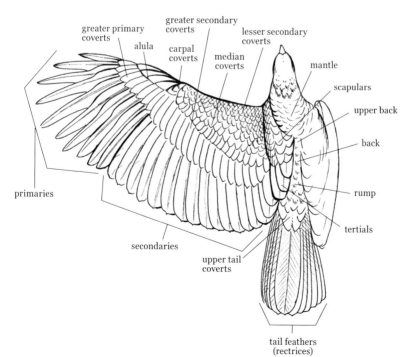

Illustrated Glossary

VAGRANTS AND HISTORICAL LISTS

Vagrants list

200	Common Quail	*Coturnix coturnix*	Probably a very uncommon vagrant. No positive records in recent past
104	Yellow-billed Duck	*Anas undulata*	Rare vagrant, could be recorded at any of the Park's waterbodies
112	Cape Shoveler	*Anas smithii*	Rare vagrant, could be recorded at any of the Park's waterbodies
489	Red-throated Wryneck	*Jynx ruficollis*	Rare vagrant
384	African Emerald Cuckoo	*Chrysococcyx cupreus*	Uncommon vagrant
390	Senegal Coucal	*Centropus senegalensis*	Rare vagrant, recorded in northern regions of the Park
348	Rock Dove	*Columba livia*	Uncommon vagrant
209	Grey Crowned Crane	*Balearica regulorum*	Rare vagrant
216	Striped Crake	*Aenigmatolimnas marginalis*	Rare vagrant on the floodplains and river margins in the northern half of the Park
223	African Purple Swamphen	*Porphyrio madagascariensis*	Rare vagrant, could be recorded at any of the Park's waterbodies
262	Ruddy Turnstone	*Arenaria interpres*	Rare vagrant, could be recorded at any of the Park's waterbodies
291	Red Phalarope	*Phalaropus fulicarius*	Rare vagrant to dams, could be found anywhere in the Park
254	Grey Plover	*Pluvialis squatarola*	Rare vagrant
247	Chestnut-banded Plover	*Charadrius pallidus*	Rare vagrant
315	Grey-headed Gull	*Larus cirrocephalus*	Rare vagrant
322	Caspian Tern	*Sterna caspia*	Rare vagrant
126	Black Kite	*Milvus migrans*	Uncommon non-br Palaearctic migrant to region
164	Western Marsh-Harrier	*Circus aeruginosus*	Uncommon non-br Palaearctic migrant to region
165	African Marsh-Harrier	*Circus ranivorus*	Rare vagrant
136	Booted Eagle	*Aquila pennatus*	Rare vagrant, could be found anywhere in the Park
178	Red-necked Falcon	*Falco chicquera*	Rare vagrant, could be found anywhere in the Park
175	Sooty Falcon	*Falco concolor*	Rare vagrant
174	African Hobby	*Falco cuvierii*	Rare vagran.
75	Rufous-bellied Heron	*Ardeola rufiventris*	Rare vagrant
50	Pink-backed Pelican	*Pelecanus rufescens*	Rare vagrant, could be recorded at any of the Park's waterbodies
544	African Golden Oriole	*Oriolus auratus*	Rare vagrant, recorded in northern regions of the Park
732	Common Fiscal	*Lanius collaris*	Uncommon vagrant
534	Banded Martin	*Riparia cincta*	Uncommon vagrant
661	Cape Grassbird	*Sphenoeacus afer*	Uncommon vagrant
642	Broad-tailed Warbler	*Schoenicola brevirostris*	Rare vagrant, probably not resident and a passage migrant
627	River Warbler	*Locustella fluviatilis*	Rare non-br Palaearctic migrant to region
620	Common Whitethroat	*Sylvia communis*	Uncommon non-br Palaearctic migrant to region
587	Capped Wheatear	*Oenanthe pileata*	Rare vagrant
774	Gurney's Sugarbird	*Promerops gurneyi*	Rare vagrant
833	Orange-winged Pytilia	*Pytilia afra*	Recorded in the Punda Maria region. One sight record for Berg-en-Dal
726	Golden Pipit	*Tmetothylacus tenellus*	Rare vagrant
872	Cape Canary	*Serinus canicollis*	Rare vagrant, recorded in south-western regions of the Park during the winter months

Historical list

198	Red-necked Spurfowl	*Pternistis afer*
117	Maccoa Duck	*Oxyura maccoa*
106	Cape Teal	*Anas capensis*
357	Blue-spotted Wood-Dove	*Turtur afer*
218	Buff-spotted Flufftail	*Sarothrura elegans*
217	Red-chested Flufftail	*Sarothrura rufa*
210	African Rail	*Rallus caerulescens*
273	Dunlin	*Calidris alpina*
257	Black-winged Lapwing	*Vanellus melanopterus*
307	Parasitic Jaeger	*Stercorarius parasiticus*
332	Sooty Tern	*Sterna fuscata*
120	Egyptian Vulture	*Neophron percnopterus*
152	Jackal Buzzard	*Buteo rufofuscus*
182	Greater Kestrel	*Falco rupicoloides*
491	African Pitta	*Pitta angolensis*
540	Grey Cuckooshrike	*Coracina caesia*
706	Fairy Flycatcher	*Stenostira scita*
629	Basra Reed-Warbler	*Acrocephalus griseldis*
677	Levaillant's Cisticola	*Cisticola tinniens*
645	Bar-throated Apalis	*Apalis thoracica*
616	Brown Scrub-Robin	*Cercotrichas signata*
586	Mountain Wheatear	*Oenanthe monticola*
768	Black-bellied Starling	*Lamprotornis corruscus*
766	Miombo Blue-eared Starling	*Lamprotornis elisabeth*
794	Plain-backed Sunbird	*Anthreptes reichenowi*
799	White-browed Sparrow-Weaver	*Plocepasser mahali*
806	Scaly-feathered Finch	*Sporopipes squamifrons*
817	Yellow Weaver	*Ploceus subaureus*
856	Red-headed Finch	*Amadina erythrocephala*
850	Swee Waxbill	*Coccopygia melanotis*
847	Black-faced Waxbill	*Estrilda erythronotos*
848	Grey Waxbill	*Estrilda perreini*
839	Red-throated Twinspot	*Hypargos niveoguttatus*
712	Mountain Wagtail	*Motacilla clara*
725	Yellow-breasted Pipit	*Anthus chloris*

ETYMOLOGY AND FOREIGN BIRD NAMES

Names for birds in languages other than English and Afrikaans are given alphabetically: French (F), German (G), North Sotho (N-So), Portuguese (Por), South Sotho (S-So), Tsonga (Tso), Xhosa (X), Zulu (Z).

Plate 1
Ostriches. Family Struthionidae. Genus *Struthio* (L) = an ostrich. **Common Ostrich** *S. camelus* (L) = pertaining to a camel. **Francolin, Spurfowl and Quail.** Family Phasianidae. Genus *Peliperdix* (Gr) = meaning uncertain; possibly from *pelios* = bruised, referring to heavily barred/blotched underparts, and *Perdix* (L) = 'the' partridge; thus 'grey-blotched partridge'. **Coqui Francolin** *P coqui* = probably imitation of its call, *ko-kwee*. Genus *Scleroptila* (Gr) = stiff feathers. **Shelley's Francolin** *S shelleyi* = after Sir E Shelley (1827-1890), English traveller in Africa. Genus *Pternistis* (Gr) = one who trips with the heel. Refers to the double spurs of the male. **Natal Spurfowl** *P natalensis* (L) = after the settlement of Port Natal (Durban), Natal, S. Africa. **Swainson's Spurfowl** *P swainsonii* = after William Swainson (1789-1855), English naturalist, artist, collector and author. Genus *Pavo* (L) = the peacock. **Common Ostrich**, Autruche d'Afrique (F), Strauß (G), Mpšhe (N-So), Avestruz (Por), Mpšhe/Mpshoe (S-So), Yinca (Tso), Inciniba (X), iNtshe (Z). **Coqui Francolin**, Francolin coqui (F), Coquifrankolin (G), Lebudiane (N-So), Francolim-das-pedras (Por), Lesogo/Letsiakarana/MmamolangwaneMantsentse (Tso), iNswempe (Z). **Shelley's Francolin**, Francolin de Shelley (F), Shelleyfrankolin (G), Francolim de Shelley (Por), Njenjele (Tso), iNtendele (Z). **Natal Spurfowl**, Francolin du Natal (F), Natalfrankolin (G), Francolim do Natal (Por), Nghwari ma ntshengwhayi (Tso), isiKhwehle (Z). **Swainson's Spurfowl**, Francolin de Swainson (F), Swainsonfrankolin (G), Francolim de Swainson (Por), Khoale (S-So), Nghwari ya xidhaka (Tso).

Plate 2
Genus *Coturnix* (L) = a quail. **Harlequin Quail** *C delegorguei* = after Adolphe Delegorgue (1814-1850). French hunter, naturalist, collector and author. **Buttonquails.** Family Turnicidae. Genus *Turnix* (L) = a quail. **Kurrichane Buttonquail.** *T sylvaticus* (L) = of woods and trees, with reference to the savanna habitat. Genus *Dendroperdix* (Gr) = tree-partridge. **Crested Francolin** *D sephaena* = uncertain, probably based on a native name (Tswana). **Guineafowls.** Family Numididae. Genus *Guttera* (L) = bearing drop-shaped spots. **Crested Guineafowl** *G edouardi* = after Edouard Verreaux (1810-1868), French natural history dealer. Genus *Numida* (L) = derived from Nubia, a territory in ancient NE Africa. **Helmeted Guineafowl** *N meleagris* = from Meleager, a mythical Greek hero. **Harlequin Quail**, Caille arlequin (F), Harlekinwachtel (G), Codorniz-arlequim (Por), Dzurhini/N'hwarixigwaqa /Xigwatla (Tso), isiGwaca (Z) **Kurrichane Buttonquail**, Turnix d'Andalousie (F), Laufhühnchen, Rostkehl-Kampfwachtel (G), Toirão-comum (Por), Mabuaneng/Mauaneng (S-So), Xitsatsana (Tso), Ingolwane (X), uNgoqo (Z). **Crested Francolin**, Francolin huppé (F), Schopffrankolin (G), Francolim-de-poupa (Por), Nghwari (Tso), isiKhwehle (Z). **Crested Guineafowl**, Pintade de Pucheran (F), Kräuselhauben-Perlhuhn (G), Pintada-de-crista (Por), Mangoko/Xiganki (Tso), iMpangele-yehlathi/iNgekle (Z). **Helmeted Guineafowl**, Pintade de Numidie (F), Helmperlhuhn (G), Kgaka (N-So), Pintada da Guiné (Por), Khaka(S-So), Mhanghela (Tso), Impangele (X), iMpangele (Z).

Plate 3
Whistling Ducks. Family Dendrocygnidae. Genus *Dendrocygna* derived from *dendron* (Gr) = a tree + *cygnus* (L) = a swan. **Fulvous Duck** *D bicolor* (L) = 2 colours, pied. **White-faced Duck** *D viduata* (L) = with a mask. **Ducks and Geese.** Family Dendrocygnidae. Genus *Thalassornis* (Gr) = a sea bird (a misnomer for a strictly fresh-water species). **White-backed Duck** *T leuconotus* (Gr) = white backed. Family Anatidae Genus *Nettapus* (Gr) = duck foot. **African Pygmy-goose** *N auritus* (L) = eared. Genus *Anas* (L) = a duck. 9 species in s Africa. **Red-billed Teal** *A erythrorhyncha* (Gr) = red bill. **Hottentot Teal** *A hottentota* (L) = of the Hottentots, a misnomer for the Khoikhoi. **Fulvous Duck**, Dendrocygne fauve (F), Gelbe Baumente (G), Pato-assobiador-arruivado (Por), Sekwa (Tso), Idada (X). **White-faced Duck**, Dendrocygne veuf (F), Witwenente (G), Pato-assobiador-de-faces-brancas (Por), Letata (S-So), Xiyahkokeni (Tso), Idada (X), iVevenyane (Z). **White-backed Duck**, Dendrocygne à dos blanc (F), Weißrückenente (G), Pato-de-dorso-branco (Por), Letata (S-So), Sekwa (Tso), Idada (X). **African Pygmy-Goose**, Anserelle naine (F), Afrikanische Zwerggans (G), Afrikaans dwerggans (Por), Pato-orelhudo (Por). **Red-billed Teal**, Canard à bec rouge (F), Rotschnabelente (G), Marreco-de-bico-vermelho (Por), Letata/Sefuli (S-So), Sekwa (Tso), iDada (Z) **Hottentot Teal**, Sarcelle hottentote (F), Hottentottenente (G), Pünktchenente (G), Marreco-hotentote (Por), Letata (S-So), Sekwa (Tso), iDada (Z).

Plate 4
Genus *Alopochen* (Gr) = fox goose; probably because of the birds colour. **Egyptian Goose** *A aegyptiaca* Genus *Tadorna* (L) = from French 'tadorne', a shelduck. Genus *Plectropterus* (Gr) = a striker or spur and wing; referring to the spur on the carpal joint. **Spur-winged Goose** *P gambensis* (L) = from Gambia. Genus *Sarkidiornis* (Gr) =

Etymology and Foreign Bird Names

Precise derivation unclear, probable reference to fleshy comb of the male. **Comb Duck** *S melanotos* (Gr) = black back. **African Black Duck** *A sparsa* (L) = scattered, referring either to the scattered white spots on the back, or to its distribution. Genus *Netta* (Gr) = a duck. **Southern Pochard** *N erythrophthalma* (Gr) = red eye. **Egyptian Goose**, Ouette d'Égypte (F), Nilgans (G), Lefaloa (N-So), Ganso do Egipto (Por), Lefaloa (S-So), Sekwa (Tso), Ilowe (X), iLongwe (Z). **Spur-winged Goose**, Oie-armée de Gambie (F), Sporengans (G), Pato-ferrão (Por), Letsikhoi/Letsikhui (S-So), Sekwagongwana/Sekwanyarhi (Tso), Ihoye (X), iHophe/iHoye (Z). **Comb Duck**, Canard à bosse (F), Höckerente (G), Pato-de-carúncula (Por), Patu ra nhova (Tso). **African Black Duck**, Canard noirâtre (F), Schwarzente (G), Pato-preto-africano (Por), Letata/Letata-la-noka (S-So), Idada (X), iDada (Z). **Southern Pochard**, Nette brune (F), Rotaugenente (G), Zarro-africano (Por), Letata (S-So), Xinyankakeni (Tso).

Plate 5
Honeyguides. Family Indicatoridae. Genus *Indicator* (L) = one who points out or directs attention to something; alludes to the bird's habit of guiding people to bee hives. **Scaly-throated Honeyguide** *I variegates* (L) = irregularly marked and alludes to the scaly throat. **Greater Honeyguide** *I indicator*. **Lesser Honeyguide** *I minor* (L) = small. Genus *Prodotiscus* (Gr) = diminutive of *prodotes*, a betrayer, alluding to the guiding habit (in *Indicator*, but not *Prodotiscus*). **Brown-backed Honeybird** *P regulus* (L) = a prince or kinglet, use in this context obscure. **Scaly-throated Honeyguide**, Indicateur varié (F), Gefleckter Honiganzeiger (G), Indicador-de-peito-escamoso (Por), Hlalala/Nhlampfu (Tso), Intakobusi (X), iNhlava (Z). **Greater Honeyguide,** Grand Indicateur (F), Großer Honiganzeiger, Schwarzkehl-Honiganzeiger (G), Tshetlo (N-So), Indicador-grande (Por), Tsehlo/Molisa-linotši (S-So), Nhlalala (generic for honeyguides) (Tso), Intakobusi (X), iNgede/iNhlavebizelayo/uNomtsheketshe (Z). **Lesser Honeyguide**, Petit Indicateur (F), Kleiner Honiganzeiger (G), Indicador-pequeno (Por), Molisa-linotši (S-So), Nhlalala (Tso), Intakobusi (X), iNhlava (Z). **Brown-backed Honeybird**, Indicateur de Wahlberg (F), Schmalschnabel-Honiganzeiger (G), Indicador-de-bico-aguçado (Por), Molisa-linotši (S-So).

Plate 6
Woodpeckers and Wrynecks. Family Picidae. Genus *Campethera* (Gr) = caterpillar hunter. **Bennett's Woodpecker** *C bennettii* probably named after Dr Edward Bennett (1797-1836), secretary of the Zoological Society of London. **Golden-tailed Woodpecker** *C abingoni,* possibly named after fifth Earl of Abingdon (1784-1854) who may have been a member of Andrew Smith's 1834 African expedition during which the type specimen was collected. Genus *Dendropicos* (Gr) = a tree pecker. **Cardinal Woodpecker** *D fuscescens* (L) = becoming brown. **Bearded Woodpecker** *D namaquus* (L) = from Namaqualand.
Bennett's Woodpecker, Pic de Bennett (F), Bennettspecht (G), Pica-pau de Bennett (Por), Ghongoswana (generic for woodpeckers) (Tso). **Golden-tailed Woodpecker**, Pic à queue dorée (F), Goldschwanzspecht (G), Pica-pau-de-rabo-dourado (Por), Ghongoswana (Tso), iSibagwebe/isiQophamuthi/uSibagwebe (Z). **Cardinal Woodpecker**, Pic cardinal (F), Kardinalspecht (G), Pica-pau-cardeal (Por), Kokomoro (S-So), Ghongoswana (Tso), Isinqolamthi (X), iNqondaqonda (Z). **Bearded Woodpecker**, Pic barbu (F), Namaspecht (G), Pica-pau-de-bigodes (Por), Ghongoswana (Tso).

Plate 7
African Barbets. Family Lybiidae. Genus *Pogoniulus* (Gr) = bearded and diminutive. **Yellow-rumped Tinkerbird** *P bilineatus* (L) = simple or undivided. **Yellow-fronted Tinkerbird** *P chrysoconus* (L) = golden cone. In Latin *conus* = cone, but should have been *comus* = hair, a mistake in the original naming. Genus *Tricholaema* (Gr) = throat hairs. **Acacia Pied Barbet** *T leucomelas* (L) = white and black. Genus *Lybius* (Gr) = bird mentioned by Aristotle and Aristophanes, possibly allied to the woodpeckers. **Black-collared Barbet** *L torquatus* (L) = (literally) adorned with a necklace (collared). Genus *Trachyphonus* (Gr) = rough voice. **Crested Barbet** *T vaillanti* (L) = after Francois Le Vaillant (1753-1824). The first real ornithologist to visit the Cape, he travelled inland as far north as the Orange R. His book *Histoire Naturelle des Oiseaux d'Afrique* in 6 volumes published between 1790 and 1808; is a classic of S African Ornithology.
Yellow-rumped Tinkerbird, Barbion à croupion jaune (F), Goldbürzel-Bartvogel (G), Barbadinho-de-rabadilha-limão (Por). **Yellow-fronted Tinkerbird**, Barbion à front jaune (F), Gelbstirn-Bartvogel (G), Barbadinho-de-testa-amarela (Por). **Acacia Pied Barbet**, Barbican pie (F), Rotstirn-Bartvogel (G), Barbaças-das-acácias (Por), Serokolo (S-So). **Black-collared Barbet**, Barbican à collier (F), Halsband-Bartvogel (G), Barbaças-de-colar-preto (Por), Kopaope (S-So), Nwagogosane (Tso), Isinagogo (X), isiKhulukhulu/isiQonqotho (Z). **Crested Barbet**, Barbican promépic (F), Haubenbartvogel (G), Barbaças-de-poupa (Por), 'Malioache (S-So), Ngoko (Tso), iMvunduna (Z).

Plate 8
Typical Hornbills. Family Bucerotidae. Genus *Tockus* (Gr) = onomatopoeic from the calls. **Red-billed Hornbill** *T erythrorhynchus* (Gr) = red-billed. **Southern Yellow-billed Hornbill** *T leucomelas* (L) = white and black.

Crowned Hornbill *T alboterminatus* (L) = white ends, with reference to the white tips to the tail feathers. **African Grey Hornbill** *T nasutus* (L) = nose, referring to the long bill of the Senegalese form (the type) that does not have an obvious pointed casque. Genus *Bycanistes* (Gr) = trumpeter. **Trumpeter Hornbill** *B bucinator* (L) = trumpeter. **Ground-hornbills.** Family Bucorvidae. Genus *Bucorvus* (L) = large, huge or great raven. **Southern Ground-Hornbill** *B leadbeateri* = named after Benjamin Leadbeater (1760-1837) who was a leading dealer in natural history material in London.
Red-billed Hornbill, Calao à bec rouge (F), Rotschnabeltoko (G), Calau-de-bico-vermelho (Por), Nkorho (Tso), umKholwane (Z). **Southern Yellow-billed Hornbill**, Calao leucomèle (F), Gelbschnabeltoko (G), Calau-de-bico-amarelo (Por), Nkorho (Tso). **Crowned Hornbill**, Calao couronné (F), Kronentoko (G), Kgoropo (N-So), Calau-coroado (Por), Nkorhonyarhi (Tso), Umkholwane (X), umKholwane (Z). **African Grey Hornbill**, Calao à bec noir (F), Grautoko (G), Calau-cinzento (Por), Nkorho (Tso). **Trumpeter Hornbill**, Calao trompette (F), Trompeter-Hornvogel (G), Calau-trombeteiro (Por), Nkorho (generic for smaller hornbills) (Tso), Ilithwa (X), iKhunatha (Z). **Southern Ground-Hornbill**, Bucorve du Sud (F), Hornrabe (G), Calau-gigante (Por), Nghututu (Tso), Intsingizi/Intsikizi (X), iNgududu/iNsingizi (Z).

Plate 9
Hoopoes. Family Upupidae. Genus *Upupa* (L) = onomatopoeic for the bird's call, as is hoopoe. **African Hoopoe** *U africana* (L) = of Africa. **Wood-Hoopoes.** Family Phoeniculidae. Genus *Phoeniculus* (Gr) = scarlet-billed. **Green Wood-Hoopoe** *P purpureus* (L) = purple, a reference to the colour of the iridescence of the feathers. **Scimitarbills.** Family Rhinopomastidae. Genus *Rhinopomastus* (Gr) = a lid or cover of the nose. Refers to the covered nares. **Common Scimitarbill** *R cyanomelas* (Gr) = dark blue-black. **Trogons.** Family Trogonidae. Genus *Apaloderma* (Gr) = weak-, or thin-skinned. **Narina Trogon** *A narina* (Khoikhoi) = flower, original name given by Levaillant (1807). Uncertain whether Levaillant's choice was because of its colourful plumage, or because the bird reminded him of a beautiful young Khoikhoi woman by the same name.
African Hoopoe, Huppe d'Afrique (F), Wiedehopf (G), Kukuku (N-So), Poupa (Por), Popopo/Khapopo/Pupupu/'Mamokete (S-So), Pupupu (Tso), Ubhobhoyi (X), umZolozolo/uZiningweni (Z). **Green Wood-Hoopoe**, Irrisor moqueur (F), Steppenbaumhopf (G), Zombeteiro-de-bico-vermelho (Por), Yokoywana (Tso), Intlekibafazi (X), iNhlekabafazi/uNukani (Z). **Common Scimitarbill**, Irrisor namaquois (F), Sichelhopf (G), Bico-de-cimitarra (Por), Yokoywana (Tso). **Narina Trogon**, Trogon narina (F), Narina-Trogon, Zügeltrogon (G), Republicano (Por), Intshatshongo (X).

Plate 10
Typical Rollers. Family Coraciidae. Genus *Coracias* (Gr) = from *korakias*, a raven, perhaps referring to crow-like calls. **European Roller** *C garrulous* (L) = chattering, babbling. **Lilac-breasted Roller** *C caudate* (L) = with a tail. **Racket-tailed Roller** *C spatulata* (L) = a broad blade or spoon. Genus *Eurystomus* (Gr) = broad mouth, beak or bill. **Purple Roller** *C naevia* (L) = purple. **Broad-billed Roller** *E glaucurus* (L and Gr) = dull greenish-blue tail.
European Roller, Rollier d'Europe (F), Blauracke (G), Rolieiro-europeu (Por), [Letlereretlere-petaputsoa](S-So), Vhevhe (generic for rollers) (Tso), iFefe (Z). **Lilac-breasted Roller**, Rollier à longs brins (F), Gabelracke (G), Matlakela (N-So), Rolieiro-de-peito-lilás (Por), Vhevhe (Tso), iFefe (Z). **Racket-tailed Roller**, Rollier à raquettes (F), Spatelracke (G), Rolieiro-cauda-de-raquete (Por), **Purple Roller**, Rollier varié (F), Strichelracke (G), Rolieiro-de-sobrancelhas-brancas (Por), Vhevhe (Tso). **Broad-billed Roller**, Rolle violet (F), Zimtroller (G), Rolieiro-de-bico-amarelo (Por).

Plate 11
Alcedinid Kingfishers. Family Alcedinidae. Genus *Alcedo* (L) = a kingfisher. **Malachite Kingfisher** *A cristata* (L) = crested. **Half-collared Kingfisher** *A semitorquata* (L) = half-collared. **Dacelonid Kingfishers** Family Dacelonidae. Genus *Halcyon* (Gr) = a mythical bird, long associated with kingfisher, which nested on the sea on calm days (hence the term halcyon days). **Grey-headed Kingfisher** *H leucocephala* (L) = white-headed. Genus *Ispidina* (L) = resembling the kingfisher (*hispida*). **African Pygmy-Kingfisher** *I picta* (L) = painted. **Woodland Kingfisher** *H senegalensis* (L) = from Senegal.
Malachite Kingfisher, Martin-pêcheur huppé (F), Malachiteisvogel, Haubenzwergfischer (G), Pica-peixe-de-poupa (Por), Seinoli (S-So), Tshololwana (Tso), Isaxwila (X), iNhlunuyamanzi/isiKhilothi/uZangozolo (Z). **Half-collared Kingfisher**, Martin-pêcheur à demi-collier (F), Kobalteisvogel, Pica-peixe-de-colar (Por), Mavungana/N'waripetani/Xitserere (Tso), Isaxwila (X), isiXula (Z). **Grey-headed Kingfisher**, Martin-chasseur à tête grise (F), Graukopfliest (G), Pica-peixe-de-barrete-cinzento (Por), Tshololwana (Tso). **African Pygmy-Kingfisher**, Martin-pêcheur pygmée (F), Natalzwergfischer (G), Pica-peixe-pigmeu (Por), Tshololwana (Tso), iNhlunuyamanzi/isiKhilothi/isiPhikeleli/uZangozolo (Z). **Woodland Kingfisher**, Martin-chasseur du Sénégal (F), Senegalliest (G), Pica-peixe-dos-bosques (Por), Tshololwana (Tso), iNkotha/uNongozolo (Z).

Plate 12
Brown-hooded Kingfisher. *H albiventris* (L) = white below. **Striped Kingfisher** *H chelicuti* (L) = of Chelicut, in Abyssinia (Ethiopia), the Type locality. **Cerylid kingfishers.** Family Cerylidae. Genus *Megaceryle* (Gr) = great kingfisher. **Giant Kingfisher** *M maxima* (L) = the largest. Genus *Ceryle* (Gr) = a Kingfisher. **Pied Kingfisher** *C rudis* (L) = a foil, for the long pointed bill.
Brown-hooded Kingfisher, Martin-chasseur à tête brune (F), Braunkopfliest (G), Pica-peixe-de-barrete-castanho (Por), Tshololwana (Tso), Undozela (X), iNdwazela/uNongobotsha/uNongozolo (Z). **Striped Kingfisher**, Martin-chasseur strié (F), Streifenliest, Gestreifter Baumliest (G), Pica-peixe-riscado (Por), Tshololwana (Tso). **Giant Kingfisher**, Martin-pêcheur géant (F), Riesenfischer, Rieseneisvogel (G), Pica-peixe-gigante (Por), Seinoli (S-So), Tshololwana (Tso), Uxomoyi (X), isiVuba (Z). **Pied Kingfisher**, Martin-pêcheur pie (F), Graufischer (G), Pica-peixe-malhado (Por), Seinoli (S-So), Tshololwana (generic for kingfishers)(Tso), Isaxwila (X), iHlabahlabane/isiQuba/isiXula (Z).

Plate 13
Bee-eaters. Family Meropidae. Genus *Merops* (Gr) = the bee-eater. **White-fronted Bee-eater** *M bullockoides* (L) = like *Merops bullocki*, a species of the n African savannas named after William Bullock (1775-1840). **Swallow-tailed Bee-eater** *M hirundineus* (L) = swallow-like. **White-throated Bee-eater** *M albicollis* = white neck. **Little Bee-eater** *M pusillus* (L) = very small. **Blue-cheeked Bee-eater** *M persicus* (L) = of Persia.
White-fronted Bee-eater, Guêpier à front blanc (F), Weißstirnspint, Weißstirn-Bienenfresser (G), Abelharuco-de-testa-branca (Por), Muhladzanhu/Muhlagambu (Tso). **Swallow-tailed Bee-eater**, Guêpier à queue d'aronde (F), Schwalbenschwanzspint Gabelschwanzspint (G), Abelharuco-andorinha (Por). **Little Bee-eater**, Guêpier nain (F), Zwergspint, Zwergbienenfresser (G), Abelharuco-nain (Por), Tinziwolana (Tso), iNkotha (Z). **Blue-cheeked Bee-eater**, Guêpier de Perse (F), Blauwangenspint (G), Abelharuco-persa (Por), Muhladzanhu/Muhlagambu (Tso), isiThwelathwela (Z).

Plate 14
European Bee-eater *M apiaster* (L) = bee-eater. **Southern Carmine Bee-eater** *M nubicoides* (L) = like Northern Carmine Bee-eater *M. nubicus*. **Mousebirds.** Family Coliidae. Genus *Colius* (L) = derivation not clear, but possibly from *koleos* (Gr) = a scabbard or sheath, with reference to the long tail. **Speckled Mousebird** *C striatus* (L) = striped, presumably referring to the fine barring on the feathers. Genus *Urocolius* (Gr), derived from *oura* = tail and *koleos* = scabbard or sheath, with reference to the relatively longer tail. **Red-faced Mousebird** *U indicus* (L) = of India.
European Bee-eater, Guêpier d'Europe (F), Europäischer Bienenfresser (G), Abelharuco-europeu (Por), Tinziwolana (Tso). **Southern Carmine Bee-eater**, Guêpier carmin (F), Scharlachspint (G), Abelharuco-róseo (Por), Nkhonyana (Tso), iNkotha-enkulu (Z). **Speckled Mousebird**, Coliou rayé (F), Braunflügel-Mausvogel (G), Rabo-de-junco-de-peito-barrado (Por), Fariki (S-So), Tshivhovo (Tso), Indlazi (X), iNdlazi (Z). **Red-faced Mousebird**, Coliou quiriva (F), Rotzügel-Mausvogel (G), Letswiyobaba (N-So), Rabo-de-junco-de-faces-vermelhas (Por), Fariki (S-So), Indlazi (Tso), Intshili (X), umTshivovo (Z).

Plate 15
Old World Cuckoos and Malkohas. Family Cuculidae. Genus *Clamator* (L) = noisy, refers to the loud chattering cries of the Genus. **Jacobin Cuckoo** *C jacobinus* (L) = from the Jacobins, French name for Dominican friars with their Paris house in the Rue St Jacques. The pied plumage of the bird resembles the traditional habit of the friars, a black mantle worn over a white scapular and tunic. **Levaillant's Cuckoo** *C levaillantii* (L) = after Francois Le Vaillant (1753-1854), French ornithologist, explorer, collector and author. **Great Spotted Cuckoo** *C glandarius* (L), derived from *glandula* or *glandium* = acorn. Refers to acorn-eating and collecting habits of Acorn Jay *Garrulus glandarius*, and probably to the noisy, jay-like behaviour of Great Spotted Cuckoo. Genus *Cuculus* (L) = a cuckoo. **Red-chested Cuckoo** *C solitarius* (L) = solitary. Genus *Pachycoccyx* (Gr) = thick, large cuckoo. In reference to large size, dense rump feathers, and stout bill. **Thick-billed Cuckoo** *P audeberti* (L) = after Father Alphonso Audebert of Madagascar, who collected the type specimen.
Jacobin Cuckoo, Coucou jacobin (F), Jakobinerkuckuck, Elsterkuckuck (G), Cuco-jacobino (Por), [Montoephatšoa] (S-So), Tihunyi (Tso), Ilunga Legwaba (X), iNkanku (Z). **Levaillant's Cuckoo**, Coucou de Levaillant (F), Kapkuckuck (G), Cuco da Cafraria (Por), Tihunyi (Tso). **Great Spotted Cuckoo**, Coucou geai (F), Häherkuckuck (G), Cuco-rabilongo (Por), **Red-chested Cuckoo**, Coucou solitaire (F), Einsiedlerkuckuck (G), Cuco-de-peito-vermelho (Por), Tlo-nke-tsoho (S-So), Ngwafalantala (Tso), Uphezukomkhono (X), uPhezukomkhono (Z). **Thick-billed Cuckoo**, Coucou d'Audebert (F), Dickschnabelkuckuck (G), Cuco-de-bico-grosso (Por).

Plate 16
Black Cuckoo *C clamosus* (L) = noisy. **Common Cuckoo** *C canorus* (L) = musical, to sing. Genus *Chrysococcyx* (Gr) = golden cuckoo, with reference to the metallic sheen of the plumage. **Klaas's Cuckoo** *C klaas* = named

after Le Vaillant's Hottentot servant. **African Cuckoo** *C gularis* (L) = of the throat, context of use here uncertain. **Diderick Cuckoo** *C caprius* (L) = goat-like, in error for *cupreus* coppery.
Black Cuckoo, Coucou criard (F), Schwarzkuckuck (G), Cuco-preto (Por), [Tetsa-kolilo](S-So), Unomntanofayo (X), iNdodosibona (Z). **Common Cuckoo**, Coucou gris (F), Kuckuck (G), Cuco-canoro (Por). **Klaas's Cuckoo**, Coucou de Klaas (F), Klaaskuckuck (G), Cuco-bronzeado-menor (Por). **African Cuckoo**, Coucou africain (F), Afrikanischer Kuckuck (G), Cuco-africano (Por). **Diderick Cuckoo**, Coucou didric (F), Diderikkuckuck, Goldkuckuck (G), Cuco-bronzeado-maior (Por), Ntetekeng (S-So), Umgcibilitshane (X), uNononekhanda (Z).

Plate 17
Coucals. Family Centropodidae. Genus *Centropus* (Gr) = spike foot, referring to the long, straight hallux claw of most coucals. **Black Coucal** *C grillii* = after the Swedish zoologist, JW Grill (1815-1864). **Burchell's Coucal** *C burchellii* = named after the naturalist William Burchell (1778-1863), who explored S Africa in 1811. **Parrots.** Family Psittacidae. Genus *Poicephalus* (Gr) = grey head. **Meyer's Parrot** *P meyeri* = after George FW Meyer, who published 'Catalogus Avium' in 1858, the catalogue of his father's Abyssinian collections. **Grey-headed Parrot** *P fuscicollis* (L) = dusky neck. **Brown-headed Parrot** *P cryptoxanthus* (Gr) = hidden yellow, referring to yellow axillaries that are only visible in flight.
Black Coucal, Coucal noir (F), Tulukuckuck, Grillkuckuck (G), Cucal-preto-africano (Por). **Burchell's Coucal**, Coucal de Burchell (F), Tiputip (G), Cucal de Burchell (Por), [Mohofa](S-So), Mfukwana (Tso), Ubikhwe (X), uFukwe/umGugwane (Z). **Meyer's Parrot**, Perroquet de Meyer (F), Goldbugpapagei (G), Palapalagae (N-So), Papagaio de Meyer (Por), Hokwe (Tso). **Grey-headed Parrot**, Perroquet à cou brun (F), Graukopfpapage (G). **Brown-headed Parrot**, Perroquet à tête brune (F), Braunkopfpapagei (G), Papagaio-de-cabeça-castanha (Por), Yhokwe (Tso).

Plate 18
Typical Swifts. Family Apodidae. Genus *Telacanthura* (Gr) = tail with spiny tip. **Mottled Spinetail** *T ussheri* = after Hebert Ussher (1836-1880), British civil servant and Governor of the Gold Coast (Ghana), who collected the Type specimen. Genus *Neafrapus* (L) = new African swift. **Böhm's Spinetail** *N boehmi* = after German naturalist Dr R Böhm (1854-1884), who visited C and E Africa in the late 1800s. Genus *Apus* (Gr) = without feet, referring to highly reduced legs and feet. **Little Swift** *A affinis* **(L)** = neighbouring; in reference to its habit of br in close association with man. **Horus Swift** *A horus* = after Ancient Egyptian falcon deity Horus. **White-rumped Swift** *A caffer* (L) = from Caffraria, ie Cape.
Mottled Spinetail, Martinet d'Ussher (F), Baobabsegler (G), Rabo-espinhoso-malhado (Por). **Böhm's Spinetail**, Martinet de Böhm (F), Fledermaussegler (G), Rabo-espinhoso de Böhm (Por). **Little Swift**, Martinet des maisons (F), Haussegler (G), Andorinhão-pequeno (Por), Lehaqasi (S-So), Nkonjana (Tso). **Horus Swift**, Martinet horus (F), Horussegler, Erdsegler (G), Andorinhão-das-barreiras (Por), Lehaqasi (S-So), Nkonjana (Tso). **White-rumped Swift**, Martinet cafre (F), Weißbürzelsegler (G), Andorinhão-cafre (Por), Lehaqasi (S-So), Nkonjana (Tso), Ihlabankomo/Ihlankomo (X), uNonqane (Z).

Plate 19
Genus *Cypsiurus* (L), derived from *kupselos* (Gr), a kind of swallow, and *ouros* (Gr), tailed. **African Palm-Swift** *C parvus* (L) = small. Genus *Tachymarptis* (Gr) = a fast seizer. **Alpine Swift** *T melba* = name given by Linnaeus without explanation; meaning unknown. **Common Swift** *A apus*. **African Black Swift** *A barbatus* (L) = bearded.
African Palm-Swift, Martinet des palmes (F), Palmensegler (G), Andorinhão-das-palmeiras (Por), Nkonjana (Tso). **Alpine Swift**, Martinet à ventre blanc (F), Alpensegler (G), Andorinhão-real (Por), Lehaqasi/Lehaqasi-lelephatsoa(S-So), Nkonjana (Tso), Ihlabankomo/Ubhantom (X). **Common Swift**, Martinet noir (F), Mauersegler (G), Andorinhão-preto-europeu (Por), Lehaqasi (S-So), Nkonjana (Tso), Ihlabankomo/Ihlankomo (X), iJankomo/uHlolamvula (Z). **African Black Swift**, Martinet du Cap (F), Kapsegler (G), Andorinhão-preto-africano (Por), Lehaqasi (S-So), Nkonjana (Tso), Ihlabankomo/Ihlankomo (X), iHlabankomo/iHlolamvula/iJankomo (Z).

Plate 20
Barn and Grass Owls. Family Tytonidae. Genus *Tyto* (Gr) = a night owl. **African Grass-Owl** *T capensis* (L) = after the Cape of Good Hope, S Africa. **Barn Owl** *T alba* (L) = white. **Typical Owls.** Family Strigidae. Genus *Otus* (Gr) = an eared owl. **African Scops-Owl** *O senegalensis* (L) = from Senegal. Genus *Ptilopsus* (Gr) = downy feathered. **Southern White-faced Scops-Owl** *P granti* = after Captain Claude HB Grant (1878-1958). British ornithologist, collector, and author. Genus *Glaucidium* (Gr) = a very small owl. **Pearl-spotted Owlet** *G perlatum* (L) = wearing pearls. **African Barred Owlet** *G capense* (L) = after the Cape of Good Hope, S Africa.
African Grass-Owl, Effraie du Cap (F), Graseule (G), Makgohlo (N-So), Coruja-do-capim (Por), Sephooko (S-So), Musoho (Tso), Isikhova (X), isiKhova/umShwelele (Z). **Barn Owl**, Effraie des clochers (F), Schleiereule (G), Leribisi (N-So), Coruja-das-torres (Por), Sephooko (S-So), Xinkhovha (Tso), Isikhova (X), isiKhova/umZwelele (Z).

African Scops-Owl, Petit-duc africain (F), Afrikanische Zwergohreule (G), Mocho-de-orelhas-africano (Por), Xikhotlwana (Tso). **Southern White-faced Scops-Owl**, Petit-duc de Grant (F), Weißgesicht-Ohreule (G), Mocho-de-faces-brancas (Por), Kurkurtavoni (Tso), umManduburu (Z). **Pearl-spotted Owl**, Chevêchette perlée (F), Perlkauz (G), Mocho-perlado (Por), Mankhudu (Tso), iNkovana (Z). **African Barred Owlet**, Chevêchette du Cap (F), Kapkauz (G), Mocho-barrado (Por).

Plate 21
Genus *Bubo* (L) = an eagle-owl, probably onomatopoeic. **Spotted Eagle-Owl** *B africanus* (L) = from Africa. **Verreaux's Eagle-Owl** *B lacteus* (L) = milk white, milky. Genus *Scotopelia* (Gr) = *scotos* (darkness) and *peli* (black), referring to large 'black' eyes of most common and widespread sp, Pel's Fishing-Owl. **Pel's Fishing-Owl** *S peli* = after H.S. Pel (died 1854), Dutch Government official, later Governor, on the Gold Coast 1840-1850. Genus *Strix* (L) = a screech-owl. **African Wood-Owl** *S woodfordii* = after Colonel EJA Woodward (1761-1825), a London collector of pictures and watercolours of birds. Genus *Asio* (Gr) = an owl. **Marsh Owl** *A capensis* (L) = after the Cape of Good Hope, S Africa.
Spotted Eagle-Owl, Grand-duc africain (F), Fleckenuhu, Berguhu (G), Bufo-malhado (Por), Makhohlo/Morubisi/Sehihi/Sephooko (S-So), Xiyinha (Tso), Ifubesi/Isihulu-hulu (X), isiKhovampondo (Z). **Verreaux's Eagle-Owl**, Grand-duc de Verreaux (F), Milchuhu, Blaßuhu (G), Bufo-leitoso (Por), Nkhunsi (Tso), Ifubesi (X), iFubesi/isiKhova (Z). **Pel's Fishing-Owl**, Chouette-pêcheuse de Pel (F), Bindenfischeule, Fischeule (G), Corujão-pesqueiro de Pel (Por). **African Wood-Owl**, Chouette africaine (F), Woodfordkauz (G), Coruja-da-floresta (Por), Mankhudu (Tso), Ibengwana (X), uMabhengwane/uNobathekeli (Z). **Marsh Owl**, Hibou du Cap (F), Kapohreule (G), Coruja-dos-pântanos (Por), Sephooko (S-So), Xikhotlwani (Tso), iNkovane/umShwelele (Z).

Plate 22
Nightjars. Family Caprimulgidae. Genus *Caprimulgus* (L) = to milk a goat; from its wide open mouth and habit of foraging around animal pens at night, hunting insects attracted to livestock. Old English name = goatsucker. **Freckled Nightjar** *C tristigma* (L) = 3 marks, referring to the (up to 3) irregular patches on scapulars. **Fiery-necked Nightjar** *C pectoralis* (Gr) = of the breast. **Square-tailed Nightjar** *C fossii* (L) = after W Fosse, German collector in Gabon. **Rufous-cheeked Nightjar** *C rufigena* (L) = red cheeks. **European Nightjar** *C europaeus* (L) = of Europe. Genus *Macrodipteryx* (L) = 2 large feathers. **Pennant-winged Nightjar** *M vexillarius* (L) = a standard-bearer.
Freckled Nightjar, Engoulevent pointillé (F), Fleckennachtschwalbe (G), Noitibó-sardento (Por), Mahulwana/Ribyatsane/Riwuvawuva (Tso). **Fiery-necked Nightjar**, Engoulevent musicien (F), Rotnacken-Nachtschwalbe (G), Leuwauwe (N-So), Noitibó-de-pescoço-dourado (Por), Kubhasti (Tso), Udebeza (X), uZavolo (Z). **Square-tailed Nightjar**, Engoulevent du Mozambique (F), Gabunnachtschwalbe (G), Noitibó de Moçambique (Por), Kubhasti (Tso). **Rufous-cheeked Nightjar**, Engoulevent à joues rousses (F), Rostwangen-Nachtschwalbe (G), Noitibó-de-faces-ruivas (Por), Semanama (S-So), Mahulwana/Ribyatsane/Riwuvawuva (Tso). **European Nightjar**, Engoulevent d'Europe (F), Ziegenmelker, Nachtschwalbe (G), Noitibó da Europa (Por), Semanama (S-So), Mahulwana/Ribyatsane (Tso), Udebeza (X), uZavolo (Z). **Pennant-winged Nightjar**, Engoulevent porte-étendard (F), Ruderflügel (G), Noitibo-de-balanceiros (Por).

Plate 24
Pigeons and Doves. Family Columbidae. Genus *Streptopelia* (Gr) = a collared dove. **Laughing Dove** *S senegalensis* (L) = from Senegal. **African Mourning Dove** *S decipiens* (L) = to deceive, probably because it closely resembles the other s African collared doves. **Cape Turtle-Dove** *S capicola* (L) = inhabitant of the Cape. **Red-eyed Dove** *S semitorquata* (L) = half-collared. Genus *Turtur* (L) = a turtle-dove. **Emerald-spotted Wood-Dove** *T chalcospilos* (Gr) = copper or brass spots. **Tambourine Dove** *T tympanistria* (Gr) = refers to a priestess of Cybele, who used drums and tambourines in ceremonies honouring the goddess. Refers to the rhythmical call.
Laughing Dove, Tourterelle maillée (F), Senegaltaube, Palmtaube (G), Rola do Senegal (Por), Leebana-khoroana/Mphubetsoana (S-So), Gugurhwana (Tso), Icelekwane/Uvelemaxhoseni (X), uKhonzane (Z). **African Mourning Dove**, Tourterelle pleureuse (F), Angolaturteltaube, Angolalachtaube, Brillentaube (G), Rola-gemedora (Por), Tuba (Tso). **Cape Turtle-Dove**, Tourterelle du Cap (F), Kapturteltaube, Gurrtaube, Kaplachtaube (G), Leaba Kgorwana (N-So), Rola do Cabo (Por), Leebana-khoroana/Lekunkuroane (S-So), Tuva (Tso), Ihobe/Untamnyama (X), iHophe/uSamdokwe (Z). **Red-eyed Dove**, Tourterelle à collier (F), Halbmondtaube (G), Rola-de-olhos-vermelhos (Por), Leebamosu/Leebana-khoroana/Leebana (S-So), Khopola/Nyakopo (Tso), Indlasidudu/Umakhulu (X), iHophe (Z). **Emerald-spotted Wood-Dove**, Tourtelette émeraudine (F), Bronzeflecktaube (G), Rola-esmeraldina (Por), Xivhambalana (Tso), Ivukazana (X), isiKhombazane-sehlanza (Z). **Tambourine Dove**, Tourtelette tambourette (F), Tamburintaube (G), Rola-de-papo-branco (Por), Xiwambalane (Tso), Isavu (X), isiBhelu/isiKhombazane-sehlathi (Z).

Etymology and Foreign Bird Names

Plate 25
Genus *Columba* (L) = a pigeon or dove. **Speckled Pigeon** *C guinea* (L) = of Guinea. Genus *Oena* (Gr) = a wild pigeon. **Namaqua Dove** *O capensis* (L) = after the Cape of Good Hope, South Africa. Genus *Treron* (Gr) = trembling, timid, shy, easily alarmed. **African Green-Pigeon** *T calva* (L) = hair-less, bare. **Turacos.** Family Musophagidae. Genus *Corythaixoides* (L) = like *Tauraco corythaix*, Knysna Turaco. **Grey Go-away-bird** *C concolor* (L) = uniformly coloured. Genus *Gallirex* (L)= from *gallus* (cock) and *rex* (king), referring to large crest and bare red skin around eye. **Purple-crested Turaco** *G porphyreolophus* (L) = purple-crested.
Speckled Pigeon, Pigeon roussard (F), Guineataube (G), Pombo-malhado (Por), Leeba/Lehoboi/Leeba-lathaba (S-So), Ivukuthu (X), iJuba/iVukuthu (Z). **Namaqua Dove**, Tourtelette masquée (F), Kaptäubchen (G), Rola-rabilonga (Por), Mokhoroane/Mokhorane (S-So). Xivhambalana (Tso), Ihotyazana (X), isiKhombazanesenkangala/uNkombose (Z). **African Green-Pigeon**, Colombar à front nu (F), Grüne Fruchttaube, Grüntaube (G), Pombo-verde-africano (Por), Nghwamba (Tso), Intendekwane (X), iJubantondo (Z). **Grey Go-away-bird**, Touraco concolore (F), Graulärmvogel (G), Mokowe (N-So), Turaco-cinzento (Por), Nkwenyana (Tso), umKlewu (Z). **Purple-crested Turaco**, Touraco à huppe splendide (F), Glanzhaubenturako (G), Turaco-de-crista-violeta (Por), Nkwenyana (generic for louries) (Tso), iGwalagwala (Z).

Plate 26
Bustards. Family Otididae. Genus *Ardeotis* (L) = heron bustard. **Kori Bustard** *A kori* (Tsw) = Tswana name for the bird, 'Kgôri'. Genus *Lophotis* (Gr) = crested bustard. **Red-crested Korhaan** *L ruficrista* (L) = red crest. Genus *Lissotis* (Gr) = smooth bustard. **Black-bellied Bustard** *L melanogaster* (Gr) = black belly. **Sandgrouse.** Family Pteroclidae. Genus *Pterocles* (Gr) = noted for, famous for or endowed with a wing, referring to the long primaries and rapid flight. **Double-banded Sandgrouse** *P bicinctus* (L) = double girdled, referring to the double breast-bands.
Kori Bustard, Outarde kori (F), Riesentrappe (G), Kgori (N-So), Abetarda-gigante (Por), Mithisi (Tso), Iseme (X), umNgqithi (Z). **Red-crested Korhaan**, Outarde houppette (F), Rotschopftrappe (G), Abetarda-de-poupa (Por), Xicololwana lexi tsongo (Tso). **Black-bellied Bustard**, Outarde à ventre noir (F), Schwarzbauchtrappe (G), Abetarda-de-barriga-preta (Por), Xicololwana lexi kulu (Tso), uFumba/uNofunjwa (Z). **Double-banded Sandgrouse**, Ganga bibande (F), Nachtflughuhn (G), Corticol-de-duas-golas (Por), Xighwaraghwara (Tso).

Plate 27
Rails, Crakes, Gallinules, Moorhens and Coots. Family Rallidae. Genus *Crecopsis* (Gr) = similar in appearance to Corn Crake (Genus *Crex*). **African Crake** *C egregia* (L) = without flocking. In this probably solitary. Genus *Crex* (L) = a crake or a bird with long legs; onomatopoeic from the croaking call. **Corn Crake** *C crex*. Genus *Gallinago* (L) = snipe. In Latin *Gallinago* is strictly a woodcock, from *gallina* = a hen. **African Snipe** *G nigripennis* (L) = black-winged (*nigripennis* might have been an error for *albipennis* in reference to the white outer web of the outer primary). **Painted-snipes.** Family Rostratulidae. Genus *Rostratula* (L) = having a small bill (smaller than those of the true snipes). **Greater Painted-snipe** *R benghalensis* (L) = from Bengal (Benghal = alternative spelling). **Jacanas.** Family Jacanidae. Genus *Actophilornis* (L) = seashore- or beach-loving bird. **African Jacana** *A africanus* (L) = from Africa. Genus *Microparra* (L) = small bird of ill omen. **Lesser Jacana** *M capensis* (L) = from the Cape of Good Hope, S Africa.
African Crake, Râle des prés (F), Steppenralle (G), Codornizão-africano (Por). **Corn Crake**, Râle des genêts (F), Wachtelkönig (G), Codornizão-euroasiático (Por). **African Snipe**, Bécassine africaine (F), Afrikanische Bekassine (G), Narceja-africana (Por), Koe-koe-lemao/Motjoli-matsana/Koekoe-lemao (S-So), Umnquduluthi (X), uNununde (Z). **Greater Painted-snipe**, Rhynchée peinte (F), Goldschnepfe (G), Narceja-pintada (Por). **African Jacana**, Jacana à poitrine dorée (F), Blaustirn-Blatthühnchen (G), Jacana-africana (Por), iThandaluzibo/uNondwayiza (Z). **Lesser Jacana**, Jacana nain (F), Zwergblatthühnchen (G), Jacana-pequena (Por).

Plate 28
Finfoot. Family Heliornithidae. Genus *Podica* (L) = pertaining to feet, or junction of toes, referring to the strange lobed toes. **African Finfoot** *P senegalensis* (L) = from Senegal. Genus *Amaurornis* (L) = dark, dim or obscure bird. **Black Crake** *A flavirostris* (L) = yellow bill. Genus *Gallinula* (L) = a small hen. **Common Moorhen** *G chloropus* (L) = green foot. Genus *Porphyrio* (Gr) = purple, but also means gallinule. **Allen's Gallinule** *P alleni* (L) = after Rear Admiral William Allen (1793-1864), leader of the Niger expeditions. **Lesser Moorhen** *G angulata* (L) = angular, pointed, with reference to the pointed frontal shield. Genus *Fulica* (L) = a coot, from bird's colour (*fuligo* = soot). **Red-knobbed Coot** *F cristata* (L) = crested, referring to the knobs on top of the shield.
African Finfoot, Grébifoulque d'Afrique (F), Afrikanische Binsenralle (G), Pés-de-barbatanas (Por). **Black Crake**, Râle à bec jaune (F), Mohrenralle (G), Negerralle (G), Franga-d'água-preta (Por), Hukunambu/Nkukumezane (Tso), umJekejeke/umJengejenge (Z). **Common Moorhen**, Gallinule poule-d'eau (F), Teichhuhn (G), Kgogomeetse/Kgogonoka (N-So), Galinha-d'água (Por), Khohonoka (S-So), Kukumezane (Tso). **Allen's Gallinule**, Talève d'Allen (F), Afrikanisches Sultanshuhn (G), Caimão de Allen (Por). **Lesser Moorhen**, Gallinule africaine (F),

Zwergteichhuhn (G), Galinha-d'água-pequena (Por), Kukumezani (Tso). **Red-knobbed Coot**, Foulque à crête (F), Kammbleßhuhn (G), Galeirão-de-crista (Por), Mohetle/Mohoetle/Tšumu/Boleseboko (S-So), Kukumezani (Tso), Unomkqayi/Unompemvana (X).

Plate 29
Sandpipers, Greenshanks, Stints, Ruff. Family Scolopacidae. Genus *Tringa* (Gr) = a thrush-sized, white-rumped, tail-bobbing waterbird mentioned by Aristotle; name first given to Green Sandpiper by Aldovandri in 1603. **Marsh Sandpiper** *T stagnatilis* (L) = of pools or marshes, in reference to habitat. **Common Greenshank** *T nebularia* (L) = cloudy, blended colours, referring to upper part coloration. **Green Sandpiper** *T ochropus* (Gr) = pale yellow foot. **Wood Sandpiper** *T glareola* (L) = pertaining to gravel. Genus *Actitis* (Gr) = a coastal dweller. **Common Sandpiper** *A hypoleucos* (Gr) = white below.
Marsh Sandpiper, Chevalier stagnatile (F), Teichwasserläufer (G), Perna-verde-fino (Por). **Common Greenshank**, Chevalier aboyeur (F), Grünschenkel (G), Perna-verde-comum (Por), Koe-koe-lemao/[Seealemabopo-holo](S-So), N'wantshekutsheku/Xitsatsana/Xitshekutsheku (Tso), Uphendu (X). **Green Sandpiper**, Chevalier cul-blanc (F), Waldwasserläufer (G), Maçarico-bique-bique (Por). **Wood Sandpiper**, Chevalier sylvain (F), Bruchwasserläufer (G), Maçarico-bastardo (Por), Koe-koe-lemao/[Seealemabopo-khoali](S-So), N'wantshekutsheku/Xitsatsana/Xitshekutsheku (Tso), Uthuthula (X). **Common Sandpiper**, Chevalier guignette (F), Flußuferläufer (G), Maçarico-das-rochas (Por). Koe-koe-lemao/Seealemabopo-hetlatšoeu (S-So), N'wantshekutsheku (Tso), Uthuthula (X).

Plate 30
Genus *Calidris* (Gr) = a grey, waterside bird (*ex* Aristotle); not specifically identified, but thought to be either a sandpiper or wagtail. **Sanderling** *C alba* (L) = white. **Little Stint** *C minuta* = tiny. **Curlew Sandpiper** *C ferruginea* (L) = rusty coloured, in reference to br plumage. Genus *Philomachus* (Gr) = lover of battles, in reference to elaborate lek displays. **Ruff (f = Reeve)** *P pugnax* (L) = combative, fond of fighting. Both generic and specific names refer to the lekking behaviour of the males when breeding. The name of the female, **Reeve**, is thought to be a stem or vowel mutation of the male name.
Sanderling, Bécasseau sanderling (F), Sanderling (G), Pilrito-sanderlingo (Por). **Little Stint** Bécasseau minute (F), Zwergstrandläufer (G), Pilrito-pequeno (Por), Tsititsiti-nyenyane (S-So). **Curlew Sandpiper**, Bécasseau cocorli (F), Sichelstrandläufer (G), Pilrito-de-bico-comprido (Por). **Ruff**, Combattant varié (F), Kampfläufer (G), Combatente (Por), Koe-koe-lemao/Seealemabopo-se-maroboko (S-So).

Plate 31
Thick-knees (Dikkops). Family Burhinidae. Genus *Burhinus* (L) = large or great nose. **Water Thick-knee** *B vermiculatus* (L) = vermiculated, in reference to the twisted trails left by burrowing worms. **Spotted Thick-knee** *B capensis* (L) = after the Cape of Good Hope, S Africa. **Stilts and Avocets.** Family Recurvirostridae. Genus *Himantopus* (Gr) = strap-foot, name given by Pliny to a thin-legged wading bird, assumed to be a stilt. **Black-winged Stilt** *H himantopus* (Gr), *himanto* = strap shaped; *pus* = foot. Genus *Recurvirostra* (L) = recurved bill. **Pied Avocet** *R avosetta* = Italian name for Pied Avocet.
Water Thick-knee, Oedicnème vermiculé (F), Wassertriel (G), Alcaravão-d'água (Por), Mtshikuyana (Tso), Ingqangqolo (X). **Spotted Thick-knee**, Oedicnème tachard (F), Kaptriel, Bändertriel (G), Alcaravão do Cabo (Por), Khoho-ea-lira/Khoalira (S-So), Mtshikuyana (generic for dikkops) (Tso), Ingqangqolo (X), umBangaqhwa/umJenjana (Z). **Black-winged Stilt**, Échasse blanche (F), Stelzenläufer (G), Perna-longa (Por), 'Mamenotoananala (S-So). **Pied Avocet**, Avocette élégante (F), Säbelschnäbler (G), Alfaiate (Por).

Plate 32
Plovers and Lapwings. Family Charadriidae. Genus *Charadrius* (Gr) = a plover. **Common Ringed Plover** *C hiaticula* (L) = cleft dweller, in reference to its habit of breeding among pebbles and rocks. **Kittlitz's Plover** *C pecuarius* (L) = a grazer, in reference to grassland habitat. **Three-banded Plover** *C tricollaris* (L) = three-collared. **White-fronted Plover** *C marginatus* (L) = of the edge, in this case, the seashore. **Caspian Plover** *C asiaticus* (L) = from Asia.
Common Ringed Plover, Pluvier grand-gravelot (F), Sandregenpfeifer (G), Borrelho-grande-de-coleira (Por), Unokrekre (X). **Kittlitz's Plover**, Pluvier pâtre (F), Hirtenregenpfeifer (G), Borrelho-do-gado (Por). **Three-banded Plover**, Pluvier à triple collier (F), Dreiband-Regenpfeifer (G), Borrelho-de-três-golas (Por), [Patapeta-nala](S-So), N'wantshekutsheku/Xitsekutseku (Tso), Inqatha/Unokrekre (X). **White-fronted Plover**, Pluvier à front blanc (F), Weißstirn-Regenpfeifer (G), Borrelho-de-testa-branca (Por), Unocegceya/Unotelela (X). **Caspian Plover**, Pluvier asiatique (F), Wermutregenpfeifer (G), Borrelho-asiático (Por).

Plate 33
Genus *Vanellus* (L) = Northern Lapwing *V vanellus*, named for its floppy wing action resembling a winnowing fan (*vannus*). **Blacksmith Lapwing** *V armatus* (L) = armed, referring to carpal spurs. **White-crowned Lapwing** *V albiceps* (L) = white-headed. **African Wattled Lapwing** *V senegallus* (L) = from Senegal. **Crowned Lapwing** *V coronatus* (L) = crowned. **Senegal Lapwing** *V lugubris* (L) = mournful, sorrowful.
White-crowned Lapwing, Vanneau à tête blanche (F), Langspornkiebitz (G), Abibe-de-coroa-branca (Por).
Blacksmith Lapwing, Vanneau armé (F), Waffenkiebitz, Schmiedekiebitz (G), Abibe-preto-e-branco (Por), Mo-otla-tšepe (S-So), Ghelekela (Tso), iNdudumela (Z). **African Wattled Lapwing**, Vanneau du Sénégal (F), Senegalkiebitz (G), Abibe-carunculado (Por), Nghelekele (Tso). **Crowned Lapwing**, Vanneau couronné (F), Kronenkiebitz (G), Mororwane (N-So), Abibe-coroado (Por), Lekekeruane/Letletleruane (S-So), Ghelekela (generic for plovers) (Tso), Igxiya (X), iTitihoye (Z). **Senegal Lapwing**, Vanneau terne (F), Trauerkiebitz (G), Abibe-d'asanegra-pequeno (Por), Ghelekela (Tso), iTitihoye (Z).

Plate 34
Coursers and Pratincoles. Family Glareolidae. Genus *Rhinoptilus* (Gr) = feathered nostril. **Bronze-winged Courser** *R chalcopterus* (Gr) = metallic-winged. **Three-banded Courser** *R cinctus* (L) = banded. Genus *Cursorius* (L) = swift runner. **Temminck's Courser** *C temminckii* = named after Dutch ornithologist Coenraad Jacob Temminck (1778-1858). Genus *Glareola* (Gr) = living on gravel. **Collared Pratincole** *G pratincola* (L) = inhabitant of meadows. **Terns.** Family Laridae. Genus *Chlidonias* (Gr) = swallow-like. **Whiskered Tern** *C hybridus* (L) = hybrid - a reference to doubtful status (at time of description) as a full species. Pallas considered that its characteristics suggested a hybrid between Black and Common Terns. **White-winged Tern** *C leucopterus* (Gr) = white-winged.
Bronze-winged Courser, Courvite à ailes bronzées (F), Bronzeflügel-Rennvogel, Amethystrennvogel (G), Corredor-asa-de-bronze (Por), Tshembyana (Tso). **Three-banded Courser**, Courvite à triple collier (F), Bindenrennvogel (G), Corredor-de-três-golas (Por). **Temminck's Courser**, Courvite de Temminck (F), Temminckrennvogel (G), Corredor de Temminck (Por), Ucelithafa (X), uNobulongwe (Z). **Collared Pratincole**, Glaréole à collier (F), Brachschwalbe (G), Perdiz-do-mar-d'asa-vermelha (Por), iWamba (Z). **Whiskered Tern**, Guifette moustac (F), Weißbart-Seeschwalbe (G), Gaivina-de-faces-brancas (Por). **White-winged Tern**, Guifette leucoptère (F), Weißflügel-Seeschwalbe (G), Gaivina-d'asa-branca (Por).

Plate 35
Typical Raptors, Old World Vultures and Osprey. Family Accipitridae Genus *Pernis* (L) = a kind of hawk, thought to be a corruption of the Greek *pternis* = a bird of prey. **European Honey-Buzzard** *P apivorus* (L) = a bee eater. Genus *Aviceda* (L) = a killing bird. **African Cuckoo Hawk** *A cuculoides* (L) = like a cuckoo, referring to the similarity between this species and cuckoos *Cuculus* spp. when in flight. Genus *Macheiramphus* (Gr) = hooked, dagger-like bill. **Bat Hawk** *M alcinus* (Gr) = strong or brave. Genus *Buteo* (L) = a falcon or hawk. **Steppe Buzzard** *B vulpinus* (L) = fox-like, ie tawny-coloured. Genus *Milvus* (L) = a kite, or a rapacious bird. **Yellow-billed Kite** *M m parasitus* (L) = a parasite. **Black Kite** *M migrans* (L) = to wander, a migrant.
European Honey-Buzzard, Bondrée apivore (F), Wespenbussard (G), Bútio-vespeiro (Por). **African Cuckoo Hawk**, Baza coucou (F), Kuckucksweih (G), Falcão-cuco (Por). **Bat Hawk**, Milan des chauves-souris (F), Fledermausaar (G), Gavião-morcegueiro (Por). **Steppe Buzzard**, Buse des steppes (F), Mäusebussard (G), Bútiocomum (Por). Khajoane (S-So), Isangxa (X). **Yellow-billed Kite** Mmankgôdi (N-So), 'Mankholi-kholi/Kholokholo (S-So), Mangatlu (Tso), Segôdi/Untloyiya/Untloyila (X), uNhloyile/uKholwe (Z). **Black Kite**, Milan d'Afrique (F), Schmarotzermilan (G), Milhafre-preto (Por).

Plate 36
Genus *Pandion* (Gr) = king of Attica, a figure in Greek mythology. **Osprey** *P haliaetus* (Gr) = a fishing eagle. Genus *Elanus* (Gr) = a kite. **Black-shouldered Kite** *E caeruleus* (L) = blue. Genus *Haliaeetus* (Gr) = a fishing eagle. **African Fish-Eagle** *H vocifer* (L) = vociferous. Genus *Gypohierax* (Gr) = vulturine eagle. **Palm-nut Vulture** *G angolensis* (L) = from Angola.
Osprey, Balbuzard pêcheur (F), Fischadler (G), Águia-pesqueira (Por). **Black-shouldered Kite**, Élanion blanc (F), Gleitaar (G), Peneireiro-cinzento (Por), Phakoana-mafieloana/Phakoana-tšooana/Phakoana-tšoana (S-So), Nwarikapanyana (Tso), Umdlampuku/Unongwevana (X). **African Fish-Eagle**, Pygargue vocifer (F), Schreiseeadler (G), Águia-pesqueira-africana (Por), Nghunghwa (Tso), Ingqolane/Unomakhwezana (X), iNkwazi (Z). **Palm-nut Vulture**, Palmiste africain (F), Palmengeier (G), Abutre-das-palmeiras (Por), Gungwa/Ngungwamawala (Tso).

Plate 37
Genus *Necrosyrtes* (Gr) = corpse and *surtes* (Gr) = pulling. **Hooded Vulture** *N monachus* (L) = a monk, in reference to hooded appearance. Genus *Gyps* (Gr) = vulture. **White-backed Vulture** *G africanus* (L) = from

Africa. **Rüppell's Vulture** *G rueppellii* = after Wilhelm Peter Eduard Simon Rüppell (1794-1884), German explorer, naturalist and collector. **Cape Vulture** *G coprotheres* (Gr) = from *kopros* = dung or faeces, and *theres* = hunting. Genus *Aegypius* (Gr) = a vulture. **Lappet-faced Vulture** *A tracheliotos* (Gr) = gristly ears, in reference to head and neck wattles. **White-headed Vulture** *A occipitalis* (L) = of the back of the head. **Hooded Vulture**, Vautour charognard (F), Kappengeier (G), Abutre-de-capuz (Por), Koti (generic for vultures) (Tso). **White-backed Vulture**, Vautour africain (F), Weißrückengeier (G), Grifo-de-dorso-branco (Por), Koti (Tso). **Cape Vulture**, Vautour chassefiente (F), Kapgeier (G), Grifo do Cabo (Por), Lenong/Letlaka (S-So), Khoti/ Mavalanga (Tso), Ixhalanga (X), iNqe (Z). **Lappet-faced Vulture**, Vautour oricou (F), Ohrengeier (G), Lenong le Leso (N-So), Abutre-real (Por), Letlaka-pipi (S-So), Koti (Tso), Isilwangangubo (X), iNqe (Z). **White-headed Vulture**, Vautour à tête blanche (F), Wollkopfgeier (G), Abutre-de-cabeça-branca (Por), Koti (Tso).

Plate 38
Genus *Circaetus* (Gr) from *Circus* a harrier, and *aetos* an eagle = harrier-eagle. **Black-chested Snake-Eagle** *C pectoralis* (L) = of the breast. **Brown Snake-Eagle** *C cinereus* (L) = ash grey. Genus *Terathopius* (Gr) = nimble juggler, referring to its rolling flight. **Bateleur** *T ecaudatus* (L) = lacking tail. Genus *Circus* (Gr) = from *kirkos*, a hawk that circles, alluding to acrobatic displays. **Pallid Harrier** *C macrourus* (L) = long-tailed. **Montagu's Harrier** *C pygargus* = from (Gr) *pugargos*, a raptor mentioned by Pliny, Aristotle and Hesychius, not further identified.
Black-chested Snake-Eagle, Circaète à poitrine noire (F), Schwarzbrust-Schlangenadler (G), Águia-cobreira-de-peito-preto (Por), Xithaklongwa (Tso), uKhozi (Z). **Brown Snake-Eagle**, Circaète brun (F), Brauner Schlangenadler (G), Águia-cobreira-castanha (Por), Ghama (Tso). **Bateleur**, Bateleur des savanes (F), Gaukler (G), Águia-bailarina (Por), Ximongwe (Tso), Ingqanga (X), iNgqungqulu (Z). **Pallid Harrier**, Busard pâle (F), Steppenweihe (G), Tartaranhão-pálido (Por), Seitlhoaeleli (S-So), Nghotsana (Tso), Ulubisi/Umphungeni (X). **Montagu's Harrier**, Busard cendré (F), Wiesenweihe (G), Tartaranhão-caçador (Por).

Plate 39
Genus *Kaupifalco* = after Johann Jakob von Kaup (1803-1873), German zoologist and author. **Lizard Buzzard** *K monogrammicus* (L) = with a single mark, signed, referring to the mark on the throat. Genus *Melierax* (Gr) = melodious hawk. **Gabar Goshawk** *M gabar* = name given by Levaillant; probably of Hottentot origin, meaning uncertain. Genus *Accipiter* (L) = a hawk. **African Goshawk** *A tachiro* (Gr) = swift. **Shikra** *A badius* (L) = reddish-brown. **Little Sparrowhawk** *A minullus* (L) = very small. **Ovambo Sparrowhawk** *A ovampensis* (L) = from Ovampo (= Ovamboland).
Lizard Buzzard, Autour unibande (F), Sperberbussard (G), Gavião-papa-lagartos (Por), Rikhozi (Tso), uKlebe (Z). **Gabar Goshawk**, Autour gabar (F), Gabarhabicht (G), Gavião-palrador (Por), Mamphoko (S-So), Xikwhezana (Tso). **African Goshawk**, Autour tachiro (F), Afrikanischer Sperber, Tachirosperber (G), Açor-africano (Por), [Fiolo-ea-meru](S-So), iKlebe/iMvumvuyane (Z). **Shikra**, Épervier shikra (F), Schikra (G), Gavião-chicra (Por), [Fiolo-'malisakhana] (S-So). **Little Sparrowhawk**, Épervier minule (F), Zwergsperber (G), Gavião-pequeno (Por), Ukhetshana (X), uMqwayini (Z). **Ovambo Sparrowhawk**, Épervier de l'Ovampo (F), Ovambosperber (G), Gavião do Ovambo (Por).

Plate 40
Black Sparrowhawk *A melanoleucus* (Gr) = black and white. **Dark Chanting Goshawk** *M metabates* (L) = changed, ie presumably meaning different from *M. canorus*. Genus *Aquila* (L) = an eagle. **African Hawk-Eagle** *A spilogaster* (L) = spotted underside. **Ayres's Hawk-Eagle** *A ayresii* (L) = after Thomas Ayres (1828-1918), naturalist and collector in S Africa.
Black Sparrowhawk, Autour noir (F), Mohrenhabicht, Trauerhabicht (G), Açor-preto (Por). **Dark Chanting Goshawk**, Autour sombre (F), Graubürzel-Singhabicht, Dunkler Grauflügelhabicht (G), Açor-cantor-escuro (Por). **African Hawk-Eagle**, Aigle fascié (F), Habichtsadler (G), Águia-dominó (Por), Ghama (Tso). **Ayres's Hawk-Eagle**, Aigle d'Ayres (F), Fleckenadler (G), Águia de Ayres (Por).

Plate 41
Genus *Polyboroides* (L) = like *Polyborus*; alludes to the resemblance of the African Harrier-Hawk to the Crested Caracara *Polyborus plancus* of the Neotropics. **African Harrier-Hawk** *P typus* (L) = the type, typical of the genus. Genus *Aquila* (L) = an eagle. **Tawny Eagle** *A rapax* (L) = rapacious. **Steppe Eagle** *A nipalensis* (L) = from Nepal. **Lesser Spotted Eagle** *A pomarina* (L) = after Type locality. **Wahlberg's Eagle** *A wahlbergi* (L) = after Johan Wahlberg (1810-1856), Swedish collector who worked in the Cape from 1838.
African Harrier-Hawk, Gymnogène d'Afrique (F), Schlangensperber, Höhlenweihe (G), Secretário-pequeno (Por), Seitlhoaeleli (S-So). **Tawny Eagle**, Aigle ravisseur (F), Raubadler (G), Ntshukôbôkôbô (N-So), Águia-fulva (Por), Ntsu (S-So), Ghama (Generic for eagles) (Tso), Ukhozi (X). **Steppe Eagle**, Aigle des steppes (F), Steppenadler (G), Águia-das-estepes (Por), Ukhozi (X). **Lesser Spotted Eagle**, Aigle pomarin (F), Schreiadler (G),

Águia-pomarina (Por). **Wahlberg's Eagle**, Aigle de Wahlberg (F), Wahlbergs Adler (G), Águia de Wahlberg (Por), Ghama (Tso).

Plate 42
Verreaux's Eagle *A verreauxii* (L) = after the two brothers Jules (1808–1873) and Edouard (1810–1868) Verreaux, French collectors who worked at the Cape. Genus *Polemaetus* Gr) = war-like eagle. **Martial Eagle** *P bellicosus* (L) = war-like. Genus *Lophaetus* (Gr) = a crested eagle. **Long-crested Eagle** *L occipitalis* (L) = of the back of the head. Genus *Stephanoaetus* (Gr) = a crowned eagle. **African Crowned Eagle** *S coronatus* (L) = crowned. **Secretarybird**. Family Sagittariidae. **Genus** *Sagittarius* (L) = an archer. **Secretarybird** *S serpentarius* (L) = pertaining to a snake; reference to diet.
Verreauxs' Eagle, Aigle de Verreaux (F), Felsenadler, Kaffernadler (G), Águia-preta (Por), Mojalipela/Moja-lipela/ Seoli/Ntsu (S-So), Gama (Tso), Ukhozi/Untsho (X), uKhozi (Z). **Martial Eagle**, Aigle martial (F), Kampfadler (G), Águia-marcial (Por), Rikhozi (Tso), Ukhozi (X), isiHuhwa/uKhozi (Z). **Long-crested Eagle**, Aigle huppard (F), Schopfadler (G), Águia-de-penacho (Por), Masworhimasworhi (Tso), Isiphungu-phungu/Uphungu-phungu (X), isiPhungumangathi (Z). **African Crowned Eagle**, Aigle couronné (F), Kronenadler (G), Águia-coroada (Por), Ukhozi (X), isiHuhwa (Z). **Secretarybird**, Messager sagittaire (F), Sekretär (G), Thlame (N-So), Secretário (Por), Koto-li-peli/Lekheloha/'Mamolangoane (S-So), Mampfana (Tso), Ingxangxosi (X), iNtungunono (Z).

Plate 43
Falcons. Family Falconidae. Genus *Falco* (L) = a falcon, derived from *falx* (L) = a sickle (referring to hooked talons or distinctive wing shape). **Dickinson's Kestrel** *F dickinsoni* = after Dr John Dickinson (1832-1863), British doctor and missionary to Malawi. **Amur Falcon** *F amurensis* = after Amuria or Amurland, the drainage area of the Amur River between the Russia Federation and China. **Eurasian Hobby** *F subbuteo* (L) = related to a buzzard. **Dickinson's Kestrel**, Faucon de Dickinson (F), Schwarzrückenfalke (G), Falcão de Dickinson (Por). **Amur Falcon**, Faucon de l'Amour (F), Amur-Rotfußfalke (G), Falcão-de-pés-vermelhos-oriental (Por), Seotsanyana (S-So), Kavakavana/Xikavakava (Tso). **Eurasian Hobby**, Faucon hobereau (F), Baumfalke (G), Ógea-europeia (Por), Phakoe (S-So), Rigamani/Rikhozi (Tso).

Plate 44
Lesser Kestrel *F naumanni* = after Johann Friedrich Naumann (1780-1857), ornithologist and author. **Rock Kestrel** *F rupicolus* (L) = rock-dweller. **Peregrine Falcon** *F peregrinus* (L) = stranger or wanderer. **Lanner Falcon** *F biarmicus* (L) = origin obscure; literally 'two-armed', perhaps with reference to notches on each side of bill. **Lesser Kestrel**, Faucon crécerellette (F), Rötelfalke (G), Peneireiro-das-torres (Por), Seotsanyana (S-So). **Rock Kestrel**, Faucon crécerelle (F), Turmfalke (G), Peneireiro-vulgar (Por), Seotsanyana (S-So), Kavakavana/ Xikavakava (Tso), Intambanane/Uthebe-thebana (X), uMathebeni/uTebetebana (Z). **Peregrine Falcon**, Faucon pèlerin (F), Wanderfalke (G), Falcão-peregrino (Por), Leubane/Phakoe (S-So), Rigamani/Rikhozi (Tso), Ukhetshe (X), uHeshe (Z). **Lanner Falcon**, Faucon lanier (F), Lannerfalke (G), Pekwa (N-So), Falcão-alfaneque (Por), Phakoe (S-So), Rikhozi (Tso), Ukhetshe (X), uHeshe (Z).

Plate 45
Grebes. Family Podicipedidae. Genus *Tachybaptus* (Gr) = swift dipper (diver). **Little Grebe** *T ruficollis (L)* = rufous or red necked. **Darters**. Family Anhingidae. Genus *Anhinga* (Amazonian dialect) = water turkey. **African Darter** *A rufa* (L) = ruddy, referring to chestnut back. **Cormorants**. Family Phalacrocoracidae. Genus *Phalacrocorax* (L) = cormorant, from Gr = bald raven. **Reed Cormorant** *P africanus* (L) = from Africa. **Whitebreasted Cormorant** *P lucidus* (L) = clear, bright.
Little Grebe, Grèbe castagneux (F), Zwergtaucher (G), Mergulhão-pequeno (Por), Thoboloko (S-So), Unolwilwilwi/ Unonyamembi (X). **African Darter**, Anhinga d'Afrique (F), Schlangenhalsvogel (G), Mergulhão-serpente (Por), Timeletsane-lalanoha (S-So), Gororo/Nyakolwa (Tso), Ivuzi (X). **Reed Cormorant**, Cormoran africain (F), Riedscharbe (G), Corvo-marinho-africano (Por), Timeletsane-ntšo (S-So), Ugwidi (X), iPhishamanzi/uLondo (Z). **White-breasted Cormorant**, Cormoran à poitrine blanche (F), Weißbrustkormoran (G), Corvo-marinho-defaces-brancas (Por), Timeletsane-botha (S-So), Ngulukwani (Tso), Ugwidi (X), iWonde (Z).

Plate 46
Egrets, Herons and Bitterns. Family Ardeidae. Genus *Egretta* (L) = from French *aigrette*, (L) = an egret or small heron. **Little Egret** *E garzetta* (It) = Italian name for small white egret. **Yellow-billed Egret** *E intermedia* (L) = intermediate, referring to the size, ie between Great Egret and Little Egret. **Great Egret** *E alba* (L) = white. Genus *Bubulcus* (L) = a cow-herd. **Cattle Egret** *B ibis* (Gr) = an ibis, generally a wading bird. Genus *Ardeola* (L) = a small heron. **Squacco Heron** *A ralloides* (L) = like a rail.
Little Egret, Aigrette garzette (F), Seidenreiher (G), Garça-branca-pequena (Por), Leholosiane (S-So),

216 *Etymology and Foreign Bird Names*

iNgekle (Z). **Yellow-billed Egret**, Héron à bec jaune (F), Edelreiher, Mittelreiher (G), Garça-branca-intermédia (Por), Leholosiane (S-So), iNgekle (Z). **Great Egret**, Grande Aigrette (F), Silberreiher (G), Garça-branca-grande (Por), Leholosiane (S-So), iLanda (Z). **Cattle Egret**, Héron garde-boeufs (F), Kuhreiher (G), Madšadipere (N-So), Garça-boieira (Por), Leholosiane/Leholotsiane (S-So), Dzandza/Munyangana/Muthecana/Nyonimahlopi (Tso), Ilanda (X), iLanda/inGevu/umLindankomo (Z). **Squacco Heron**, Crabier chevelu (F), Rallenreiher (G), Garça-caranguejeira (Por), Kokolofitoe (S-So).

Plate 47
Black Heron *E ardesiaca* (L) = slate-coloured. Genus *Ardea* (L) = a heron. **Grey Heron** *A cinerea* (L) = grey. **Black-headed Heron** *A melanocephala* (Gr) = black head. **Goliath Heron** *A goliath* (L) = large, giant. **Purple Heron** *A purpurea* (L) = purple.
Black Heron, Aigrette ardoisée (F), Glockenreiher (G), Garça-preta (Por), iKuwela (Z). **Grey Heron**, Héron cendré (F), Graureiher (G), Garça-real (Por), Kokolofitoe [-putsoa](S-So), Isikhwalimanzi / Ukhwalimanzi (X), uNokilonki (Z). **Black-headed Heron**, Héron mélanocéphale (F), Schwarzkopfreiher (G), Garça-de-cabeça-preta (Por), Kokolofitoe [-hloontšo](S-So), Isikhwalimanzi / Ukhwalimanzi (X), uNokilonki (Z). **Goliath Heron**, Héron goliath (F), Goliathreiher (G), Garça-gigante (Por), Kokolofitoe [-kholo](S-So), uNozalizingwenyana (Z). **Purple Heron**, Héron pourpré (F), Purpurreiher (G), Garça-vermelha (Por), Kokolofitoe [-lalatšehla](S-So), Rikolwa (Tso), Ucofuza/Undofu (X).

Plate 48
Genus *Butorides* (L) = bittern-like. **Green-backed Heron** *B striata* (L) = striped. Genus *Nycticorax* (Gr) = night raven. **Black-crowned Night-Heron** *N nycticorax* (Gr) = raven of the night. Genus *Gorsachius* = after *goi-sagi*, the Japanese name for Black-crowned Night-Heron. **White-backed Night-Heron** *G leuconotus* (Gr) = white back. Genus *Ixobrychus* (Gr) = from *ixos* = reed, and *brychus* = roar, referring to habitat and grunting calls. **Little Bittern** *I minutus* (L) = small. **Dwarf Bittern** *I sturmii* (L) = after Johan Heinrich Christian Sturm (1805-1862), bird artist and naturalist.
Green-backed Heron, Héron strié (F), Mangrovereiher (G), Garça-de-dorso-verde (Por). **Black-crowned Night-Heron**, Bihoreau gris (F), Nachtreiher (G), Garça-nocturna (Por), Kokolofitoe (S-So), uSiba (Z). **White-backed Night-Heron**, Bihoreau à dos blanc (F), Weißrücken-Nachtreiher (G), Garça-nocturna-de-dorso-branco (Por). **Little Bittern**, Blongios nain (F), Zwergrohrdommel (G), Garçote-comum (Por), Khoitinyane (S-So), Ihashe (X). **Dwarf Bittern**, Blongios de Sturm (F), Sturms Zwergrohrdommel (G), Garçote-anão (Por).

Plate 49
Ibises and Spoonbills. Family Threskiornithidae. Genus *Plegadis* (Gr) = a sickle. **Glossy Ibis** *P falcinellus* (L) = small and curved. **Hamerkop.** Family Scopidae**.** Genus *Scopus* (L) = a broom or brush and alludes to the bird's crest or (Gr) = to search, examine or consider. *S umbretta* (L) = shade, or shady, for the dusky brown colour, or for the crest and bill which give the head an umbrella-like appearance. **Genus** *Bostrychia* (Gr) = curved, referring to the curved beak. **Hadeda Ibis** *B hagedash* (L) = onamatopoeic rendering of the bird's repetitive call. Genus *Threskiornis* (Gr) = a religious bird. **African Sacred Ibis** *T aethiopicus* (L) = of Ethiopia, ie Africa south of the Sahara.
Glossy Ibis, Ibis falcinelle (F), Brauner Sichler (G), Ibis-preto (Por). **Hamerkop**, Ombrette africaine (F), Hammerkopf (G), Pássaro-martelo (Por), Mamasianoke/Masianoke (S-So), Mandonzwana (Tso), Uqhimngqoshe/Uthekwane (X), uThekwane (Z). **Hadeda Ibis**, Ibis hagedash (F), Hagedasch-Ibis (G), Singanga (Por), Lengaangane/Lengangane (S-So), Xikohlwa hi jambo (Tso), Ing'ang'ane (X), iNkankane (Z). **African Sacred Ibis**, Ibis sacré (F), Heiliger Ibis (G), Ibis-sagrado (Por), Lehalanyane/Leholotsoane (S-So), N'wafayaswitlangi (Tso), umXwagele (Z).

Plate 50
Flamingos. Family Phoenicopteridae. Genus *Phoenicopterus* (Gr) = bright red or scarlet wing. **Greater Flamingo** *P ruber* (L) = red. **Lesser Flamingo** *P minor* (L) = small. Genus *Platalea* (L) = the Spoonbill *P. leucorodia*. **African Spoonbill** *P alba* (L) = white. **Pelicans.** Family Pelecanidae. Genus *Pelecanus* (L and Gr) = a pelican. **Great White Pelican** *P onocrotalus* (L) = a pelican.
Greater Flamingo, Flamant rose (F), Flamingo (G), Flamingo-comum (Por), uKholwase/uNondwebu (Z). **Lesser Flamingo**, Flamant nain (F), Zwergflamingo (G), Flamingo-pequeno (Por), uNondwebu (Z). **African Spoonbill**, Spatule d'Afrique (F), Afrikanischer Löffler (G), Colhereiro-africano (Por), [Molomo-khaba](S-So), iNkenkane/isiXulamasele (Z). **Great White Pelican**, Pélican blanc (F), Rosapelikan (G), Pelicano-branco (Por), Gumbula/Khungulu/Manawavembe/Xilandzaminonga (Tso), Ingcwanguba (X), iFuba/iKhungula/iVuba (Z).

Plate 51
Storks. Family Ciconiidae. Genus *Anastomus* (Gr) = a coming together, or bringing to a point, referring to bill. **African Openbill** *A lamelligerus* (L) = a small thin metal plate; in reference to flattened feather shafts, especially

on underparts. Genus *Ciconia* (L) = the White Stork *C ciconia*. **Black Stork** *C nigra* (L) = black. **Abdim's Stork** *C abdimii* = after El Arnaut Abdim Bey (1780-1827), Egyptian governor of the Wadi Halfa area of Sudan, 1821-1825. **Woolly-necked Stork** *C episcopus* (L) = a bishop, in reference to black 'skullcap'. **African Openbill**, Bec-ouvert africain (F), Klaffschnabel (G), Bico-aberto (Por), Mukyindlopfu (Tso), isiQhophamnenke (Z). **Black Stork**, Cigogne noire (F), Schwarzstorch (G), Cegonha-preta (Por), Mokoroane (S-So), Unocofu (X). **Abdim's Stork**, Cigogne d'Abdim (F), Abdimsstorch, Regenstorch (G), Cegonha de Abdim (Por), Lekololoane/Mokoroane/Roba-re-bese (S-So). **Woolly-necked Stork**, Cigogne épiscopale (F), Wollhalsstorch (G), Cegonha-episcopal (Por), isiThandamanzi (Z).

Plate 52
White Stork *C ciconia* (L) = the stork. Genus *Mycteria* (Gr) = nose, or snout. **Yellow-billed Stork** *M ibis* (Gr) = an ibis. Genus *Ephippiorhynchus* (Gr) = saddle-billed, referring to frontal shield. **Saddle-billed Stork** *E sengalensis* (L) = from Senegal. Genus *Leptoptilos* (Gr) = thin, slender plumes. **Marabou Stork** *L crumeniferus* (L) = a leather pouch, referring to inflatable air sac.
White Stork, Cigogne blanche (F), Weißstorch (G), Leakaboswana le Lešweu (N-So), Cegonha-branca (Por), Mokoroane/Mokotatsie (S-So), Gumba/Ntsavila/Xaxari (Tso), Ingwamza/Unowanga (X), uNogolantethe/ uNowanga (Z). **Yellow-billed Stork**, Tantale ibis (F), Nimmersatt (G), Cegonha-de-bico-amarelo (Por), Mokotatsie (S-So). **Saddle-billed Stork**, Jabiru d'Afrique (F), Sattelstorch (G), Jabiru (Por), Ngwamhlanga (Tso). **Marabou Stork**, Marabout d'Afrique (F), Marabu (G), Mmakaitšimeletša (N-So), Marabu (Por), Ghumba (Tso).

Plate 53
Old World Orioles. Family Oriolidae. Genus *Oriolus* (L) = golden or yellow. **Eurasian Golden Oriole** *O oriolus*. **Black-headed Oriole** *O larvatus* (L) = masked. **Crested Flycatchers and Paradise-Flycatchers**. Family Monarchidae. Genus *Terpsiphone* (Gr) = delightful voice. **African Paradise-Flycatcher** *T viridis* (L) = green, presumably after greenish-tinged head. **Drongos**. Family Dicruridae. Genus *Dicrurus* (Gr) = two tailed, ie forked. **Fork-tailed Drongo** *D adsimilis* (L) = like or similar, refers to the all black plumage. Genus *Trochocercus* (Gr) derived from *trochos* = a wheel or disk, and *cercus* = tail, with reference to habit of fanning tail. **Blue-mantled Crested-Flycatcher** *T cyanomelas* (Gr) = dark blue-black.
Eurasian Golden Oriole, Loriot d'Europe (F), Europäischer Pirol (G), Papa-figos-europeu (Por), Sebabole-hetlantšo (S-So), Ndukuzani (Tso). **Black-headed Oriole**, Loriot masqué (F), Maskenpirol (G), Papa-figos-de-cabeça-preta (Por), Phamahumu/Phamanyarhi (Tso), Umkro/Umqokolo (X), umBhicongo/umQoqongo (Z). **African Paradise-Flycatcher**, Tchitrec d'Afrique (F), Paradiesschnäpper (G), Mmakgwadi (N-So), Papa-moscas-do-paraíso (Por), [Kapantsi-ea-meru](S-So), Nglhazi (Tso), Ujejane/Unomaphelana (X), iNzwece/uVe (Z). **Fork-tailed Drongo**, Drongo brillant (F), Trauerdrongo, Gabelschwanzdrongo (G), Theko (N-So), Drongo-de-cauda-forcada (Por), Mantengu (Tso), Intengu (X), iNtengu (Z). **Blue-mantled Crested-Flycatcher**, Tchitrec du Cap (F), Blaumantel-Schopfschnäpper (G), Papa-moscas-de-poupa (Por), Igotyi (X).

Plate 54
Bush-shrikes, Puffbacks, Tchagras, Boubous, Helmet-shrikes, Batises and Wattle-eyes. Family Malaconotinae. Genus *Nilaus* = anagram of *Lanius*, genus of true shrikes. **Brubru** *N afer* (L) = of Africa. Genus *Dryoscopus* (Gr) = a watcher from trees. **Black-backed Puffback** *D cubla* = named by Le Vaillant; of Bantu or Hottentot origin, onomatopoeic, the 'c' pronounced with a click. Genus *Tchagra* = a Levaillant name, onomatopoeic for grating call. **Black-crowned Tchagra** *T senegalus* (L) = from Senegal. **Brown-crowned Tchagra** *T australis* (L) = southern.
Brubru, Brubru africain (F), Brubru (G), Brubru (Por). **Black-backed Puffback**, Cubla boule-de-neige (F), Schneeballwürger (G), Picanço-de-almofadinha (Por), Phavomu (Tso), Intakembila/Unomaswana (X), iBhoboni (Z). **Black-crowned Tchagra**, Tchagra à tête noire (F), Senegaltschagra (G), Picanço-assobiador-de-coroa-preta (Por), Mghubhana lowu kulu (Tso), Umnguphane (X), umNguphane (Z). **Brown-crowned Tchagra**, Tchagra à tête brune (F), Damaratschagra (G), Picanço-assobiador-de-coroa-castanha (Por), Mghubhana lowu tsongo (Tso).

Plate 55
Genus *Laniarius* (L) = a butcher. **Tropical Boubou** *L aethiopicus* (L) = from Ethiopia. **Southern Boubou** *L ferrugineus* (L) = rusty. **Crimson-breasted Shrike** *L atrococcineus* (Gr) = black and scarlet. Genus *Telophorus* (Gr) = to carry far, referring to loud ringing calls of Bokmakierie. **Gorgeous Bush-Shrike** *T viridis* (L) = green.
Tropical Boubou, Gonolek d'Abyssinie (F), Orgelwürger, Tropischer Flötenwürger (G), Picanço-tropical (Por). **Southern Boubou**, Gonolek boubou (F), Flötenwürger (G), Picanço-ferrugíneo (Por), Hwilo/Samjukwa/ Xighigwa (Tso), Igqubusha (X), iBhoboni/iGqumasha (Z). **Crimson-breasted Shrike**, Gonolek rouge et noir (F), Rotbauchwürger, Reichsvogel, Kaiservogel (G), Picanço-preto-e-vermelho (Por), **Gorgeous Bush-Shrike**, Gladiateur quadricolore (F), Vierfarbenwürger (G), Picanço-quadricolor (Por), iNgongoni (Z).

Plate 56
Orange-breasted Bush-Shrike *T sulphureopectus* (L) = yellow breast. Genus *Malaconotus* (Gr) = refers to soft, fluffy back and rump feathers. **Grey-headed Bush-Shrike** *M blanchoti* = after M or P Blanchot, French Governor of Senegal ca 1790, but details sketchy. Genus *Prionops* (Gr) = from *priōn* = a saw, and *ops* = appearance; refers to ragged fleshy eye wattles. **White-crested Helmet-Shrike** *P plumatus* (L) = plumed, with reference to the crested forehead. **Retz's Helmet-Shrike** *P retzii* = after Anders Adolf Retzius (1796-1860), Swedish anatomist.
Orange-breasted Bush-Shrike, Gladiateur soufré (F), Orangewürger, Orangebrustwürger (G), Picanço-de-peito-laranja (Por), uHlaza (Z). **Grey-headed Bush-Shrike**, Gladiateur de Blanchot (F), Graukopfwürger, Riesenbuschwürger (G), Picanço-de-cabeça-cinzenta (Por), Umbhankro (X), uHlaza (Z). **White-crested Helmet-Shrike**, Bagadais casqué (F), Brillenwürger (G), Atacador-de-poupa-branca (Por), Urhiana (Tso). iPhemvu/uThimbakazane (Z). **Retz's Helmet-Shrike**, Bagadais de Retz (F), Dreifarbenwürger, Dreifarb-Brillenwürger (G), Atacador-de-poupa-preta (Por), Urhiana (Tso).

Plate 57
Genus *Batis* (Gr) = unidentified worm-eating bird mentioned by Aristotle. **Cape Batis** *B capensis* (L) = from Cape of Good Hope, S Africa. **Chinspot Batis** *B molitor* (L) = a miller, in reference to the bird's 'stone-rubbing' call. Genus *Platysteira* (Gr) = broad or flat keel of a ship. Refers to the shape of the bird's bill. **Black-throated Wattle-eye** *P peltata* (Gr and L) = small shield of light leather; refers to wattle around eye. **Typical Shrikes.** Family Laniidae. Genus *Lanius* (L) = a butcher. **Red-backed Shrike** *L collurio* (Gr) = an ancient name (*kollyrion*) for a thrush-like bird. **Lesser Grey Shrike** *L minor* (L) = small.
Cape Batis, Pririt du Cap (F), Kapschnäpper (G), Batis do Cabo (Por), Ingedle/Unongedle (X), uDokotela/umNqube (Z). **Chinspot Batis**, Pririt molitor (F), Weißflankenschnäpper (G), Batis-comum (Por), Ximgenngwamangwami (Tso), Undyola/Unondyola (X), umNqube (Z). **Black-throated Wattle-eye**, Pririt à gorge noire (F), Schwarzkehl-Lappenschnäpper (G), Olho-carunculado (Por). **Red-backed Shrike**, Pie-grièche écorcheur (F), Neuntöter (G), Picanço-de-dorso-ruivo (Por), Tšemeli (S-So), Mghubhana lokhulu (Tso), Ihlolo (X). **Lesser Grey Shrike**, Pie-grièche à poitrine rose (F), Schwarzstirnwürger (G), Picanço-pequeno (Por), Tšemeli (S-So), Juka/Rhiyani (Tso).

Plate 58
Crows and Ravens. Family Corvidae. Genus *Corvus* (L) = a raven or crow. **Pied Crow** *C albus* (L) = white. **White-necked Raven** *C albicollis* (L) = white neck. Genus *Corvinella* (L) = a small crow. **Magpie Shrike** *C melanoleuca* (Gr) = black and white. Genus *Eurocephalus* (Gr) = broad-headed. **Southern White-crowned Shrike** *E anguitimens* (L) = from *Anguitinens*, the constellation Ophiuchus, the serpent-holder. Refers to supposed prey.
Pied Crow, Corbeau pie (F), Schildrabe (G), Legokobu (N-So), Gralha-seminarista (Por), Mohakajane (S-So), Qigwana (Tso), Igwangwa/Igwarhube (X), iGwababa/uGwabayi (Z). **White-necked Raven**, Corbeau à nuque blanche (F), Geierrabe (G), Corvo-das-montanhas (Por), Lekhoaba/Moqukubi S-So), Gwavava/Ukuuku (Tso), Ihlungulu/Igrwababa/Umfundisi (X), iHubulu/iWabayi (Z). **Magpie Shrike**, Corvinelle noir et blanc (F), Elsterwürger (G), Picanço-rabilongo (Por), Ncilongi (Tso), umQonqotho (Z). **Southern White-crowned Shrike**, Eurocéphale à couronne blanche (F), Weißscheitelwürger (G), Picanço-de-coroa-branca (Por).

Plate 59
Cuckooshrikes. Family Campephagidae. Genus *Coracina* (Gr) = crow-like, or a little raven. **White-breasted Cuckooshrike** *C pectoralis* (Gr) = belonging to, or pertaining to the breast. Genus *Campephaga* (Gr) = caterpillar eating. **Black Cuckooshrike** *C flava* (L) = yellow. **Tits and Penduline-Tits.** Family Paridae. Genus *Anthoscopus* (Gr) = to look at or examine a flower. **Grey Penduline-Tit** *A caroli* = a corruption of 'Charles'; after Charles Andersson (1827-1867), Swedish naturalist and collector in Namibia. Genus *Parus* (L) = a titmouse. **Southern Black Tit** *P niger* (L) = black.
White-breasted Cuckooshrike, Échenilleur à ventre blanc (F), Weißbrust-Raupenfänger (G), Lagarteiro-cinzento-e-branco (Por). **Black Cuckooshrike**, Échenilleur à épaulettes jaunes (F), Kuckuckswürger (G), Rankwitšidi (N-So), Lagarteiro-preto (Por), Umthethi/Usinga Olumnyama (X), iNhlangu (Z). **Grey Penduline-Tit**, Rémiz minute (F), Kapbeutelmeise (G), Pássaro-do-algodão-cinzento (Por), Unogushana/Unothoyi (X). **Southern Black T**it, Mésange nègre (F), Mohrenmeise (G), Chapim-preto-meridional (Por), Xidzhavadzhava (Tso), Isicukujeje (X).

Plate 60
Swallows and Martins. Family Hirundinidae. Genus *Riparia* (L) = from *ripa*, a stream bank, ie frequenting stream or river-bank. **Sand Martin** *R riparia*. **Brown-throated Martin** *R paludicola* (L) = marsh-dweller. Genus *Hirundo* (L) = a swallow. **Barn Swallow** *H rustica* (L) = of the country. **Rock Martin** *H fuligula* (L) = dull brown or sooty.
Sand Martin, Hirondelle de rivage (F), Europäische Uferschwalbe (G), Andorinha-das-barreiras-europeia (Por).

Etymology and Foreign Bird Names

Brown-throated Martin, Hirondelle paludicole (F), Braunkehl-Uferschwalbe, Afrikanische Uferschwalbe (G), Andorinha-das-barreiras-africana (Por), Lekabelane/Sekatelane (S-So), Mbawulwana/Nyenga (Tso). **Barn Swallow**, Hirondelle rustique (F), Rauchschwalbe (G), Andorinha-das-chaminés (Por), Lefokotsane/'Malinakana/ Lekabelane (S-So), Nyengha (generic for swallows) (Tso), Inkonjane/Ucelizapholo/Udlihashe (X), iNkonjane (Z). **Rock Martin**, Hirondelle isabelline (F), Felsenschwalbe, Steinschwalbe (G), Andorinha-das-rochas-africana (Por), Lekabelane (S-So), Mbawulwana/Nyenga (Tso), Inkonjane/Unongubendala/Unongubende (X), iNhlolamvula (Z).

Plate 61
Genus *Pseudhirundo* (Gr & L) = false swallow. **Grey-rumped Swallow** *P griseopyga* (Gr) = grey rump. Genus *Delichon* = anagram of Greek *chelidon* = a swallow. **Common House-Martin** *D urbicum* (L) = from the city. **Wire-tailed Swallow** *H smithii* = after Sir Andrew Smith (1797-1872), Scottish herpetologist and ornithologist, and first Director of the South African Museum, Cape Town. **Pearl-breasted Swallow** *H dimidiata* (L) = divided through the middle; refers to incomplete throat band.
Grey-rumped Swallow, Hirondelle à croupion gris (F), Graubürzelschwalbe (G), Andorinha-de-rabadilha-cinzenta (Por). **Common House-Martin**, Hirondelle de fenêtre (F), Mehlschwalbe (G), Andorinha-dos-beirais (Por), Lekabelane (S-So). **Wire-tailed Swallow**, Hirondelle à longs brins (F), Rotkappenschwalbe (G), Andorinha-cauda-de-arame (Por), Mbawulwana/Nyenga (Tso), iNkonjane (Z). **Pearl-breasted Swallow** Hirondelle à gorge perlée (F), Perlbrustschwalbe (G), Andorinha-de-peito-pérola (Por), Lefokotsane (S-So), Mbawulwana/Nyenga (Tso), iNkonjane (Z).

Plate 62
Lesser Striped Swallow *H abyssinica* (L) = of Abyssinia (= Ethiopia). **Red-breasted Swallow** *H semirufa* (L) = half red. **Mosque Swallow** *H senegalensis* (L) = from Senegal. Genus *Psalidoprocne*, derived from *psalis* or *psalidos* (Gr) = pair of shears, and *progne* (L) = a swallow; reference to barbed edges of outer primaries. **Black Saw-wing** *P holomelaena* (Gr) = black all over.
Lesser Striped Swallow, Hirondelle striée (F), Kleine Streifenschwalbe (G), Andorinha-estriada-pequena (Por), Nyengha (Tso), Inkonjane (X), iNkonjane (Z). **Red-breasted Swallow**, Hirondelle à ventre roux (F), Rotbauchschwalbe (G), Andorinha-de-peito-ruivo (Por), Lekabelane (S-So), Nyengha leyi kulu (Tso). **Mosque Swallow**, Hirondelle des mosquées (F), Senegalschwalbe (G), Andorinha-das-mesquitas (Por), Nyengha (Tso). **Black Saw-wing**, Hirondelle du Ruwenzori (F), Sundevalls Sägeflügelschwalbe (G), Andorinha-preta (Por), Inkonjane/Unomalahlana (X).

Plate 63
Bulbuls and Nicators. Family Pycnonotidae. Genus *Pycnonotus* (Gr) = literally 'thick back', in reference to their thickly feathered backs. **Dark-capped Bulbul** *P tricolor* (L) = tri-coloured. Genus *Andropadus* (Gr) from *andros* = male, and *padus* = tree, = 'man of the trees'. **Sombre Greenbul** *A importunus* (L) = troublesome, referring to loud, persistent song. Genus *Chlorocichla* (Gr) = green thrush. **Yellow-bellied Greenbul** *C flaviventris* (L) = yellow belly. Genus *Nicator* (Gr) = a conqueror. **Eastern Nicator** *N gularis* (L) = of the throat. Genus *Phyllastrephus* (Gr) = from *phullon* = a leaf, and *strepho* = to bend or toss, referring to habit of scratching about in leaf litter. **Terrestrial Brownbul** *P terrestris* (L) = terrestrial, referring to ground-dwelling habit.
Dark-capped Bulbul, Bulbul tricolore (F), Gelbsteißbülbül, Graubülbül (G), Rankgwetšhe (N-So), Tuta-negra (Por), Bwoto/Chigwenhure/MugwetureHlakahlotoana (S-So), Bhokota (Tso), Ikhwebula (X), iPhothwe/iPogota (Z). **Sombre Greenbul**, Bulbul importun (F), Kap-Grünbülbül (G), Tuta-sombria (Por), Inkwili (X), iWili (Z). **Yellow-bellied Greenbul**, Bulbul à poitrine jaune (F), Gelbbrustbülbül (G), Tuta-amarela (Por), iBhada (Z). **Eastern Nicator**, Bulbul à tête brune (F), Bülbülwürger (G), Tuta-malhada (Por). **Terrestrial Brownbul**, Bulbul jaboteur (F), Laubbülbül (G), Tuta-da-terra (Por), Ikhalakandla/Ugwegwegwe/Umnqu (X).

Plate 64
Leaf-Warblers, Babblers and Warblers. Family Sylviidae. Genus *Sylvietta* (L) = diminutive of *Sylvia*, a warbler. **Long-billed Crombec** *S rufescens* (L) = reddish. Genus *Eremomela* (Gr) = a wilderness or desert song. **Yellow-bellied Eremomela** *E icteropygialis* (Gr) = yellow-rumped. **Green-capped Eremomela** *E scotops* (Gr) = dark eye or face, referring to dusky lores. **Burnt-necked Eremomela** *E usticollis* = *usti*, from *usticius* (L) = brown, produced by burning, and *collis*, from *collum* (L) = neck.
Long-billed Crombec, Crombec à long bec (F), Langschnabel-Sylvietta, Kurzschwanz-Sylvietta (G), Rabicurta-de-bico-comprido (Por), Nqcunu (Tso), iNdibilitshe (Z). **Yellow-bellied Eremomela**, Érémomèle à croupion jaune (F), Gelbbauch-Eremomela (G), Eremomela-de-barriga-amarela (Por). **Green-capped Eremomela**, Érémomèle à calotte verte (F), Grünkappen-Eremomela (G), Eremomela-de-barrete-verde (Por). **Burnt-necked Eremomela**, Érémomèle à cou roux (F), Rostkehl-Eremomela, Rostband-Eremomela (G), Eremomela-de-garganta-castanha (Por).

Plate 65
Genus *Bradypterus* (Gr) = slow wing, presumably referring to reluctance to flush. **Little Rush-Warbler** *B baboecala* = unknown, possibly from *babul* (L) = a babbler, probably with reference to call, or from *babax* (Gr) = a chatterer. Genus *Acrocephalus* (Gr) = pointed head, referring to head shape of the male when singing. **Sedge Warbler** *A schoenobaenus* (L) = reed-dwelling. **African Reed-Warbler** *A baeticatus* (L) = having brown (plumage). **Lesser Swamp-Warbler** *A gracilirostris* (L) = slender-billed. **Great Reed-Warbler** *A arundinaceus* (L) = reed-like.
Little Rush-Warbler, Bouscarle caqueteuse (F), Sumpfbuschsänger (G), Felosa-dos-juncos-africana (Por), Unomakhwane (X). **Sedge Warbler**, Phragmite des joncs (F), Schilfrohrsänger (G), Felosa-dos-juncos (Por). **African Reed-Warbler**, Rousserolle africaine (F), Gartenrohrsänger (G), Rouxinol-dos-caniços-africano (Por), Soamahlaka-sa-mefero (S-So). **Lesser Swamp-Warbler**, Rousserolle à bec fin (F), Kaprohrsänger (G), Rouxinol-pequeno-dos-pântanos (Por), Soamahlaka-ntšitšoeu (S-So). **Great Reed-Warbler**, Rousserolle turdoïde (F), Drosselrohrsänger (G), Rouxinol-grande-dos-caniços (Por), Soamahlaka-kholo (S-So).

Plate 66
Marsh Warbler *A palustris* (L) = swampy or marshy. Genus *Sylvia* (L) = from *sylvaticus*, woodland, referring to typical habitat. **Garden Warbler** *S borin* (It) = Italian name for the bird. Genus *Hippolais* (Gr) = a small warbler-like bird. **Olive-tree Warbler** *H olivetorum* (L) = of the olive trees. **Icterine Warbler** *H icterina* (L) = yellow. Genus *Phylloscopus* (Gr) = leaf-watcher, referring to leaf-gleaning behaviour. **Willow Warbler** *P trochilus* (Gr) = a small bird mentioned by Aristotle, identified by later writers with wren *Troglodytes*. *Trochilus* also means to twist or swivel, which would be appropriate, the birds twisting round thin twigs as they feed.
Marsh Warbler, Rousserolle verderolle (F), Sumpfrohrsänger (G), Felosa-palustre (Por). **Garden Warbler**, Fauvette des jardins (F), Gartengrasmücke (G), Felosa-das-figueiras (Por), Soamahlaka-sa-jarete (S-So). **Olive-tree Warbler**, Hypolaïs des oliviers (F), Olivenspötter (G), Felosa-das-oliveiras (Por). **Icterine Warbler**, Hypolaïs ictérine (F), Gelbspötter (G), Felosa-icterina (Por). **Willow Warbler**, Pouillot fitis (F), Fitis (G), Felosa-musical (Por), Timba Pilipili-sa-mabelete (S-So), Unothoyi (X).

Plate 67
Genus *Hyliota* (Gr) = living in woodland. **Southern Hyliota** *H australis* (L) = southern. Genus *Turdoides* (L) = like a thrush. **Arrow-marked Babbler** *T jardineii* = named after Sir William Jardine (1800-1874), naturalist and the editor of The Naturalist's Library from 1833-1845. Genus *Parisoma* derived from *Parus* (L) = body of a tit, ie tit-like in appearance. **Chestnut-vented Tit-Babbler** *P subcaeruleum* (L) = blue-grey coloured below. **White-eyes**. Family Zosteropidae. Genus *Zosterops* (Gr) = girdle and eye, referring to eye rings. **African Yellow White-eye** *Z senegalensis* (L) = from Senegal. **Cape White-eye** *Z virens* = green.
Southern Hyliota, Hyliote australe (F), Maschona-Hyliota (G), Papa-moscas-austral (Por). **Arrow-marked Babbler**, Cratérope fléché (F), Braundroßling (G), Zaragateiro-castanho (Por), Tlekedhwana (Tso), iHelkehle (Z). **Chestnut-vented Tit-Babbler**, Parisome grignette (F), Meisensänger (G), Felosa-chapim-dos-bosques (Por). **African Yellow White-eye**, Zostérops jaune (F), Senegalbrillenvogel (G), Olho-branco-amarelo (Por), Manqiti (Tso), umBicini/ uMehlwane (Z). **Cape White-eye**, Zostérops du Cap (F), Oranjebrillenvogel (G), Olho-branco do Cabo (Por), Setona-mahloana (S-So), Manqiti (Tso), Intukwane (X), umBicini/ uMehlwane (Z).

Plate 68
African warblers. Family Cisticolidae. Genus *Cisticola* (Gr) = a flowering shrub inhabitant, or (L) = a water tank or reservoir inhabitant, perhaps referring to Zitting Cisticola's bottle-shaped nest. *Cista* is also a basket of woven twigs, possibly also referring to their woven nests. **Red-faced Cisticola** *C erythrops* (Gr) = red eye. **Lazy Cisticola** *C aberrans* (L) = aberrant, departing from the usual. **Rattling Cisticola** *C chinianus* (Tsw) = Chue or Choo which was one of Burchell's collecting localities, or may have been derived from the name of the Cheyane Mt in NW Province. **Rufous-winged Cisticola** *C galactotes* (L) = milk white.
Red-faced Cisticola, Cisticole à face rousse (F), Rotgesicht-Zistensänger (G), Fuinha-de-faces-vermelhas (Por), **Lazy Cisticola**, Cisticole paresseuse (F), Smiths Zistensänger (G), Fuinha-preguiçosa (Por).). **Rattling Cisticola**, Cisticole grinçante (F), Rotscheitel-Zistensänger (G), Fuinha-chocalheira (Por), Mantsiyana (Tso), iNqoba (Z). **Rufous-winged Cisticola**, Cisticole roussâtre (F), Schwarzrücken-Zistensänger (G), Fuinha-de-dorso-preto (Por).

Etymology and Foreign Bird Names

Plate 69
Croaking Cisticola *C natalensis* (L) = after the settlement of Port Natal (Durban), Natal, South Africa. **Neddicky** *C fulvicapillus* (L) = tawny hair. **Zitting Cisticola** *C juncidis* (L) = of rushes, probably referring to some of the features of the habitat used by this species. **Desert Cisticola** *C aridulus* (L) = of dry places or desert, referring to the habitat used by this species.

Croaking Cisticola, Cisticole striée (F), Strichelzistensänger (G), Fuinha do Natal (Por), Matinti (Tso), Igabhoyi/ Ubhoyi-bhoyi (X), iBhoyi (Z), Matinti (Tso). **Neddicky**, Cisticole à couronne rousse (F), Brauner Zistensänger (G), Fuinha-de-cabeça-ruiva (Por), Motintinyane (S-So), Matinti (Tso), Incede (X), iNcede/uQoyi (Z). **Zitting Cisticola**, Cisticole des joncs (F), Zistensänger (G), Tangtang (N-So), Fuinha-dos-juncos (Por), Motintinyane (S-So), Matinti (Tso), Unonzwi (X), uNcede (Z). **Desert Cisticola**, Cisticole du désert (F), Kalahari-Zistensänger (G), Fuinha-do-deserto (Por), Motintinyane (S-So), Matinti (Tso).

Plate 70
Genus *Prinia* = the Javanese name for the Tawny-flanked Prinia *Prinia subflava blythi*. **Tawny-flanked Prinia** *P subflava* (L) = almost yellow. Genus *Camaroptera*, derived from *kamara* (Gr) = arched, vaulted, and *ptera* (Gr) = wings, ie arched wings. **Green-backed Camaroptera** *C brachyura* (Gr) = short tail. **Grey-backed Camaroptera** *C brevicaudata* (L) = short-tailed. Genus *Apalis* (Gr) = soft. **Yellow-breasted Apalis** *A flavida* (L) = pale yellow. **Rudd's Apalis** *A ruddi* = after Charles Rudd (1844-1916), who financed Grant's collecting trips in S Africa. Genus *Calamonastes* (Gr) = a reed singer. **Stierling's Wren-Warbler** *C stierlingi* = after N. Stierling (fl. 1901), German traveller and collector in Malawi and Tanzania.

Tawny-flanked Prinia, Prinia modeste (F), Rahmbrustprinie (G), Prínia-de-flancos-castanhos (Por), Matsinyani (Tso), Ungcuze (X). **Green-backed Camaroptera**, Camaroptère à tête grise (F), Meckergrasmücke (Grünrücken-Camaroptera, Graurücken-Camaroptera) (G), Felosa-de-dorso-verde (Por), Unomanyuku/Unome (X), iMbuzi-yehlathi/ umBuzana (Z). **Grey-backed Camaroptera**, Camaroptère à dos gris (F), Graurücken-Grasmücke (G). **Yellow-breasted Apalis**, Apalis à gorge jaune (F), Gelbbrust-Feinsänger (G), Apalis-de-peito-amarelo (Por), N'walanga/Xinyamukhwarani (Tso). **Rudd's Apalis**, Apalis de Rudd (F), Rudds Feinsänger (G), Apalis de Rudd (Por). **Stierling's Wren-Warbler**, Camaroptère de Stierling (F), Stierling-Bindensänger (G), Felosa de Stierling (Por), Xingede (Tso).

Plate 71
Larks and Sparrowlarks. Family Alaudidae Genus *Mirafra* = etymology undescribed. **Monotonous Lark** *M passerina* (L) = sparrow-like. **Rufous-naped Lark** *M africana* (L) = of Africa. **Flappet Lark** *M rufocinnamomea* (L) = reddish-cinnamon. Genus *Calendulauda* (L) = combination of 2 lark Genera, *Calendula* and *Alauda*. **Sabota Lark** *C sabota* (Tsw) = *sebotha* or *sebothê*, generic name for larks.

Monotonous Lark, Alouette monotone (F), Sperlingslerche (G), Cotovia-monótona (Por), Mapuluhweni (Tso). **Rufous-naped Lark**, Alouette à nuque rousse (F), Rotnackenlerche (G), Cotovia-de-nuca-vermelha (Por), Tsiroane (S-So), Mapuluhweni (Tso), Igwangqa/Iqabathule (X). **Flappet Lark**, Alouette bourdonnante (F), Baumklapperlerche, Zimtbaumlerche (G), Cotovia-das-castanholas (Por), uQaqashe (Z). **Sabota Lark**, Alouette sabota (F), Sabotalerche (G), Cotovia-sabota (Por), Urimakutata/Vhumakutata (Tso).

Plate 72
Genus *Pinarocorys* (Gr) = a dirty, lark-like bird. **Dusky Lark** *P nigricans* (L) = black, or swarthy. **Fawn-coloured Lark** *C africanoides* (New L) = like *C. africana*, Rufous-naped Lark. Genus *Eremopterix* (Gr) = desert bird, referring to arid habitats of most sparrowlarks. **Chestnut-backed Sparrowlark** *E leucotis* (Gr) = white ears. **Grey-backed Sparrowlark** *E verticalis* (L) = perpendicular, upright. Genus *Calandrella* (Gr) = a small lark. **Red-capped Lark** *C cinerea* (L) = ash-coloured.

Dusky Lark, Alouette brune (F), Drossellerche (G), Cotovia-sombria (Por), Xihelagadzi (Tso). **Fawn-coloured Lark**, Alouette fauve (F), Steppenlerche (G), Cotovia-cor-d'areia (Por), Mapuluhweni (Tso). **Chestnut-backed Sparrowlark**, Moinelette à oreillons blancs (F), Weißwangenlerche (G), Cotovia-pardal-de-dorso-castanho (Por), 'Maliberoane/'Mamphemphe (S-So). **Grey-backed Sparrowlark**, Moinelette à dos gris (F), Nonnenlerche (G), Cotovia-pardal-de-dorso-cinzento (Por). **Red-capped Lark**, Alouette cendrille (F), Rotscheitellerche (G), Cotovia-de-barrete-vermelho (Por), Thesta-balisana/Tsiroane (S-So), Intibane/Intutyane (X), umNtoli (Z).

Plate 73

Thrushes, Robins, Chats and Old World Flycatchers. Family Muscicapidae. Genus *Turdus* (L) = a thrush. **Kurrichane Thrush** *T libonyanus* (Tsw) = Lebonyana is a personal name in Tswana. Genus *Psophocichla* (Gr) = a noisy thrush. **Groundscraper Thrush** *P litsitsirupa* (Tswana) = ground scraper, onomatopoeic from Ietshutshuroopoo. Genus *Bradornis* (Gr) = a slow or sluggish bird. **Pale Flycatcher** *B pallidus* (L) = pallid, pale. Genus *Melaenornis* (Gr) = black bird. **Southern Black Flycatcher** *M pammelaina* (Gr) = altogether olive, for the entirely black plumage with a greenish gloss. **Marico Flycatcher** *B mariquensis* (L) = from Marico.
Kurrichane Thrush, Merle kurrichane (F), Rotschnabeldrossel (G), Tordo-chicharrio (Por), Mbyhiyoni (Tso), umuNswi (Z). **Groundscraper Thrush**, Merle litsipsirupa (F), Akaziendrossel (G), Tordo-de-peito-malhado (Por). **Pale Flycatcher**, Gobemouche pâle (F), Fahlschnäpper (G), Papa-moscas-pálido (Por). **Southern Black Flycatcher**, Gobemouche sud-africain (F), Drongoschnäpper (G), Papa-moscas-preto-meridional (Por), umMbesi (Z). **Marico Flycatcher**, Gobemouche du Marico (F), Maricoschnäpper (G), Papa-moscas do Marico (Por).

Plate 74

Genus *Sigelus* (L) = mute or silent, from Sigalion, Egyptian god of silence. **Fiscal Flycatcher** *S silens* (L) = silent. Genus *Muscicapa* (L) = fly catcher. **African Dusky Flycatcher** *M adusta* (L) = blackened or scorched, hence dusky. **Spotted Flycatcher** *M striata* (L) = striped. **Ashy Flycatcher** *M caerulescens* (L) = bluish or becoming blue. Genus *Myioparus*, derived from *muia* (Gr) a fly, and *parus* (L) a tit. **Grey Tit-Flycatcher** *M plumbeus* (L) = leaden grey.
Fiscal Flycatcher, Gobemouche fiscal (F), Würgerschnäpper (G), Papa-moscas-fiscal (Por), Icola (X). **African Dusky Flycatcher**, Gobemouche sombre (F), Dunkelschnäpper (G), Papa-moscas-sombrio (Por), Unomaphelana (X). **Spotted Flycatcher**, Gobemouche gris (F), Grauschnäpper (G), Papa-moscas-cinzento (Por), [Kapantsi-tubatubi](S-So). **Ashy Flycatcher**, Gobemouche à lunettes (F), Schieferschnäpper (G), Papa-moscas-azulado (Por). **Grey Tit-Flycatcher**, Gobemouche mésange (F), Meisenschnäpper (G), Para-moscas-de-leque (Por).

Plate 75

Genus *Cossypha* (Gr) = the European Blackbird *Turdus merula*, a singing bird. **Cape Robin-Chat** *C caffra* (L) = from Caffraria, ie Cape. **White-throated Robin-Chat** *C humeralis* (L) = of the shoulders, referring to the white wing coverts. **White-browed Robin-Chat** *C heuglini* = after Theodor von Heuglin (1824-1876). An ornithologist who made several journeys to NE Africa between 1850 and 1875. **Red-capped Robin-Chat** *C natalensis* (L) = after the settlement of Port Natal (Durban), Natal, S Africa. Genus *Cercotrichas* (Gr) = a long-tailed thrush. **Bearded Scrub-Robin** *C quadrivirgata* (L) = four-streaked, a reference to the alternating dark and light stripes on the face. **White-browed Scrub-Robin** *C leucophrys* (L) = white eyebrow.
Cape Robin-Chat, Cossyphe du Cap (F), Kaprötel (G), Cossifa do Cabo (Por), Sethoenamoru/Sethoena-moru/Setholo-moru (S-So), Ugaga (X), uGaga/umBhekle (Z). **White-throated Robin-Chat**, Cossyphe à gorge blanche (F), Weißkehlrötel (G), Cossifa-de-peito-branco (Por). **White-browed Robin-Chat**, Cossyphe de Heuglin (F), Weißbrauenrötel (G), Cossifa de Heuglin (Por). **Red-capped Robin-Chat**, Cossyphe à calotte rousse (F), Natalrötel (G), Cossifa do Natal (Por), Nyarhututu (Tso). **Bearded Scrub-Robin**, Agrobate à moustaches (F), Brauner Bartheckensänger (G), Rouxinol-do-mato-de-bigodes (Por). **White-browed Scrub-Robin**, Agrobate à dos roux (F), Weißbrauen-Heckensänger (G), Rouxinol-do-mato-estriado (Por), Mtsherhitani (Tso).

Plate 76

Genus *Cichladusa* (Gr) = a thrush. **Collared Palm-Thrush** *C arquata* (L) = arched (arcuate), referring to the black mark on the chest. Genus *Luscinia* (L) = the nightingale. **Thrush Nightingale** *L luscinia*. Genus *Saxicola* (L) = dwelling among stones. **African Stonechat** *S torquata* (L) = collared. Genus *Cercomela* (Gr) = a black tail. **Familiar Chat** *C familiaris* (L) = familiar. Genus *Myrmecocichla* (Gr) = an ant thrush. **Arnot's Chat** *M arnoti*. Named after David Arnot a law agent at Colesberg and advisor to the Griqua Chief N Waterboer, who collected fossil reptiles, mammals, birds and insects for the South African Museum between 1858 and 1868. Genus *Thamnolaea* (Gr) = a bush, and a thrush. **Mocking Cliff-Chat** *T cinnamomeiventris* (New L) = cinnamon-coloured belly.
Collared Palm-Thrush, Cichladuse à collier (F), Morgenrötel (G), Tordo-das-palmeiras-de-colar (Por). **Thrush Nightingale**, Rossignol progné (F), Sprosser (G), Rouxinol-russo (Por). **African Stonechat**, Tarier pâtre (F), Schwarzkehlchen (G), Cartaxo-comum (Por). Hlatsinyane/Tlhatsinyane (S-So), Ingcaphe/Isangcaphe (X), isAncaphela/isAnqawane/isiChegu (Z). **Familiar Chat**, Traquet familier (F), Rostschwanzschmätzer, Rostschwanz (G), Chasco-familiar (Por), Letlerenyane/Letleretsane (S-So), Isikretyane/Unongungu (X), umBexe (Z). **Arnot's Chat**, Traquet d'Arnott (F), Arnotschmätzer (G), Chasco de Arnot (Por), Mandlakeni (Tso). **Mocking Cliff-Chat**, Traquet à ventre roux (F), Rotbauchschmätzer (G), Chasco-poliglota (Por), iQumutsha-lamawa (Z).

Plate 77
Starlings, Mynas and Oxpeckers. Family Sturnidae. Genus *Lamprotornis* (Gr) = a bright or shining bird. **Cape Glossy Starling** *L nitens* (L) = shining, bright. **Meves's Starling** *L mevesii* = after Friederich Wilhelm Meves (1814-1891). Curator of the Royal Museum, Stockholm, under whom Wahlberg studied. **Greater Blue-eared Starling** *L chalybaeus* (L) = steely, or steel blue. The Chalybes were an ancient nation of Asia Minor, famous for their ability in working iron. **Burchell's Starling** *L australis* (L) = southern.
Cape Glossy Starling, Choucador à épaulettes rouges (F), Rotschulter-Glanzstar (G), Legodi (N-So), Estorninho-metálico (Por), Leholi-piloane(S-So), Kwezu leri tsongo (Tso), Inyakrili/Inyakrini (X), iKhwezi/iKhwinsi (Z).
Meves's Starling, Choucador de Meves (F), Meves-Glanzstar (G), Estorninho-rabilongo (Por). **Greater Blue-eared Starling**, Choucador à oreillons bleus (F), Grünschwanz-Glanzstar (G), Estorninho-grande-d'orelha-azul (Por), Kwezu leri tsongo (Tso). **Burchell's Starling**, Choucador de Burchell (F), Riesenglanzstar, Glanzelstar (G), Estorninho de Burchell (Por), Kwezu leri kulu (Tso).

Plate 78
Genus *Onychognathus* (Gr) = a finger nail, talon, claw or hoof and the jaw. The bill resembles a claw or nail. **Red-winged Starling** *O morio* (Gr) = black; or (L) = contraction of 'mormorion' a dark brown stone, perhaps alluding to the red-brown wings. Genus *Cinnyricinclus* (Gr) = a shining thrush. **Violet-backed Starling** *C leucogaster* (Gr) = white belly. Genus *Buphagus* (L) = ox-eating. **Yellow-billed Oxpecker** *B africanus* (L) = from Africa. **Red-billed Oxpecker** *B erythrorhynchus* (Gr) = red bill. Genus *Acridotheres* (Gr) = locust or grasshopper hunting. Genus *Creatophora* (Gr) = to carry flesh. For the loose, fleshy wattles that the males 'carry' in the breeding season. **Wattled Starling** *C cinerea* (L) = grey, describing the plumage.
Red-winged Starling, Rufipenne morio (F), Rotschwingenstar (G), Estorninho-d'asa-castanha (Por), Letšoemila/ Letšomila (S-So), Isomi (X), iNsomi/iSomi (Z). **Violet-backed Starling**, Spréo améthyste (F), Amethystglanzstar (G), Estorinho-de-dorso-violeta (Por), Xinwavulombe (Tso). **Yellow-billed Oxpecker**, Piqueboeuf à bec jaune (F), Gelbschnabel-Madenhacker (G), Pica-bois-de-bico-amarelo (Por), iHlalankomo/iHlalanyathi (Z). **Red-billed Oxpecker**, Piqueboeuf à bec rouge (F), Rotschnabel-Madenhacker (G), Pica-bois-de-bico-vermelho (Por), Yandhana (Tso), Ihlalanyathi (X), iHlalankomo/iHlalanyathi (Z). **Wattled Starling**, Étourneau caronculé (F), Lappenstar (G), Estorninho-carunculado (Por), Leholi (S-So), Kwezu elimhlope (Tso), Unowambu/Uwambu (X), iMpofazana (Z).

Plate 79
Sunbirds. Family Nectariniidae. Genus *Chalcomitra* (Gr) = copper or bronze head-band or cap. **Scarlet-chested Sunbird** *C senegalensis* (L) = from Senegal. **Amethyst Sunbird** *C amethystina* (L) = amethyst-coloured. Genus *Hedydipna* (Gr) = sweet feeding or sweet-eating. **Collared Sunbird** *H collaris* (L) = collared. Genus *Cinnyris* (Gr) = shining. **White-bellied Sunbird** *C talatala* (Tsw) = greenish, referring to colour of head and mantle. **Marico Sunbird** *C mariquensis* = Marico district, NW Province, by inference. *mariquensis* = from Marico district.
Scarlet-chested Sunbird, Souimanga à poitrine rouge (F), Rotbrust-Glanzköpfchen, Rotbrust-Nektarvogel (G), Beija-flor-de-peito-escarlate (Por), Nwapyopyamhanya (Tso). **Amethyst Sunbird**, Souimanga améthyste (F), Amethyst-Glanzköpfchen (G), Beija-flor-preto (Por), Nwapyopyamhanya (Tso), Ingcungcu (X). **Collared Sunbird**, Souimanga à collier (F), Waldnektarvogel (G), Stahlnektarvogel (G), Beija-flor-de-colar (Por), Nwapyopyamhanya (Tso), Inqathane (X), iNgqwathane/iNqwathane/iNtonso (Z). **White-bellied Sunbird**, Souimanga à ventre blanchâtre (F), Weißbauch-Nektarvogel (G), Beija-flor-de-barriga-branca (Por), Nwapyopyamhanya (Tso). **Marico Sunbird**, Souimanga de Mariqua (F), Bindennektarvogel (G), Beija-flor de Marico (Por), Nwapyopyamhanya (generic for sunbirds) (Tso).

Plate 80
Weavers, Queleas and Widowbirds. Family Ploceidae. Genus *Ploceus* (Gr) = a weaver. **Lesser Masked-Weaver** *P intermedius* (L) = intermediate in size. **Spectacled Weaver** *P ocularis* (L) = of the eye, referring to distinctive markings around eye. **Golden Weaver** *P xanthops* (Gr) = golden eye. **Southern Masked-Weaver** *P velatus* (L) = masked, ie covered or partly concealed from view. **Village Weaver** *P cucullatus* (L) = hooded.
Lesser Masked-Weaver, Tisserin intermédiaire (F), Cabanisweber (G), Tecelão-pequeno-de-mascarilha (Por), Ndzheyana (Tso), umZwingili (Z). **Spectacled Weaver**, Tisserin à lunettes (F), Brillenweber (G), Tecelão-de-lunetas (Por), Sowa (Tso), Ikreza (X), iGelegekle/iGeleja (Z). **Golden Weaver**, Tisserin safran (F), Großer Goldweber, Safranweber (G), Tecelão-dourado (Por), Sowa (Tso), iHlokohloko (Z). **Southern Masked-Weaver**, Tisserin à tête rousse (F), Maskenweber, Schwarzstirnweber (G), Tecelão-de-máscara (Por), Letolopje/Thaha (S-So), Ndzheyana (Tso), Ihobo-hobo (X), iHlokohloko (Z). **Village Weaver**, Tisserin gendarme (F), Textor, Dorfweber (G), Thaga (N-So), Tecelão-malhado (Por), Letholopje (S-So), Ndzheyana (generic for weavers) (Tso), Ihobo-hobo (X), iHlokohloko (Z).

Etymology and Foreign Bird Names

Plate 81

Genus *Bubalornis* (L) = a buffalo bird. **Red-billed Buffalo-Weaver** *B niger* (L) = black. Genus *Quelea* (L) = from *qualea* = quail, perhaps referring to streaked upper parts and buzzing flight. **Red-billed Quelea** *Q quelea* = probably from the English quail, which has the variant spellings *qualia* and *qualea*. Genus *Euplectes* woven, referring to woven nests. Genus *Amblyospiza* (Gr) = from *amblus* = blunt, and *spiza* = finch, referring to heavy bill. **Thick-billed Weaver** *A albifrons* (L) = white forehead. Genus *Anaplectes* (L) = like a true weaver. **Red-headed Weaver** *A melanotis* = black ear.
Red-billed Buffalo-Weaver, Alecto à bec rouge (F), Büffelweber (G), Tecelão-de-bico-vermelho (Por), Xighonyombha (Tso). **Red-billed Quelea**, Travailleur à bec rouge (F), Blutschnabelweber (G), Lerwerwe (N-So), Quelea-de-bico-vermelho (Por), Thaha (S-So), Ndzheyana (Tso). **Thick-billed Weaver**, Amblyospize à front blanc (F), Weißstirnweber (G), Tecelão-de-bico-grosso (Por). **Red-headed Weaver**, Tisserin écarlate (F), Scharlachweber (G), Tecelão-de-cabeça-vermelha (Por), Ndzheyana ya nhloko ya ka phsuku (Tso).

Plate 82

Genus *Euplectes* woven, referring to woven nests. **Yellow-crowned Bishop** *E afer* (L) = of Africa. **Southern Red Bishop** *E orix* (L) = 'oryza'= rice, presumably referring to habit of feeding in rice paddies. **Red-collared Widowbird** *E ardens* L) = glowing (Type description based on an inaccurate illustration which shows a red belly spot, described as 'like a glowing coal'). **White-winged Widowbird** *E albonotatus* (L) = white marked. **Fan-tailed Widowbird** *E axillaries* (L) = of the shoulder.
Yellow-crowned Bishop, Euplecte vorabé (F), Tahaweber, Napoleonweber (G), Cardeal-tecelão-amarelo (Por), Thaha-pinyane/Thaha-tsehle/Tsehle/Thaha (S-So), Mantunje/Xikhungumala (Tso). **Southern Red Bishop**, Euplecte ignicolore (F), Oryxweber (G), Thagalehlaka (N-So), Cardeal-tecelão-vermelho (Por), Khube/Thaha-khube/Thaha-khubelu (S-So), Intakomlilo/Ucumse/Umlilo (X), iBomvana/iNtakansinsi/isiGwe (Z). **Red-collared Widowbird**, Euplecte veuve-noire (F), Schildwida (G), Viúva-de-colar-vermelho (Por), Molepe/Thaha/Tjobolo (S-So), Intakazana/Ujobela (X), iNtaka/uJojo (Z). **White-winged Widowbird**, Euplecte à épaules blanches (F), Spiegelwida (G), Viúva-d'asa-branca (Por), iNtakansinsi (Z). **Fan-tailed Widowbird**, Euplecte à épaules orangées (F), Stummelwida (G), Viúva-de-espáduas-vermelhas (Por), Isahomba/Isakhomba (X), iNtaka/uMahube/uMangube (Z).

Plate 83

Waxbills, Firefinches and Twinspots. Family Estrildidae. Genus *Sporaeginthus* (Gr), from *sporos* = a seed, and *Aegintha* = waxbill Genus. **Orange-breasted Waxbill** *S subflavus* (L) = yellowish below. Genus *Ortygospiza* (Gr) = a quail finch. **African Quailfinch** *O atricollis* (L) = black-necked. Genus *Amadina* = corrupt diminutive of *Ammodramus*, a Genus of Nearctic 'sparrows' (Emberizinae, Fringillidae). **Cut-throat Finch** *A fasciata* (L) = broadly, transversely striped. Genus *Mandingoa* = after Mandingo tribe from Niger R valley in W Africa. **Green Twinspot** *M nitidula* (L) = bright, trim, glittering; alluding to bird's colourful appearance.
Orange-breasted Waxbill, Bengali zébré (F), Goldbrüstchen (G), Bico-de-lacre-de-peito-laranja (Por), Borahane/Borane (S-So), Xidzingirhi (Tso). **African Quailfinch**, Astrild-caille à lunettes (F), Wachtelastrild (G), Bico-de-lacre-codorniz (Por), Lekolikotoana/Lekolukotoana (S-So), Unonkxwe (X), iNxenge/uNonklwe (Z). **Cut-throat Finch**, Amadine cou-coupé (F), Bandfink (G), Degolado (Por). **Green Twinspot**, Sénégali vert (F), Grüner Tropfenastrild (G), Pintadinha-verde (Por).

Plate 84

Genus *Estrilda* derived from *astra* (L) = a star, starred. **Common Waxbill** *E astrild* (L) = a star, starred. Genus *Granatina* (L) = a garnet, in reference to the violet-purple cheeks and ear coverts, or (L) = seedy, referring to diet. **Violet-eared Waxbill** *G granatina* (L) = a garnet, in reference to the violet-purple cheeks and ear coverts. Genus *Uraeginthus* (Gr) = tail like a finch. **Blue Waxbill** *U angolensis* (L) = from Angola. Genus *Hypargos* (Gr) = below Argos, the mythological 100-eyed guardian, ie having 100 eyes below. **Pink-throated Twinspot** *H margaritatus* (L) = pearly or pearl-like, referring to spotting on underparts. Genus *Pytilia* (Gr) = diminutive of Genus *Pitylus*, grosbeak. **Green-winged Pytilia** *P melba* = name used without explanation by Linnaeus; derivation unknown.
Common Waxbill, Astrild ondulé (F), Wellenastrild (G), Bico-de-lacre-comum (Por), Borahane/Borane (S-So), Xindzingiri bhanga (Tso), Intshiyane (X), iNtiyane (Z). **Violet-eared Waxbill**, Cordonbleu grenadin (F), Granatastrild, Blaubäckchen (G), Monsenhor (Por), Xindzingiri bhanga (Tso). **Blue Waxbill**, Cordonbleu de l'Angola (F), Angola-Schmetterlingsfink (G), Peito-celeste (Por), Xindzingiri (Tso). **Pink-throated Twinspot**, Sénégali de Verreaux (F), Perlastrild (G), Pintadinho-de-peito-rosado (Por). **Green-winged Pytilia**, Beaumarquet melba (F), Buntastrild (G), Maracachão-d'asa-verde (Por), Xindzingiri bhanga (Tso).

Plate 85
Genus *Lagonosticta* (Gr) = spotted flanks. **Red-billed Firefinch** *L senegala* (L) = from Senegal. **African Firefinch** *L rubricata* (L) = red. **Jameson's Firefinch** *L rhodopareia* (Gr) = red or rose cheek. Genus *Spermestes* (Gr) = a seed eater. **Bronze Mannikin** *S cucullatus* (L) = hooded. **Red-backed Mannikin** *S bicolor* (L) = 2 colours, pied.
Red-billed Firefinch, Amarante du Sénégal (F), Senegal-Amarant, Amarant (G), Peito-de-fogo-de-bico-vermelho (Por), Borane (S-So), Xidzingirhi (Tso). **African Firefinch**, Amarante foncé (F), Dunkelroter Amarant (G), Peito-de-fogo-de-bico-azul (Por), Isicibilili (X), ubuCubu (Z). **Jameson's Firefinch**, Amarante de Jameson (F), Rosenamarant, Jamesons Amarant (G), Peito-de-fogo de Jameson (Por), Xidzingirhi (Tso). **Bronze Mannikin**, Capucin nonnette (F), Kleinelsterchen (G), Freirinha-bronzeada (Por), Rijajani (Tso), Ingxenge/Ungxenge (X). **Red-backed Mannikin**, Capucin à dos brun (F), Glanzelsterchen (G), Freirinha-de-dorso-vermelho (Por), Rijajani (Tso).

Plate 86
Whydahs, Indigobirds and Cuckoo Finch. Family Viduidae. Genus *Vidua* (L) = a widow, referring to black plumage or widow's train (long tail of some spp.). **Long-tailed Paradise-Whydah** *V paradisaea* (L) = of paradise. **Shaft-tailed Whydah** *V regia* (L) = royal; use in this context obscure. **Pin-tailed Whydah** *V macroura* (Gr) = large tail. **Village Indigobird** *V chalybeata* (Gr) = steely, referring to plumage colour; the Chalybes were an ancient nation of Asia Minor, famed for their ability to work iron. **Dusky Indigobird** *V funereal* (L) = dark, belonging to a funeral, referring to black plumage. **Purple Indigobird** *V purpurascens* (L) = purplish, referring to glossy plumage.
Long-tailed Paradise-Whydah, Veuve de paradis (F), Spitzschwanz-Paradieswitwe (G), Viúva-do-paraíso-oriental (Por), Mitikahincila (Tso), uJojokhaya (Z). **Shaft-tailed Whydah**, Veuve royale (F), Königswitwe (G), Viúva-seta (Por). **Pin-tailed Whydah**, Veuve dominicaine (F), Dominikanerwitwe (G), Viuvinha (Por),'Mamarungoana/Selahlamarungoana/Selahlamarumo(S-So), N'waminungu (Tso), Uhlakhwe/Ujobela (X), uHlekwane (Z). **Village Indigobird**, Combassou du Sénégal (F), Rotschnabel-Atlaswitwe (G), Viúva-azul (Por). **Dusky Indigobird**, Combassou noir (F), Purpur-Atlaswitwe (G), Viúva-negra (Por). **Purple Indigobird**, Combassou violacé (F), Weißfuß-Atlaswitwe (G), Viúva-púrpura (Por).

Plate 87
Genus *Anomalospiza* (L) = unusual finch. **Cuckoo Finch** *A imberbis* (L) = beardless; use in this context obscure. **Sparrows and Petronias.** Family Passeridae. Genus *Passer* (L) = a sparrow. **House Sparrow** *P domesticus* (Gr) = belonging to the household. Genus *Petronia* (L) = to do with rocks, with reference to the habitat of the Rock Sparrow *Petronia petronia*. **Yellow-throated Petronia** *P superciliarus* (L) = with an eyebrow. **Southern Grey-headed Sparrow** *P diffuses* (L) = diffuse, extensive, referring to grey-washed head and neck.
Cuckoo Finch, Anomalospize parasite (F), Kuckucksweber (G), Tecelão-parasita (Por). **House Sparrow**, Moineau domestique (F), Haussperling (G), Pardal-dos-telhados (Por), Serobele (S-So), **Yellow-throated Petronia**, Moineau bridé (F), Gelbkehlsperling (G), Pardal-de-garganta-amarela (Por), **Southern Grey-headed Sparrow**, Moineau sud-africain (F), Graukopfsperling (G), Pardal-de-cabeça-cinzenta-meridional (Por), Serobele (S-So).

Plate 88
Wagtails, Pipits and Longclaws. Family Motacillinae. Genus *Motacilla* (L) = little mover. **African Pied Wagtail** *M aguimp* (Fr) = with a wimple, referring to black hood on head, neck and sides of face. **Cape Wagtail** *M capensis* (L) = from Cape of Good Hope, S Africa. **Yellow Wagtail** *M flava* (L) = yellow. Genus *Macronyx* (Gr) = large talon or claw, referring to large hind claws. **Yellow-throated Longclaw** *M croceus* (L) = yellow.
African Pied Wagtail, Bergeronnette pie (F), Witwenstelze (G), Alvéola-preta-e-branca (Por), Motjoli (S-So), Umcelu/Umvemve/Umventshana (X), umVemve (Z). **Cape Wagtail**, Bergeronnette du Cap (F), Kapstelze (G), Moletašaka (N-So), Alvéola do Cabo (Por), Motjoli (S-So), Mandzedzerekundze/Matsherhani/N'wapesupesu (Tso), Umcelu/Umvemve/Umventshana (X), umVemve (Z). **Yellow Wagtail**, Bergeronnette printanière (F), Schafstelze (G), Alvéola-amarela (Por). **Yellow-throated Longclaw**, Sentinelle à gorge jaune (F), Gelbkehlgroßsporn, Safrangroßsporn (G), Unha-longa-amarelo (Por), Holiyo/Hwiyo (Tso), iGwili/iNqomfi (Z).

Plate 89
Genus *Anthus* (Gr) = flower. **Striped Pipit** *A lineiventris* (L) = streaked underparts, referring to ventral striping. **African Pipit** *A cinnamomeus* (L) = cinnamon coloured. **Plain-backed Pipit** *A leucophrys* (Gr) = white eyebrow. **Buffy Pipit** *A vaalensis* (L) = from the Vaal R. **Bushveld Pipit** *A caffer* (L) = of Kaffraria, after historical name for tribes in E Cape.
Striped Pipit, Pipit de Sundevall (F), Streifenpieper (G), Petinha-estriada (Por), Intsasana (X). **African Pipit**, Pipit africain (F), Weidelandpieper, (Spornpieper) (G), Petinha-do-capim (Por), Tšase/Tšaase-ea-lithota (S-So), Icelu/Icetshu (X), umNgcelekeshu/umNgcelu (Z). **Plain-backed Pipit**, Pipit à dos uni (F), Braunrückenpieper (G), Petinha-de-dorso-liso (Por), Tšase (S-So), Xihitagadzi (Tso), Icelu/Icetshu (X), umNgcelekeshu/umNgcelu (Z). **Buffy Pipit**, Pipit du Vaal (F), Vaalpieper (G), Petinha do Vaal (Por), Tšase (S-So). **Bushveld Pipit**, Pipit des arbres (F), Buschpieper (G), Petinha-do-mato (Por).

Plate 90
Canaries and Buntings. Family Fringillidae. Genus *Crithagra* (Gr) = derivation obscure, possibly from *krith* = barley, and *agra* = hunting, ie barley-eater. **Yellow-fronted Canary** *C mozambicus* (L) = of Mozambique. **Lemon-breasted Canary** *C citrinipectus* (L) = lemon breast. **Brimstone Canary** *C sulphuratus* (L) = sulphur-coloured. **Streaky-headed Seedeater** *C gularis* (L) = of the throat.
Yellow-fronted Canary, Serin du Mozambique (F), Mossambikgirlitz (G), Canário de Moçambique (Por), Tšoere (S-So), Manswikidyani/Risunyani/ Ritswiri (Tso), Unyileyo (X), umBhalane (Z). **Lemon-breasted Canary**, Serin à poitrine citron (F), Gelbbrustgirlitz (G), Canário-de-peito-limão (Por). **Brimstone Canary**, Serin soufré (F), Schwefelgirlitz (G), Canário-girassol (Por), Indweza/Indweza Eluhlaza (X). **Streaky-headed Seedeater**, Serin gris (F), Brauengirlitz (G), Canário-de-cabeça-estriada (Por), Tsoere (S-So), Indweza (X), umBhalane/ umDendeliswe (Z).

Plate 91
Lark-like Bunting *E impetuani;* the derivation of the name is unknown, but probably of Tswana origin. **Cinnamon-breasted Bunting** *E tahapisi*. Derivation of the name unknown, but probably from the Tswana name "tahapitsi" for various finches and indigobirds. *Pietsi* may be onomatopoeic, but *pietsi* in Tswana is a zebra. So the striped head may be referred to. **Golden-breasted Bunting** *E flaviventris* (L) = yellow belly.
Lark-like Bunting, Bruant des rochers (F), Lerchenammer (G), Escrevedeira-cotovia (Por). **Cinnamon-breasted Bunting**, Bruant cannelle (F), Bergammer, Siebenstreifenammer (G), Escrevedeira-das-pedras (Por), 'Maborokoane (S-So), Undenjenje/Undenzeni (X), umDinasibula (Z). **Golden-breasted Bunting**, Bruant à poitrine dorée (F), Gelbbauchammer (G), Escrevedeira-de-peito-dourado (Por), Rhakweni/rhanciyoni (Tso), Intsasa (X), umNdweza (Z).

KEY: (445) = Roberts' number, 209 = Etymology, 13 = Plate number

INDEX TO PORTUGUESE NAMES

Abelharuco-andorinha (445), 209, **13**
-de-testa-branca (443), 209, **13**
-dourado (444), 209, **13**
-europeu (438), 209, **14**
-persa (440), 209, **13**
-róseo (441), 209, **14**
Abetarda-de-barriga-preta (238), 212, **26**
-de-poupa (237), 212, **26**
-gigante (230), 212, **26**
Abibe-carunculado (260), 214, **33**
-coroado (255), 214, **33**
-d'asa-negra-pequeno (256), 214, **33**
-de-coroa-branca (259), 214, **33**
-preto-e-branco (258), 214, **33**
Abutre-das-palmeiras (147), 214, **36**
-de-cabeça-branca (125), 215, **37**
-de-capuz (121), 215, **37**
-real (124), 215, **37**
Açor-africano (160), 215, **39**
-cantor-escuro (163), 215, **40**
-preto (158), 215, **40**
Águia de Ayres (138), 215, **40**
de Wahlberg (135), 216, **41**
-bailarina (146), 215, **38**
-cobreira-castanha (142), 215, **38**
-cobreira-de-peito-preto (143), 215, **38**
-coroada (141), 216, **42**
-das-estepes (133), 215, **41**
-de-penacho (139), 216, **42**
-dominó (137), 215, **40**
-fulva (132), 215, **41**
-marcial (140), 216, **42**
-pesqueira (170), 214, **36**
-pesqueira-africana (148), 214, **36**
-pomarina (134), 216, **41**
-preta (131), 216, **42**
Alcaravão do Cabo (297), 213, **31**
-d'água (298), 213, **31**
Alfaiate (294), 213, **31**
Alvéola do Cabo (713), 226, **88**
-amarela (714), 226, **88**
-preta-e-branca (711), 226, **88**
Andorinha-cauda-de-arame (522), 220, **61**
-das-barreiras-africana (533), 220, **60**
-das-barreiras-europeia (532), 219, **60**
-das-chaminés (518), 220, **60**
-das-mesquitas (525), 220, **62**
-das-rochas-africana (529), 220, **60**
-de-peito-pérola (523), 220, **61**
-de-peito-ruivo (524), 220, **62**
-de-rabadilha-cinzenta (531), 220, **61**
-dos-beirais (530), 220, **61**

-estriada-pequena (527), 220, **62**
Andorinhão-cafre (415), 210, **18**
-das-barreiras (416), 210, **18**
-das-palmeiras (421), 210, **19**
-pequeno (417), 210, **18**
-preto-africano (412), 210, **19**
-preto-europeu (411), 210, **19**
-real (418), 210, **19**
Andorinha-preta (536), 220, **62**
Apalis de Rudd (649), 222, **70**
-de-peito-amarelo (648), 222, **70**
Atacador-de-poupa-branca (753), 219, **56**
-de-poupa-preta (754), 219, **56**
Avestruz (1), 206, **1**
Barbaças-das-acácias (465), 207, **7**
-de-colar-preto (464), 207, **7**
-de-poupa (473), 207, **7**
Barbadinho-de-rabadilha-limão (471), 207, 7
-de-testa-amarela (470), 207, **7**
Batis do Cabo (700), 219, **57**
-comum (701), 219, **57**
Beija-flor de Marico (779), 224, **79**
-flor-de-barriga-branca (787), 224, **79**
-flor-de-colar (793), 224, **79**
-flor-de-peito-escarlate (791), 224, **79**
-flor-preto (792), 224, **79**
Bico-aberto (87), 218, **51**
-de-cimitarra (454), 208, **9**
-de-lacre-codorniz (852), 225, **83**
-de-lacre-comum (846), 225, **84**
-de-lacre-de-peito-laranja (854), 225, **83**
Borrelho-asiático (252), 213, **32**
-de-testa-branca (246), 213, **32**
-de-três-golas (249), 213, **32**
-do-gado (248), 213, **32**
-grande-de-coleira (245), 213, **32**
Brubru (741), 218, **54**
Bufo-leitoso (402), 211, **21**
-malhado (401), 211, **21**
Bútio-comum (149), 214, **35**
-vespeiro (130), 214, **35**
Caimão de Allen (224), 212, **28**
Calau-cinzento (457), 208, **8**
-coroado (460), 208, **8**
-de-bico-amarelo (459), 208, **8**
-de-bico-vermelho (458), 208, **8**
-gigante (463), 208, **8**
-trombeteiro (455), 208, **8**
Canário de Moçambique (869), 227, **90**
-de-cabeça-estriada (881), 227, **90**
-de-peito-limão (871), 227, **90**
-girassol (877), 227, **90**

228   *Index to Portuguese Names*

Cardeal-tecelão-amarelo (826), 225, **82**
  -tecelão-vermelho (824), 225, **82**
Cartaxo-comum (596), 223, **76**
Cegonha de Abdim (85), 218, **51**
  -branca (83), 218, **52**
  -de-bico-amarelo (90), 218, **52**
  -episcopal (86), 218, **51**
  -preta (84), 218, **51**
Chapim-preto-meridional (554), 219, **59**
Chasco de Arnot (594), 223, **76**
  -familiar (589), 223, **76**
  -poliglota (593), 223, **76**
Codornizão-africano (212), 212, **27**
  -euroasiático (211), 212, **27**
Colhereiro-africano (95), 217, **50**
Combatente (284), 213, **30**
Corredor de Temminck (300), 214, **34**
  -asa-de-bronze (303), 214, **34**
  -de-três-golas (302), 214, **34**
Corticol-de-duas-golas (347), 212, **26**
Coruja-da-floresta (394), 211, **21**
  -das-torres (392), 210, **20**
  -do-capim (393), 210, **20**
  -dos-pântanos (395), 211, **21**
Corujão-pesqueiro de Pel (403), 211, **21**
Corvo-das-montanhas (550), 219, **58**
  -marinho-africano (58), 216, **45**
  -marinho-de-faces-brancas (55), 216, **45**
Cossifa de Heuglin (599), 223, **75**
  do Cabo (601), 223, **75**
  do Natal (600), 223, **75**
  -de-peito-branco (602), 223, **75**
Cotovia-cor-d'areia (497), 222, **72**
  -das-castanholas (496), 222, **71**
  -de-barrete-vermelho (507), 222, **72**
  -de-nuca-vermelha (494), 222, **71**
  -monótona (493), 222, **71**
  -pardal-de-dorso-castanho (515), 222, **72**
  -pardal-de-dorso-cinzento (516), 222, **72**
  -sabota (498), 222, **71**
  -sombria (505), 222, **72**
Cucal de Burchell (391), 210, **17**
  -preto-africano (388), 210, **17**
Cuco da Cafraria (381), 209, **15**
  -africano (375), 210, **16**
  -bronzeado-maior (386), 210, **16**
  -bronzeado-menor (385), 210, **16**
  -canoro (374), 210, **16**
  -de-peito-vermelho (377), 209, **15**
  -jacobino (382), 209, **15**
  -preto (378), 210, **16**
  -rabilongo (380), 209, **15**
Degolado (855), 225, **83**
Drongo-de-cauda-forcada (541), 218, **53**

Eremomela-de-barrete-verde (655), 220, **64**
  -de-barriga-amarela (653), 220, **64**
  -de-garganta-castanha (656), 220, **64**
Escrevedeira-cotovia (887), 227, **91**
  -das-pedras (886), 227, **91**
  -de-peito-dourado (884), 227, **91**
Estorinho-de-dorso-violeta (761), 224, **78**
  de Burchell (762), 224, **77**
  -carunculado (760), 224, **78**
  -d'asa-castanha (769), 224, **78**
  -grande-d'orelha-azul (765), 224, **77**
  -metálico (764), 224, **77**
  -rabilongo (763), 224, **77**
Falcão de Dickinson (185), 216, **43**
  -alfaneque (172), 216, **44**
  -cuco (128), 214, **35**
  -de-pés-vermelhos-oriental (180), 216, **43**
  -peregrino (171), 216, **44**
Felosa de Stierling (659), 222, **70**
  -chapim-dos-bosques (621), 221, **67**
  -das-figueiras (619), 221, **66**
  -das-oliveiras (626), 221, **66**
  -de-dorso-verde (657), 222, **70**
  -dos-juncos (634), 221, **65**
  -dos-juncos-africana (638), 221, **65**
  -icterina (625), 221, **66**
  -musical (643), 221, **66**
  -palustre (633), 221, **66**
Flamingo-comum (96), 217, **50**
  -pequeno (97), 217, **50**
Francolim de Shelley (191), 206, **1**
  de Swainson (199), 206, **1**
  do Natal (196), 206, **1**
  -das-pedras (188), 206, **1**
  -de-poupa (189), 206, **2**
Franga-d'água-preta (213), 212, **28**
Freirinha-bronzeada (857), 226, **85**
  -de-dorso-vermelho (858), 226, **85**
Fuinha do Natal (678), 222, **69**
  -chocalheira (672), 221, **68**
  -de-cabeça-ruiva (681), 222, **69**
  -de-dorso-preto (675), 221, **68**
  -de-faces-vermelhas (674), 221, **68**
  -do-deserto (665), 222, **69**
  -dos-juncos (664), 222, **69**
  -preguiçosa (679), 221, **68**
Gaivina-d'asa-branca (339), 214, **34**
  -de-faces-brancas (338), 214, **34**
Galeirão-de-crista (228), 213, **28**
Galinha-d'água (226), 212, **28**
  -d'água-pequena (224), 213, **28**
Ganso do Egipto (102), 207, **4**
Garça-boieira (71), 217, **46**
  -branca-grande (66), 217, **46**

-branca-intermédia (68), 217, **46**
-branca-pequena (67), 216, **46**
-caranguejeira (72), 217, **46**
-de-cabeça-preta (63), 217, **47**
-de-dorso-verde (74), 217, **48**
-gigante (64), 217, **47**
-nocturna (76), 217, **48**
-nocturna-de-dorso-branco (77), 217, **48**
-preta (69), 217, **47**
-real (62), 217, **47**
-vermelha (65), 217, **47**
Garçote-anão (79), 217, **48**
-comum (78), 217, **48**
Gavião do Ovambo (156), 215, **39**
-chicra (159), 215, **39**
-morcegueiro (129), 214, **35**
-palrador (161), 215, **39**
-papa-lagartos (154), 215, **39**
-pequeno (157), 215, **39**
Gralha-seminarista (548), 219, **58**
Grifo do Cabo (122), 215, **37**
-de-dorso-branco (123), 215, **37**
Ibis-preto (93), 217, **49**
-sagrado (91), 217, **49**
Indicador-de-bico-aguçado (478), 207, **5**
-de-peito-escamoso (475), 207, **5**
-grande (474), 207, **5**
-pequeno (476), 207, **5**
Jabiru (88), 218, **52**
Jacana-africana (240), 212, **27**
-pequena (241), 212, **27**
Lagarteiro-cinzento-e-branco (539), 219, **59**
-preto (538), 219, **59**
Maçarico-bastardo (266), 213, **29**
-bique-bique (265), 213, **29**
-das-rochas (264), 213, **29**
Marabu (89), 218, **52**
Maracachão-d'asa-verde (834), 225, **84**
Marreco-de-bico-vermelho (108), 206, **3**
-hotentote (107), 206, **3**
Mergulhão-pequeno (8), 216, **45**
-serpente (60), 216, **45**
Milhafre-preto (126), 214, **35**
Mocho-barrado (399), 211, **20**
-de-faces-brancas (397), 211, **20**
-de-orelhas-africano (396), 211, **20**
-perlado (398), 211, **20**
Monsenhor (845), 225, **84**
Narceja-africana (286), 212, **27**
-pintada (242), 212, **27**
Noitibó da Europa (404), 211, **22**
   de Moçambique (409), 211, **22**
-de-balanceiros (410), 211, **22**
-de-faces-ruivas (406), 211, **22**

-de-pescoço-dourado (405), 211, **22**
-sardento (408), 211, **22**
Ógea-europeia (173), 216, **43**
Olho-branco do Cabo (796), 221, **67**
-branco-amarelo (797), 221, **67**
-carunculado (705), 219, **57**
Papa-figos-de-cabeça-preta (545), 218, **53**
-figos-europeu (543), 218, **53**
-moscas do Marico (695), 223, **73**
-moscas-austral (624), 221, **67**
-moscas-azulado (691), 223, **74**
-moscas-cinzento (689), 223, **74**
-moscas-de-poupa (708), 218, **53**
-moscas-do-paraíso (710), 218, **53**
-moscas-fiscal (698), 223, **74**
-moscas-pálido (696), 223, **73**
-moscas-preto-meridional (694), 223, **73**
-moscas-sombrio (690), 223, **74**
Papagaio de Meyer (364), 210, **17**
-de-cabeça-castanha (363), 210, **17**
Para-moscas-de-leque (693), 223, **74**
Pardal-de-cabeça-cinzenta-meridional (804), 226, **87**
-de-garganta-amarela (805), 226, **87**
-dos-telhados (801), 226, **87**
Pássaro-do-algodão-cinzento (558), 219, **59**
-martelo (81), 217, **49**
Pato-assobiador-arruivado (100), 206, **3**
-assobiador-de-faces-brancas (99), 206, **3**
-de-carúncula (115), 207, **4**
-de-dorso-branco (101), 206, **3**
-ferrão (116), 207, **4**
-orelhudo (114), 206, **3**
-preto-africano (105), 207, **4**
Peito-celeste (844), 225, **84**
-de-fogo de Jameson (841), 226, **85**
-de-fogo-de-bico-azul (840), 226, **85**
-de-fogo-de-bico-vermelho (842), 226, **85**
Pelicano-branco (49), 217, **50**
Peneireiro-cinzento (127), 214, **36**
-das-torres (183), 216, **44**
-vulgar (181), 216, **44**
Perdiz-do-mar-d'asa-vermelha (304), 214, **34**
Perna-longa (295), 213, **31**
-verde-comum (270), 213, **29**
-verde-fino (269), 213, **29**
Pés-de-barbatanas (229), 212, **28**
Petinha do Vaal (719), 227, **89**
-de-dorso-liso (718), 227, **89**
-do-capim (716), 227, **89**
-do-mato (723), 227, **89**
-estriada (720), 227, **89**
Pica-bois-de-bico-amarelo (771), 224, **78**
-bois-de-bico-vermelho (772), 224, **78**
-pau de Bennett (481), 207, **6**

-pau-cardeal (486), 207, **6**
-pau-de-bigodes (487), 207, **6**
-pau-de-rabo-dourado (483), 207, **6**
-peixe-de-barrete-castanho (435), 209, **12**
-peixe-de-barrete-cinzento (436), 208, **11**
-peixe-de-colar (430), 208, **11**
-peixe-de-poupa (431), 208, **11**
-peixe-dos-bosques (433), 208, **11**
-peixe-gigante (429), 209, **12**
-peixe-malhado (428), 209, **12**
-peixe-pigmeu (432), 208, **11**
-peixe-riscado (437), 209, **12**
Picanço-assobiador-de-coroa-castanha (743), 218, **54**
-assobiador-de-coroa-preta (744), 218, **54**
-de-almofadinha (740), 218, **54**
-de-cabeça-cinzenta (751), 219, **56**
-de-coroa-branca (756), 219, **58**
-de-dorso-ruivo (733), 219, **57**
-de-peito-laranja (748), 219, **56**
-ferrugíneo (736), 218, **55**
-pequeno (731), 219, **57**
-preto-e-vermelho (739), 218, **55**
-quadricolor (747), 218, **55**
-rabilongo (735), 219, **58**
-tropical (737), 218, **55**
Pilrito-de-bico-comprido (272), 213, **30**
-pequeno (274), 213, **30**
-sanderlingo (281), 213, **30**
Pintada da Guiné (203), 206, **2**
-de-crista (204), 206, **2**
Pintadinha-verde (835), 225, **83**
-de-peito-rosado (838), 225, **84**
Pombo-malhado (349), 212, **25**
-verde-africano (361), 212, **25**
Poupa (451), 208, **9**
Prínia-de-flancos-castanhos (683), 222, **70**
Quelea-de-bico-vermelho (821), 225, **81**
Rabicurta-de-bico-comprido (651), 220, **64**
Rabo-de-junco-de-faces-vermelhas (426), 209, **14**
-de-junco-de-peito-barrado (424), 209, **14**
-espinhoso de Böhm (423), 210, **18**
-espinhoso-malhado (422), 210, **18**
Republicano (427), 208, **9**
Rola do Cabo (354), 211, **24**
do Senegal (355), 211, **24**
-de-olhos-vermelhos (352), 211, **24**
-de-papo-branco (359), 211, **24**
-esmeraldina (358), 211, **24**
-rabilonga (356), 212, **25**

Rolieiro-cauda-de-raquete (448), 208, **10**
-de-bico-amarelo (450), 208, **10**
-de-peito-lilás (447), 208, **10**
-de-sobrancelhas-brancas (449), 208, **10**
-europeu (446), 208, **10**
Rouxinol-do-mato-de-bigodes (617), 223, **75**
-do-mato-estriado (613), 223, **75**
-dos-caniços-africano (631), 221, **65**
-grande-dos-caniços (628), 221, **65**
-pequeno-dos-pântanos (635), 221, **65**
-russo (609), 223, **76**
Secretário (118), 216, **42**
-pequeno (169), 215, **41**
Singanga (94), 217, **49**
Tartaranhão-caçador (166), 215, **38**
-pálido (167), 215, **38**
Tecelão-de-bico-grosso (807), 225, **81**
-de-bico-vermelho (798), 225, **81**
-de-cabeça-vermelha (819), 225, **81**
-de-lunetas (810), 224, **80**
-de-máscara (814), 224, **80**
-dourado (816), 224, **80**
-malhado (811), 224, **80**
-parasita (820), 226, **87**
-pequeno-de-mascarilha (815), 224, **80**
Toirão-comum (205), 206, **2**
Tordo-chicharrio (576), 223, **73**
-das-palmeiras-de-colar (603), 223, **76**
-de-peito-malhado (580), 223, **73**
Turaco-cinzento (373), 212, **25**
-de-crista-violeta (371), 212, **25**
Tuta-amarela (574), 220, **63**
-da-terra (569), 220, **63**
-malhada (575), 220, **63**
-negra (568), 220, **63**
-sombria (572), 220, **63**
Unha-longa-amarelo (728), 226, **88**
Viúva-azul (867), 226, **86**
-d'asa-branca (829), 225, **82**
-de-colar-vermelho (831), 225, **82**
-de-espáduas-vermelhas (828), 225, **82**
-do-paraíso-oriental (862), 226, **86**
-negra (864), 226, **86**
-púrpura (865), 226, **86**
-seta (861), 226, **86**
Viuvinha (860), 226, **86**
Zaragateiro-castanho (560), 221, **67**
Zarro-africano (113), 207, **4**
Zombeteiro-de-bico-vermelho (452), 208, **9**

KEY: (85) = Roberts' number, 218 = Etymology, **51** = Plate number

INDEX TO GERMAN NAMES

Abdimsstorch, Regenstorch (85), 218, **51**
Afrikanische Bekassine (286), 212, **27**
    Binsenralle (229), 212, **28**
    Zwerggans (114), 206, **3**
    Zwergohreule (396), 211, **20**
    Kuckuck (375), 210, **16**
    Löffler (95), 217, **50**
    Sperber, Tachirosperber (160), 215, **39**
Afrikanisches Sultanshuhn (224), 212, **28**
Akaziendrossel (580), 223, **73**
Alpensegler (418), 210, **19**
Amethyst-Glanzköpfchen (792), 224, **79**
Amethystglanzstar (761), 224, **78**
Amur-Rotfußfalke (180), 216, **43**
Angola-Schmetterlingsfink (844), 225, **84**
Angolaturteltaube, Angolalachtaube, Brillentaube (353), 211, **24**
Arnotschmätzer (594), 223, **76**
Bandfink (855), 225, **83**
Baobabsegler (422), 210, **18**
Baumfalke (173), 216, **43**
Baumklapperlerche, Zimtbaumlerche (496), 222, **71**
Bennettspecht (481), 207, **6**
Bergammer, Siebenstreifenammer (886), 227, **91**
Bindenfischeule, Fischeule (403), 211, **21**
Bindennektarvogel (779), 224, **79**
Bindenrennvogel (302), 214, **34**
Blaumantel-Schopfschnäpper (708), 218, **53**
Blauracke (446), 208, **10**
Blaustirn-Blatthühnchen, Jacana (240), 212, **27**
Blauwangenspint (440), 209, **13**
Blutschnabelweber (821), 225, **81**
Brachschwalbe (304), 214, **34**
Brauengirlitz (881), 227, **90**
Braundroßling (560), 221, **67**
Brauner Bartheckensänger (617), 223, **75**
    Schlangenadler (142), 215, **38**
    Sichler (93), 217, **49**
    Zistensänger (681), 222, **69**
Braunflügel-Mausvogel (424), 209, **14**
Braunkehl-Uferschwalbe, Afrikanische Uferschwalbe (533), 220, **60**
Braunkopfliest (435), 209, **12**
Braunkopfpapagei (363), 210, **17**
Braunrückenpieper (718), 227, **89**
Brillenweber (810), 224, **80**
Brillenwürger (753), 219, **56**
Bronzeflecktaube (358), 211, **24**
Bronzeflügel-Rennvogel, Amethystrennvogel (303), 214, **34**
Brubru (741), 218, **54**
Bruchwasserläufer (266), 213, **29**

Büffelweber (798), 225, **81**
Bülbülwürger (575), 220, **63**
Buntastrild (834), 225, **84**
Buschpieper (723), 227, **89**
Cabanisweber (815), 224, **80**
Coquifrankolin (188), 206, **1**
Damaratschagra (743), 218, **54**
Dickschnabelkuckuck (383), 209, **15**
Diderikkuckuck, Goldkuckuck (386), 210, **16**
Dominikanerwitwe (860), 226, **86**
Dreiband-Regenpfeifer (249), 213, **32**
Dreifarbenwürger, Dreifarb-Brillenwürger (754), 219, **56**
Drongoschnäpper (694), 223, **73**
Drossellerche (505), 222, **72**
Drosselrohrsänger (628), 221, **65**
Dunkelroter Amarant (840), 226, **85**
Dunkelschnäpper (690), 223, **74**
Edelreiher, Mittelreiher (68), 217, **46**
Einsiedlerkuckuck (377), 209, **15**
Elsterwürger (735), 219, **58**
Europäische Uferschwalbe (532), 219, **60**
Europäischer Bienenfresser (438), 209, **14**
    Pirol (543), 218, **53**
Fahlschnäpper (696), 223, **73**
Felsenadler, Kaffernadler (131), 216, **42**
Felsenschwalbe, Steinschwalbe (529), 220, **60**
Fischadler (170), 214, **36**
Fitis (643), 221, **66**
Flamingo (96), 217, **50**
Fleckenadler (138), 215, **40**
Fleckennachtschwalbe (408), 211, **22**
Fleckenuhu, Berguhu (401), 211, **21**
Fledermausaar (129), 214, **35**
Fledermaussegler (423), 210, **18**
Flötenwürger (736), 218, **55**
Flußuferläufer (264), 213, **29**
Gabarhabicht (161), 215, **39**
Gabelracke (447), 208, **10**
Gabunnachtschwalbe (409), 211, **22**
Gartengrasmücke (619), 221, **66**
Gartenrohrsänger (631), 221, **65**
Gaukler (146), 215, **38**
Gefleckter Honiganzeiger (475), 207, **5**
Geierrabe (550), 219, **58**
Gelbbauchammer (884), 227, **91**
Gelbbauch-Eremomela (653), 220, **64**
Gelbbrustbülbül (574), 220, **63**
Gelbbrust-Feinsänger (648), 222, **70**
Gelbbrustgirlitz (871), 227, **90**
Gelbe Baumente (100), 206, **3**
Gelbkehlgroßsporn, Safrangroßsporn (728), 226, **88**

Gelbkehlsperling (805), 226, **87**
Gelbschnabel-Madenhacker (771), 224, **78**
Gelbschnabeltoko (459), 208, **8**
Gelbspötter (625), 221, **66**
Gelbsteißbülbül, Graubülbül (568), 220, **63**
Gelbstirn-Bartvogel (470), 207, **7**
Glanzelsterchen (858), 226, **85**
Glanzhaubenturako (371), 212, **25**
Gleitaar (127), 214, **36**
Glockenreiher (69), 217, **47**
Goldbrüstchen (854), 225, **83**
Goldbugpapagei (364), 210, **17**
Goldbürzel-Bartvogel (471), 207, **7**
Goldschnepfe (242), 212, **27**
Goldschwanzspecht (483), 207, **6**
Goliathreiher (64), 217, **47**
Granatastrild, Blaubäckchen (845), 225, **84**
Graseule (393), 210, **20**
Graubürzelschwalbe (531), 220, **61**
Graubürzel-Singhabicht, Dunkler Grauflügelhabicht (163), 215, **40**
Graufischer (428), 209, **12**
Graukopfliest (436), 208, **11**
Graukopfpapage (-), 210, **17**
Graukopfsperling (804), 226, **87**
Graukopfwürger, Riesenbuschwürger (751), 219, **56**
Graulärmvogel (373), 212, **25**
Graureiher (62), 217, **47**
Graurücken-Grasmücke (-), 222, **70**
Grauschnäpper (689), 223, **74**
Grautoko (457), 208, **8**
Großer Goldweber, Safranweber (816), 224, **80**
  Honiganzeiger, Schwarzkehl-Honiganzeiger (474), 207, **5**
Grüne Fruchttaube, Grüntaube (361), 212, **25**
  Tropfenastrild (835), 225, **83**
Grünkappen-Eremomela (655), 220, **64**
Grünschenkel (270), 213, **29**
Grünschwanz-Glanzstar (765), 224, **77**
Guineataube (349), 212, **25**
Habichtsadler (137), 215, **40**
Hagedasch-Ibis (94), 217, **49**
Häherkuckuck (380), 209, **15**
Halbmondtaube (352), 211, **24**
Halsband-Bartvogel (464), 207, **7**
Hammerkopf (81), 217, **49**
Harlekinwachtel (201), 206, **2**
Haubenbartvogel (473), 207, **7**
Haussegler (417), 210, **18**
Haussperling (801), 226, **87**
Heiliger Ibis (91), 217, **49**
Helmperlhuhn (203), 206, **2**
Hirtenregenpfeifer (248), 213, **32**
Höckerente (115), 207, **4**

Hornrabe (463), 208, **8**
Horussegler, Erdsegler (416), 210, **18**
Hottentottenente, Pünktchenente (107), 206, **3**
Jakobinerkuckuck, Elsterkuckuck (382), 209, **15**
Kalahari-Zistensänger (665), 222, **69**
Kammbleßhuhn (228), 213, **28**
Kampfadler (140), 216, **42**
Kampfläufer (284), 213, **30**
Kapbeutelmeise (558), 219, **59**
Kapgeier (122), 215, **37**
Kap-Grünbülbül (572), 220, **63**
Kapkauz (399), 211, **20**
Kapkuckuck (381), 209, **15**
Kapohreule (395), 211, **21**
Kappengeier (121), 215, **37**
Kaprohrsänger (635), 221, **65**
Kaprötel (601), 223, **75**
Kapschnäpper (700), 219, **57**
Kapsegler (412), 210, **19**
Kapstelze (713), 226, **88**
Kaptäubchen (356), 212, **25**
Kaptriel, Bändertriel (297), 213, **31**
Kapturteltaube, Gurrtaube, Kaplachtaube (354), 211, **24**
Kardinalspecht (486), 207, **6**
Klaaskuckuck (385), 210, **16**
Klaffschnabel (87), 218, **51**
Kleine Streifenschwalbe (527), 220, **62**
Kleinelsterchen (857), 226, **85**
Kleiner Honiganzeiger (476), 207, **5**
Kobalteisvogel (430), 208, **11**
Königswitwe (861), 226, **86**
Kräuselhauben-Perlhuhn (204), 206, **2**
Kronenadler (141), 216, **42**
Kronenkiebitz (255), 214, **33**
Kronentoko (460), 208, **8**
Kuckuck (374), 210, **16**
Kuckucksweber (820), 226, **87**
Kuckucksweih (128), 214, **35**
Kuckuckswürger (538), 219, **59**
Kuhreiher (71), 217, **46**
Langschnabel-Sylvietta, Kurzschwanz-Sylvietta (651), 220, **64**
Langspornkiebitz (259), 214, **33**
Lannerfalke (172), 216, **44**
Lappenstar (760), 224, **78**
Laubbülbül (569), 220, **63**
Laufhühnchen, Rostkehl-Kampfwachtel (205), 206, **2**
Lerchenammer (887), 227, **91**
Malachiteisvogel, Haubenzwergfischer (431), 208, **11**
Mangrovereiher (74), 217, **48**
Marabu (89), 218, **52**
Maricoschnäpper (695), 223, **73**
Maschona-Hyliota (624), 221, **67**

Index to German Names 233

Maskenpirol (545), 218, **53**
Maskenweber, Schwarzstirnweber (814), 224, **80**
Mauersegler (411), 210, **19**
Mäusebussard (149), 214, **35**
Meckergrasmücke (Grünrücken-Camaroptera, Graurücken-Camaroptera) (657), 222, **70**
Mehlschwalbe (530), 220, **61**
Meisensänger (621), 221, **67**
Meisenschnäpper (693), 223, **74**
Meves-Glanzstar (763), 224, **77**
Milchuhu, Blaßuhu (402), 211, **21**
Mohrenhabicht, Trauerhabicht (158), 215, **40**
Mohrenmeise (554), 219, **59**
Mohrenralle, Negerralle (213), 212, **28**
Morgenrötel (603), 223, **76**
Mossambikgirlitz (869), 227, **90**
Nachtflughuhn (347), 212, **26**
Nachtreiher (76), 217, **48**
Namaspecht (487), 207, **6**
Narina-Trogon, Zügeltrogon (427), 208, **9**
Natalfrankolin (196), 206, **1**
Natalrötel (600), 223, **75**
Natalzwergfischer (432), 208, **11**
Neuntöter (733), 219, **57**
Nilgans (102), 207, **4**
Nimmersatt (90), 218, **52**
Nonnenlerche (516), 222, **72**
Ohrengeier (124), 215, **37**
Olivenspötter (626), 221, **66**
Orangewürger, Orangebrustwürger (748), 219, **56**
Oranjebrillenvogel (796), 221, **67**
Orgelwürger, Tropischer Flötenwürger (737), 218, **55**
Oryxweber (824), 225, **82**
Ovambosperber (156), 215, **39**
Palmengeier (147), 214, **36**
Palmensegler (421), 210, **19**
Paradiesschnäpper (710), 218, **53**
Perlastrild (838), 225, **84**
Perlbrustschwalbe (523), 220, **61**
Perlkauz (398), 211, **20**
Purpur-Atlaswitwe (864), 226, **86**
Purpurreiher (65), 217, **47**
Rahmbrustprinie (683), 222, **70**
Rallenreiher (72), 217, **46**
Raubadler (132), 215, **41**
Rauchschwalbe (518), 220, **60**
Riedscharbe (58), 216, **45**
Riesenfischer, Rieseneisvogel (429), 209, **12**
Riesenglanzstar, Glanzelstar (762), 224, **77**
Riesentrappe (230), 212, **26**
Rosapelikan (49), 217, **50**
Rosenamarant, Jamesons Amarant (841), 226, **85**
Rostkehl-Eremomela, Rostband-Eremomela (656), 220, **64**

Rostschwanzschmätzer, Rostschwanz (589), 223, **76**
Rostwangen-Nachtschwalbe (406), 211, **22**
Rotaugenente (113), 207, **4**
Rotbauchschmätzer (593), 223, **76**
Rotbauchschwalbe (524), 220, **62**
Rotbauchwürger, Reichsvogel, Kaiservogel (739), 218, **55**
Rotbrust-Glanzköpfchen, Rotbrust-Nektarvogel (791), 224, **79**
Rötelfalke (183), 216, **44**
Rotgesicht-Zistensänger (674), 221, **68**
Rotkappenschwalbe (522), 220, **61**
Rotnackenlerche (494), 222, **71**
Rotnacken-Nachtschwalbe (405), 211, **22**
Rotscheitellerche (507), 222, **72**
Rotscheitel-Zistensänger (672), 221, **68**
Rotschnabel-Atlaswitwe (867), 226, **86**
Rotschnabeldrossel (576), 223, **73**
Rotschnabelente (108), 206, **3**
Rotschnabel-Madenhacker (772), 224, **78**
Rotschnabeltoko (458), 208, **8**
Rotschopftrappe (237), 212, **26**
Rotschulter-Glanzstar (764), 224, **77**
Rotschwingenstar (769), 224, **78**
Rotstirn-Bartvogel (465), 207, **7**
Rotzügel-Mausvogel (426), 209, **14**
Rudds Feinsänger (649), 222, **70**
Ruderflügel (410), 211, **22**
Säbelschnäbler (294), 213, **31**
Sabotalerche (498), 222, **71**
Sanderling (281), 213, **30**
Sandregenpfeifer (245), 213, **32**
Sattelstorch (88), 218, **52**
Schafstelze (714), 226, **88**
Scharlachspint (441), 209, **14**
Scharlachweber (819), 225, **81**
Schieferschnäpper (691), 223, **74**
Schikra (159), 215, **39**
Schildrabe (548), 219, **58**
Schildwida (831), 225, **82**
Schilfrohrsänger (634), 221, **65**
Schlangenhalsvogel (60), 216, **45**
Schlangensperber, Höhlenweihe (169), 215, **41**
Schleiereule (392), 210, **20**
Schmalschnabel-Honiganzeiger (478), 207, **5**
Schmarotzermilan (126), 214, **35**
Schneeballwürger (740), 218, **54**
Schopfadler (139), 216, **42**
Schopffrankolin (189), 206, **2**
Schreiadler (134), 215, **41**
Schreiseeadler (148), 214, **36**
Schwalbenschwanzspint Gabelschwanzspint (445), 209, **13**
Schwarzbauchtrappe (238), 212, **26**

Schwarzbrust-Schlangenadler (143), 215, **38**
Schwarzente (105), 207, **4**
Schwarzkehlchen (596), 223, **76**
Schwarzkehl-Lappenschnäpper (705), 219, **57**
Schwarzkopfreiher (63), 217, **47**
Schwarzkuckuck (378), 210, **16**
Schwarzrückenfalke (185), 216, **43**
Schwarzrücken-Zistensänger (675), 221, **68**
Schwarzstirnwürger (731), 219, **57**
Schwarzstorch (84), 218, **51**
Schwefelgirlitz (877), 227, **90**
Seidenreiher (67), 216, **46**
Sekretär (118), 216, **42**
Senegal-Amarant, Amarant (842), 226, **85**
Senegalbrillenvogel (797), 221, **67**
Senegalkiebitz (260), 214, **33**
Senegalliest (433), 208, **11**
Senegalschwalbe (525), 220, **62**
Senegaltaube, Palmtaube (355), 211, **24**
Senegaltschagra (744), 218, **54**
Shelleyfrankolin (191), 206, **1**
Sichelhopf (454), 208, **9**
Sichelstrandläufer (272), 213, **30**
Silberreiher (66), 217, **46**
Smiths Zistensänger (679), 221, **68**
Spatelracke (448), 208, **10**
Sperberbussard (154), 215, **39**
Sperlingslerche (493), 222, **71**
Spiegelwida (829), 225, **82**
Spitzschwanz-Paradieswitwe (862), 226, **86**
Sporengans (116), 207, **4**
Sprosser (609), 223, **76**
Stelzenläufer (295), 213, **31**
Steppenbaumhopf (452), 208, **9**
Steppenlerche (497), 222, **72**
Steppenralle (212), 212, **27**
Steppenweihe (167), 215, **38**
Stierling-Bindensänger (659), 222, **70**
Strauß (1), 206, **1**
Streifenliest, Gestreifter Baumliest (437), 209, **12**
Streifenpieper (720), 227, **89**
Strichelracke (449), 208, **10**
Strichelzistensänger (678), 222, **69**
Stummelwida (828), 225, **82**
Sturms Zwergrohrdommel (79), 217, **48**
Sumpfbuschsänger (638), 221, **65**
Sumpfrohrsänger (633), 221, **66**
Sundevalls Sägeflügelschwalbe (536), 220, **62**
Swainsonfrankolin (199), 206, **1**
Tahaweber, Napoleonweber (826), 225, **82**
Tamburintaube (359), 211, **24**
Teichhuhn (226), 212, **28**
Teichwasserläufer (269), 213, **29**
Temminckrennvogel (300), 214, **34**

Textor, Dorfweber (811), 224, **80**
Tiputip (391), 210, **17**
Trauerdrongo, Gabelschwanzdrongo (541), 218, **53**
Trauerkiebitz (256), 214, **33**
Trompeter-Hornvogel (455), 208, **8**
Tulukuckuck, Grillkuckuck (388), 210, **17**
Turmfalke (181), 216, **44**
Vaalpieper (719), 227, **89**
Vierfarbenwürger (747), 218, **55**
Wachtelastrild (852), 225, **83**
Wachtelkönig (211), 212, **27**
Waffenkiebitz, Schmiedekiebitz (258), 214, **33**
Wahlbergs Adler (135), 216, **41**
Waldnektarvogel, Stahlnektarvogel (793), 224, **79**
Waldwasserläufer (265), 213, **29**
Wanderfalke (171), 216, **44**
Wassertriel (298), 213, **31**
Weidelandpieper, (Spornpieper) (716), 227, **89**
Weißbart-Seeschwalbe (338), 214, **34**
Weißbauch-Nektarvogel (787), 224, **79**
Weißbrauen-Heckensänger (613), 223, **75**
Weißbrauenrötel (599), 223, **75**
Weißbrustkormoran (55), 216, **45**
Weißbrust-Raupenfänger (539), 219, **59**
Weißbürzelsegler (415), 210, **18**
Weißflankenschnäpper (701), 219, **57**
Weißflügel-Seeschwalbe (339), 214, **34**
Weißfuß-Atlaswitwe (865), 226, **86**
Weißgesicht-Ohreule (397), 211, **20**
Weißkehlrötel (602), 223, **75**
Weißrückenente (101), 206, **3**
Weißrückengeier (123), 215, **37**
Weißrücken-Nachtreiher (77), 217, **48**
Weißscheitelwürger (756), 219, **58**
Weißstirn-Regenpfeifer (246), 213, **32**
Weißstirnspint, Weißstirn-Bienenfresser (443), 209, **13**
Weißstirnweber (807), 225, **81**
Weißstorch (83), 218, **52**
Weißwangenlerche (515), 222, **72**
Wellenastrild (846), 225, **84**
Wermutregenpfeifer (252), 213, **32**
Wespenbussard (130), 214, **35**
Wiedehopf (451), 208, **9**
Wiesenweihe (166), 215, **38**
Witwenente (99), 206, **3**
Witwenstelze (711), 226, **88**
Wollhalsstorch (86), 218, **51**
Wollkopfgeier (125), 215, **37**
Woodfordkauz (394), 211, **21**
Würgerschnäpper (698), 223, **74**
Ziegenmelker, Nachtschwalbe (404), 211, **22**
Zimtroller (450), 208, **10**
Zistensänger (664), 222, **69**

Index to German Names

Zwergblatthühnchen (241), 212, **27**
Zwergflamingo (97), 217, **50**
Zwergrohrdommel (78), 217, **48**
Zwergsperber (157), 215, **39**

Zwergspint, Zwergbienenfresser (444), 209, **13**
Zwergstrandläufer (274), 213, **30**
Zwergtaucher (8), 216, **45**
Zwergteichhuhn (224), 213, **28**

KEY: (613) = Roberts' number, 223 = Etymology, **75** = Plate number

INDEX TO FRENCH NAMES

Agrobate à dos roux (613), 223, **75**
  à moustaches (617), 223, **75**
Aigle couronné (141), 216, **42**
  d'Ayres (138), 215, **40**
  de Verreaux (131), 216, **42**
  de Wahlberg (135), 216, **41**
  des steppes (133), 215, **41**
  fascié (137), 215, **40**
  huppard (139), 216, **42**
  martial (140), 216, **42**
  pomarin (134), 215, **41**
  ravisseur (132), 215, **41**
Aigrette ardoisée (69), 217, **47**
  garzette (67), 216, **46**
Alecto à bec rouge (798), 225, **81**
Alouette à nuque rousse (494), 222, **71**
  bourdonnante (496), 222, **71**
  brune (505), 222, **72**
  cendrille (507), 222, **72**
  fauve (497), 222, **72**
  monotone (493), 222, **71**
  sabota (498), 222, **71**
Amadine cou-coupé (855), 225, **83**
Amarante de Jameson (841), 226, **85**
  du Sénégal (842), 226, **85**
  foncé (840), 226, **85**
Amblyospize à front blanc (807), 225, **81**
Anhinga d'Afrique (60), 216, **45**
Anomalospize parasite (820), 226, **87**
Anserelle naine (114), 206, **3**
Apalis à gorge jaune (648), 222, **70**
  de Rudd (649), 222, **70**
Astrild ondulé (846), 225, **84**
  -caille à lunettes (852), 225, **83**
Autour gabar (161), 215, **39**
  noir (158), 215, **40**
  sombre (163), 215, **40**
  tachiro (160), 215, **39**
  unibande (154), 215, **39**
Autruche d'Afrique (1), 206, **1**
Avocette élégante (294), 213, **31**
Bagadais casqué (753), 219, **56**
  de Retz (754), 219, **56**
Balbuzard pêcheur (170), 214, **36**

Barbican à collier (464), 207, **7**
  pie (465), 207, **7**
  promépic (473), 207, **7**
Barbion à croupion jaune (471), 207, **7**
  à front jaune (470), 207, **7**
Bateleur des savanes (146), 215, **38**
Baza coucou (128), 214, **35**
Beaumarquet melba (834), 225, **84**
Bécasseau cocorli (272), 213, **30**
  minute (274), 213, **30**
  sanderling (281), 213, **30**
Bécassine africaine (286), 212, **27**
Bec-ouvert africain (87), 218, **51**
Bengali zébré (854), 225, **83**
Bergeronnette du Cap (713), 226, **88**
  pie (711), 226, **88**
  printanière (714), 226, **88**
Bihoreau à dos blanc (77), 217, **48**
  gris (76), 217, **48**
Blongios de Sturm (79), 217, **48**
  nain (78), 217, **48**
Bondrée apivore (130), 214, **35**
Bouscarle caqueteuse (638), 221, **65**
Bruant à poitrine dorée (884), 227, **91**
  cannelle (886), 227, **91**
  des rochers (887), 227, **91**
Brubru africain (741), 218, **54**
Bucorve du Sud (463), 208, **8**
Bulbul à poitrine jaune (574), 220, **63**
  à tête brune (575), 220, **63**
  importun (572), 220, **63**
  jaboteur (569), 220, **63**
  tricolore (568), 220, **63**
Busard cendré (166), 215, **38**
  pâle (167), 215, **38**
Buse des steppes (149), 214, **35**
Caille arlequin (201), 206, **2**
Calao à bec noir (457), 208, **8**
  à bec rouge (458), 208, **8**
  couronné (460), 208, **8**
  leucomèle (459), 208, **8**
  trompette (455), 208, **8**
Camaroptère à dos gris (-), 222, **70**
  à tête grise (657), 222, **70**

236   *Index to French Names*

de Stierling (659), 222, **70**
Canard à bec rouge (108), 206, **3**
 à bosse (115), 207, **4**
 noirâtre (105), 207, **4**
Capucin à dos brun (858), 226, **85**
 nonnette (857), 226, **85**
Chevalier aboyeur (270), 213, **29**
 cul-blanc (265), 213, **29**
 guignette (264), 213, **29**
 stagnatile (269), 213, **29**
 sylvain (266), 213, **29**
Chevêchette perlée (398), 211, **20**
Choucador à épaulettes rouges (764), 224, **77**
 à oreillons bleus (765), 224, **77**
 de Burchell (762), 224, **77**
 de Meves (763), 224, **77**
Chouette africaine (394), 211, **21**
 -pêcheuse de Pel (403), 211, **21**
Cichladuse à collier (603), 223, **76**
Cigogne blanche (83), 218, **52**
 d'Abdim (85), 218, **51**
 épiscopale (86), 218, **51**
 noire (84), 218, **51**
Circaète à poitrine noire (143), 215, **38**
 brun (142), 215, **38**
Cisticole à couronne rousse (681), 222, **69**
 à face rousse (674), 221, **68**
 des joncs (664), 222, **69**
 du désert (665), 222, **69**
 grinçante (672), 221, **68**
 paresseuse (679), 221, **68**
 roussâtre (675), 221, **68**
 striée (678), 222, **69**
Coliou quiriva (426), 209, **14**
 rayé (424), 209, **14**
Colombar à front nu (361), 212, **25**
Combassou du Sénégal (867), 226, **86**
 noir (864), 226, **86**
 violacé (865), 226, **86**
Combattant varié (284), 213, **30**
Corbeau à nuque blanche (550), 219, **58**
 pie (548), 219, **58**
Cordonbleu de l'Angola (844), 225, **84**
 grenadin (845), 225, **84**
Cormoran à poitrine blanche (55), 216, **45**
 africain (58), 216, **45**
Corvinelle noir et blanc (735), 219, **58**
Cossyphe à calotte rousse (600), 223, **75**
 de Heuglin (599), 223, **75**
 du Cap (601), 223, **75**
Coucal de Burchell (391), 210, **17**
 noir (388), 210, **17**
Coucou africain (375), 210, **16**
 criard (378), 210, **16**
 d'Audebert (383), 209, **15**
 de Klaas (385), 210, **16**

de Levaillant (381), 209, **15**
 didric (386), 210, **16**
 geai (380), 209, **15**
 gris (374), 210, **16**
 jacobin (382), 209, **15**
 solitaire (377), 209, **15**
Courvite à ailes bronzées (303), 214, **34**
 à triple collier (302), 214, **34**
 de Temminck (300), 214, **34**
Crabier chevelu (72), 217, **46**
Cratérope fléché (560), 221, **67**
Crombec à long bec (651), 220, **64**
Cubla boule-de-neige (740), 218, **54**
 boule-de-neige (740), 218, **54**
Dendrocygne à dos blanc (101), 206, **3**
 fauve (100), 206, **3**
 veuf (99), 206, **3**
Drongo brillant (541), 218, **53**
Échasse blanche (295), 213, **31**
Échenilleur à épaulettes jaunes (538), 219, **59**
 à ventre blanc (539), 219, **59**
Effraie des clochers (392), 210, **20**
 du Cap (393), 210, **20**
Élanion blanc (127), 214, **36**
Engoulevent à joues rousses (406), 211, **22**
 d'Europe (404), 211, **22**
 du Mozambique (409), 211, **22**
 musicien (405), 211, **22**
 pointillé (408), 211, **22**
 porte-étendard (410), 211, **22**
Épervier de l'Ovampo (156), 215, **39**
 minule (157), 215, **39**
 shikra (159), 215, **39**
Érémomèle à calotte verte (655), 220, **64**
 à cou roux (656), 220, **64**
 à croupion jaune (653), 220, **64**
Étourneau caronculé (760), 224, **78**
Euplecte à épaules blanches (829), 225, **82**
 à épaules orangées (828), 225, **82**
 ignicolore (824), 225, **82**
 veuve-noire (831), 225, **82**
 vorabé (826), 225, **82**
Eurocéphale à couronne blanche (756), 219, **58**
Faucon crécerelle (181), 216, **44**
 crécerellette (183), 216, **44**
 de Dickinson (185), 216, **43**
 de l'Amour (180), 216, **43**
 hobereau (173), 216, **43**
 lanier (172), 216, **44**
 pèlerin (171), 216, **44**
Fauvette des jardins (619), 221, **66**
Flamant nain (97), 217, **50**
 rose (96), 217, **50**
Foulque à crête (228), 213, **28**
Francolin coqui (188), 206, **1**
 de Shelley (191), 206, **1**

Index to French Names 237

de Swainson (199), 206, **1**
du Natal (196), 206, **1**
huppé (189), 206, **2**
Gallinule africaine (227), 212, **28**
  poule-d'eau (226), 212, **28**
Ganga bibande (347), 212, **26**
Gladiateur de Blanchot (751), 219, **56**
  quadricolore (747), 218, **55**
  soufré (748), 219, **56**
Glaréole à collier (304), 214, **34**
Gobemouche à lunettes (691), 223, **74**
  du Marico (695), 223, **73**
  fiscal (698), 223, **74**
  gris (689), 223, **74**
  mésange (693), 223, **74**
  pâle (696), 223, **73**
  sombre (690), 223, **74**
  sud-africain (694), 223, **73**
Gonolek boubou (736), 218, **55**
  d'Abyssinie (737), 218, **55**
  rouge et noir (739), 218, **55**
Grand Indicateur (474), 207, **5**
  -duc africain (401), 211, **21**
  -duc de Verreaux (402), 211, **21**
Grande Aigrette (66), 217, **46**
Grèbe castagneux (8), 216, **45**
Grébifoulque d'Afrique (229), 212, **28**
Guêpier à front blanc (443), 209, **13**
  à queue d'aronde (445), 209, **13**
  carmin (441), 209, **14**
  d'Europe (438), 209, **14**
  de Perse (440), 209, **13**
  nain (444), 209, **13**
Guifette leucoptère (339), 214, **34**
  moustac (338), 214, **34**
Gymnogène d'Afrique (169), 215, **41**
Héron à bec jaune (68), 217, **46**
  cendré (62), 217, **47**
  garde-boeufs (71), 217, **46**
  goliath (64), 217, **47**
  pourpré (65), 217, **47**
  strié (74), 217, **48**
Hibou du Cap (395), 211, **21**
Hirondelle à croupion gris (531), 220, **61**
  à gorge perlée (523), 220, **61**
  à longs brins (522), 220, **61**
  à ventre roux (524), 220, **62**
  de fenêtre (530), 220, **61**
  de rivage (532), 219, **60**
  des mosquées (525), 220, **62**
  du Ruwenzori (536), 220, **62**
  isabelline (529), 220, **60**
  paludicole (533), 220, **60**
  rustique (518), 220, **60**
  striée (527), 220, **62**
Huppe d'Afrique (451), 208, **9**

Hyliote australe (624), 221, **67**
Hypolaïs des oliviers (626), 221, **66**
  ictérine (625), 221, **66**
Ibis falcinelle (93), 217, **49**
  hagedash (94), 217, **49**
  sacré (91), 217, **49**
Indicateur de Wahlberg (478), 207, **5**
  varié (475), 207, **5**
Irrisor moqueur (452), 208, **9**
  namaquois (454), 208, **9**
Jabiru d'Afrique (88), 218, **52**
Jacana à poitrine dorée (240), 212, **27**
  nain (241), 212, **27**
Loriot d'Europe (543), 218, **53**
  masqué (545), 218, **53**
Marabout d'Afrique (89), 218, **52**
Martin-chasseur à tête brune (435), 209, **12**
  -chasseur à tête grise (436), 208, **11**
  -chasseur du Sénégal (433), 208, **11**
  -chasseur strié (437), 209, **12**
Martinet à ventre blanc (418), 210, **19**
  cafre (415), 210, **18**
  d'Ussher (422), 210, **18**
  de Böhm (423), 210, **18**
  des maisons (417), 210, **18**
  des palmes (421), 210, **19**
  du Cap (412), 210, **19**
  horus (416), 210, **18**
  noir (411), 210, **19**
Martin-pêcheur à demi-collier (430), 208, **11**
  -pêcheur géant (429), 209, **12**
  -pêcheur huppé (431), 208, **11**
  -pêcheur pie (428), 209, **12**
  -pêcheur pygmée (432), 208, **11**
Merle kurrichane (576), 223, **73**
  litsipsirupa (580), 223, **73**
Mésange nègre (554), 219, **59**
Messager sagittaire (118), 216, **42**
Milan d'Afrique (126), 214, **35**
  des chauves-souris (129), 214, **35**
Moineau bridé (805), 226, **87**
  domestique (801), 226, **87**
  sud-africain (804), 226, **87**
Moinelette à dos gris (516), 222, **72**
  à oreillons blancs (515), 222, **72**
Nette brune (113), 207, **4**
Oedicnème tachard (297), 213, **31**
  vermiculé (298), 213, **31**
Oie-armée de Gambie (116), 207, **4**
Ombrette africaine (81), 217, **49**
Outarde à ventre noir (238), 212, **26**
  houppette (237), 212, **26**
  kori (230), 212, **26**
Palmiste africain (147), 214, **36**
Parisome grignette (621), 221, **67**
Pélican blanc (49), 217, **50**

Perroquet à cou brun (-), 210, **17**
  à tête brune (363), 210, **17**
  de Meyer (364), 210, **17**
Petit Indicateur (476), 207, **5**
  -duc africain (396), 211, **20**
  -duc de Grant (397), 211, **20**
Phragmite des joncs (634), 221, **65**
Pic à queue dorée (483), 207, **6**
  barbu (487), 207, **6**
  cardinal (486), 207, **6**
  de Bennett (481), 207, **6**
Pie-grièche à poitrine rose (731), 219, **57**
  -grièche écorcheur (733), 219, **57**
Pigeon roussard (349), 212, **25**
Pintade de Numidie (203), 206, **2**
  de Pucheran (204), 206, **2**
Pipit à dos uni (718), 227, **89**
  africain (716), 227, **89**
  de Sundevall (720), 227, **89**
  des arbres (723), 227, **89**
  du Vaal (719), 227, **89**
Piqueboeuf à bec jaune (771), 224, **78**
  à bec rouge (772), 224, **78**
Pluvier à front blanc (246), 213, **32**
  à triple collier (249), 213, **32**
  asiatique (252), 213, **32**
  grand-gravelot (245), 213, **32**
  pâtre (248), 213, **32**
Pouillot fitis (643), 221, **66**
Prinia modeste (683), 222, **70**
Pririt à gorge noire (705), 219, **57**
  du Cap (700), 219, **57**
  molitor (701), 219, **57**
Pygargue vocifer (148), 214, **36**
Râle à bec jaune (213), 212, **28**
  des genêts (211), 212, **27**
  des prés (212), 212, **27**
Rémiz minute (558), 219, **59**
Rhynchée peinte (242), 212, **27**
Rolle violet (450), 208, **10**
Rollier à longs brins (447), 208, **10**
  à raquettes (448), 208, **10**
  d'Europe (446), 208, **10**
  varié (449), 208, **10**
Rossignol progné (609), 223, **76**
Rousserolle à bec fin (635), 221, **65**
  africaine (631), 221, **65**
  turdoïde (628), 221, **65**
  verderolle (633), 221, **66**
Rufipenne morio (769), 224, **78**
Sarcelle hottentote (107), 206, **3**
Sénégali de Verreaux (838), 225, **84**
  vert (835), 225, **83**
Sentinelle à gorge jaune (728), 226, **88**
Serin à poitrine citron (871), 227, **90**
  du Mozambique (869), 227, **90**

  gris (881), 227, **90**
  soufré (877), 227, **90**
Souimanga à collier (793), 224, **79**
  à poitrine rouge (791), 224, **79**
  à ventre blanchâtre (787), 224, **79**
  améthyste (792), 224, **79**
  de Mariqua (779), 224, **79**
Spatule d'Afrique (95), 217, **50**
Spréo améthyste (761), 224, **78**
Talève d'Allen (224), 212, **28**
Tantale ibis (90), 218, **52**
Tarier pâtre (596), 223, **76**
Tchagra à tête brune (743), 218, **54**
  à tête noire (744), 218, **54**
Tchitrec d'Afrique (710), 218, **53**
  du Cap (708), 218, **53**
Tisserin à lunettes (810), 224, **80**
  à tête rousse (814), 224, **80**
  écarlate (819), 225, **81**
  gendarme (811), 224, **80**
  intermédiaire (815), 224, **80**
  safran (816), 224, **80**
Touraco à huppe splendide (371), 212, **25**
  concolore (373), 212, **25**
Tourtelette émeraudine (358), 211, **24**
  masquée (356), 212, **25**
  Tourtelette tambourette (359), 211, **24**
Tourterelle à collier (352), 211, **24**
  du Cap (354), 211, **24**
  maillée (355), 211, **24**
  pleureuse (353), 211, **24**
Traquet à ventre roux (593), 223, **76**
  d'Arnott (594), 223, **76**
  familier (589), 223, **76**
Travailleur à bec rouge (821), 225, **81**
Trogon narina (427), 208, **9**
Turnix d'Andalousie (205), 206, **2**
Vanneau à tête blanche (259), 214, **33**
  armé (258), 214, **33**
  couronné (255), 214, **33**
  du Sénégal (260), 214, **33**
  terne (256), 214, **33**
Vautour à tête blanche (125), 215, **37**
  africain (123), 215, **37**
  charognard (121), 215, **37**
  chassefiente (122), 215, **37**
  oricou (124), 215, **37**
Veuve de paradis (862), 226, **86**
  dominicaine (860), 226, **86**
  royale (861), 226, **86**
Zostérops du Cap (796), 221, **67**
  jaune (797), 221, **67**

Index to French Names

KEY: (574) = Roberts' number, 220 = Etymology, **63** = Plate number

INDEX TO ZULU NAMES

iBhada (574), 220, **63**
iBhoboni (740), 218, **54**
iBhoboni/iGqumusha (736), 218, **55**
iBhoyi (678), 222, **69**
iBomvana/iNtakansinsi/isiGwe (824), 225, **82**
iDada (Z) (108), 206, **3**
 (105), 207, **4**
 (107), 206, **3**
iFefe (446), 208, **10**
 (447), 208, **10**
iFuba/iKhungula/iVuba (49), 217, **50**
iFubesi/isiKhova (402), 211, **21**
iGelegekle/iGeleja (810), 224, **80**
iGwababa/uGwabayi (548), 219, **58**
iGwalagwala (371), 212, **25**
iGwili/iNqomfi (728), 226, **88**
iHelkehle (560), 221, **67**
iHlabahlabane/isiQuba/isiXula (428), 209, **12**
iHlabankomo/iHlolamvula/iJankomo (412), 210, **19**
iHlalankomo/iHlalanyathi (771), 224, **78**
iHlalankomo/iHlalanyathi (772), 224, **78**
iHlokohloko (811), 224, **80**
 (814), 224, **80**
 (816), 224, **80**
iHophe (352), 211, **24**
iHophe/iHoye (116), 207, **4**
iHophe/uSamdokwe (354), 211, **24**
iHubulu/iWabayi (550), 219, **58**
iJankomo/uHlolamvula (411), 210, **19**
iJubantondo (361), 212, **25**
iKhunatha (455), 208, **8**
iKhwezi/iKhwinsi (764), 224, **77**
iKlebe/iMvumvuyane (160), 215, **39**
iKuwela (69), 217, **47**
iLanda (66), 217, **46**
iLanda/inGevu/umLindankomo (71), 217, **46**
iLongwe (102), 207, **4**
iMbuzi-yehlathi/ umBuzana (657), 222, **70**
iMpangele (203), 206, **2**
iMpangele-yehlathi/iNgekle (204), 206, **2**
iMpofazana (760), 224, **78**
iMvunduna (473), 207, **7**
iNcede/uQoyi (681), 222, **69**
iNdibilitshe (651), 220, **64**
iNdlazi (424), 209, **14**
iNdodosibona (378), 210, **16**
iNdudumela (258), 214, **33**
iNdwazela/uNongobotsha/uNongozolo (435), 209, **12**
iNgede/iNhlavebizelayo/uNomtsheketshe (474), 207, **5**
iNgekle (67), 217, **46**
 (68), 217, **46**
iNgongoni (747), 218, **55**

iNgqungqulu (146), 215, **38**
iNgqwathane/iNqwathane/iNtonso (793), 224, **79**
iNgududu/iNsingizi (463), 208, **8**
iNhlangu (538), 219, **59**
iNhlava (475), 207, **5**
 (476), 207, **5**
iNhlekabafazi/uNukani (452), 208, **9**
iNhlolamvula (529), 220, **60**
iNhlunuyamanzi/isiKhilothi/isiPhikeleli/uZangozolo (432), 208, **11**
iNhlunuyamanzi/isiKhilothi/uZangozolo (431), 208, **11**
iNkankane (94), 217, **49**
iNkanku (382), 209, **15**
iNkenkane/isiXulamasele (95), 217, **50**
iNkonjane (518), 220, **60**
 (522), 220, **61**
 (523), 220, **61**
 (527), 220, **62**
iNkotha (444), 209, **13**
iNkotha/uNongozolo (433), 208, **11**
iNkotha-enkulu (441), 209, **14**
iNkovana (398), 211, **20**
iNkovane/umShwelele (395), 211, **21**
iNkwazi (148), 214, **36**
iNqe (122), 215, **37**
 (124), 215, **37**
iNqoba (672), 221, **68**
iNqondaqonda (486), 207, **6**
iNsomi/iSomi (769), 224, **78**
iNswempe (188), 206, **1**
iNtaka/uJojo (831), 225, **82**
iNtaka/uMahube/uMangube (828), 225, **82**
iNtakansinsi (829), 225, **82**
iNtendele (191), 206, **1**
iNtengu (541), 218, **53**
iNtiyane (846), 225, **84**
iNtshe (1), 206, **1**
iNtungunono (118), 216, **42**
iNxenge/uNonklwe (852), 225, **83**
iNzwece/uVe (710), 218, **53**
iPhemvu/uThimbakazane (753), 219, **56**
iPhishamanzi/uLondo (58), 216, **45**
iPhothwe/iPogota (568), 220, **63**
iQumutsha-lamawa (593), 223, **76**
isAncaphela/isAnqawane/isiChegu (596), 223, **76**
iSibagwebe/isiQophamuthi/uSibagwebe (483), 207, **6**
isiBhelu/isiKhombazane-sehlathi (359), 211, **24**
isiGwaca (Z) (201), 206, **2**
isiHuhwa (141), 216, **42**
isiHuhwa/uKhozi (140), 216, **42**
isiKhombazane-sehlanze (358), 211, **24**
isiKhombazane-senkangala/uNkombose (356), 212, **25**

isiKhova/umShwelele (393), 210, **20**
isiKhova/umZwelele (392), 210, **20**
isiKhovampondo (401), 211, **21**
isiKhulukhulu/isiQonqotho (464), 207, **7**
isiKhwehle (189), 206, **2**
(196), 206, **1**
isiPhungumangathi (139), 216, **42**
isiQhophamnenke (87), 218, **51**
isiThandamanzi (86), 218, **51**
isiThwelathwela (440), 209, **13**
isiVuba (429), 209, **12**
isiXula (430), 208, **11**
iThandaluzibo/uNondwayiza (240), 212, **27**
iTitihoye (255), 214, **33**
(256), 214, **33**
iVevenyane (99), 206, **3**
Ivukuthu iJuba/iVukuthu (349), 212, **25**
iWamba (304), 214, **34**
iWili (572), 220, **63**
iWonde (55), 216, **45**
ubuCubu (840), 226, **85**
uDokotela/umNqube (700), 219, **57**
uFukwe/umGugwane (391), 210, **17**
uFumba/uNofunjwa (238), 212, **26**
uGaga/umBhekle (601), 223, **75**
uHeshe (171), 216, **44**
(172), 216, **44**
uHlaza (748), 219, **56**
(751), 219, **56**
uHlekwane (860), 226, **86**
uJojokhaya (862), 226, **86**
uKholwase/uNondwebu (96), 217, **50**
uKhonzane (355), 211, **24**
uKhozi (131), 216, **42**
(143), 215, **38**
uKlebe (154), 215, **39**
uMabhengwane/uNobathekeli (394), 211, **21**
uMandubulu (397), 211, **20**
uMathebeni/uTebetebana (181), 216, **44**
umBangaqhwa/umJenjana (297), 213, **31**
umBexe (589), 223, **76**
umBhalane (869), 227, **90**
umBhalane/ umDendeliswe (881), 227, **90**
umBhicongo/umQoqongo (545), 218, **53**

umBicini/ uMehlwane (796), 221, **67**
(797), 221, **67**
umDinasibula (886), 227, **91**
umJekejeke/umJengejenge (213), 212, **28**
umKholwane (458), 208, **8**
(460), 208, **8**
umKlewu (373), 212, **25**
umMbesi (694), 223, **73**
umNdweza (884), 227, **91**
umNgcelekeshu/umNgcelu (716), 227, **89**
(718), 227, **89**
umNgqithi (230), 212, **26**
umNguphane (744), 218, **54**
umNqube (701), 219, **57**
umNtoli (507), 222, **72**
umQonqotho (735), 219, **58**
uMqwayini (157), 215, **39**
umTshivovo (426), 209, **14**
umuNswi (576), 223, **73**
umVemve (711), 226, **88**
(713), 226, **88**
umXwagele (91), 217, **49**
umZolozolo/uZiningweni (451), 208, **9**
umZwingili (815), 224, **80**
uNcede (664), 222, **69**
uNgcede (678), 222, **69**
uNgoqo (205), 206, **2**
uNhloyile/uKholwe (126b), 214, **35**
uNobulongwe (300), 214, **34**
uNogolantethe/uNowanga (83), 218, **52**
uNokilonki (62), 217, **47**
(63), 217, **47**
uNondwebu (97), 217, **50**
uNononekhanda (386), 210, **16**
uNonqane (415), 210, **18**
uNozalizingwenyana (64), 217, **47**
uNununde (286), 212, **27**
uPhezukomkhono (377), 209, **15**
uQaqashe (496), 222, **71**
uSiba (76), 217, **48**
uThekwane (81), 217, **49**
uZavolo (404), 211, **22**
(405), 211, **22**

Index to Zulu Names

KEY: (394) = Roberts' number, 211 = Etymology, **21** = Plate number

INDEX TO XHOSA NAMES

Ibengwana (394), 211, **21**
Icelekwane/Uvelemaxhoseni (355), 211, **24**
Icelu/Icetshu (716), 227, **89**
   (718), 227, **89**
Icola (698), 223, **74**
Idada (105), 207, **4**
   (99), 206, **3**
   (100), 206, **3**
   (101), 206, **3**
Ifubesi (402), 211, **21**
Ifubesi/Isihulu-hulu (401), 211, **21**
Igabhoyi/Ubhoyi-bhoyi (678), 222, **69**
Igotyi (708), 218, **53**
Igqubusha (736), 218, **55**
Igwangqa/Iqabathule (494), 222, **71**
   (548), 219, **58**
Igxiya (255), 214, **33**
Ihashe (78), 217, **48**
Ihlabankomo/Ihlankomo (411), 210, **19**
   (412), 210, **19**
   (415), 210, **18**
Ihlabankomo/Ubhantom (418), 210, **19**
Ihlalanyathi (772), 224, **78**
Ihlolo (733), 219, **57**
Ihlungulu/Igrwababa/Umfundisi (550), 219, **58**
Ihobe/Untamnyama (354), 211, **24**
Ihobo-hobo (811), 224, **80**
   (814), 224, **80**
Ihotyazana (356), 212, **25**
Ihoye (116), 207, **4**
Ikhwebula (568), 220, **63**
Ikreza (810), 224, **80**
Ilanda (71), 217, **46**
Ilithwa (455), 208, **8**
Ilowe (102), 207, **4**
Ilunga Legwaba (382), 209, **15**
Impangele (203), 206, **2**
Incede (681), 222, **69**
Inciniba (1), 206, **1**
Indlasidudu/Umakhulu (352), 211, **24**
Indlazi (424), 209, **14**
Indweza (881), 227, **90**
Indweza/Indweza Eluhlaza (877), 227, **90**
Ing'ang'ane (94), 217, **49**
Ingcaphe/Isangcaphe (596), 223, **76**
Ingcungcu (792), 224, **79**
Ingcwanguba (49), 217, **50**
Ingedle/Unongedle (700), 219, **57**
Ingolwane (205), 206, **2**
Ingqanga (146), 215, **38**

Ingqangqolo (297), 213, **31**
   (298), 213, **31**
Ingqolane/Unomakhwezana (148), 214, **36**
Ingwamza/Unowanga (83), 218, **52**
Ingxangxosi (118), 216, **42**
Ingxenge/Ungxenge (857), 226, **85**
Inkonjane (527), 220, **62**
Inkonjane/Ucelizapholo/Udlihashe (518), 220, **60**
Inkonjane/Unomalahlana (536), 220, **62**
Inkonjane/Unongubendala/Unongubende (529), 220, **60**
Inkwili (572), 220, **63**
Inqatha/Unokrekre (249), 213, **32**
Inqathane (793), 224, **79**
Intakazana/Ujobela (831), 225, **82**
Intakembila/Unomaswana (740), 218, **54**
Intakobusi (474), 207, **5**
   (475), 207, **5**
   (476), 207, **5**
Intakomlilo/Ucumse/Umlilo (824), 225, **82**
Intambanane/Uthebe-thebana (181), 216, **44**
Intendekwane (361), 212, **25**
Intengu (541), 218, **53**
Intibane/Intutyane (507), 222, **72**
Intlekibafazi (452), 208, **9**
Intsasa (884), 227, **91**
Intsasana (720), 227, **89**
Intshatshongo (427), 208, **9**
Intshili (426), 209, **14**
Intshiyane (846), 225, **84**
Intsingizi/Intsikizi (463), 208, **8**
Intukwane (796), 221, **67**
Inyakrili/Inyakrini (764), 224, **77**
Isahomba/Isakhomba (828), 225, **82**
Isangxa (149), 214, **35**
Isavu (359), 211, **24**
Isaxwila (428), 209, **12**
   (430), 208, **11**
   (431), 208, **11**
Iseme (230), 212, **26**
Isicibilili (840), 226, **85**
Isicukujeje (554), 219, **59**
Isikhova (392), 210, **20**
   (393), 210, **20**
Isikhwalimanzi/Ukhwalimanzi (62), 217, **47**
   (63), 217, **47**
Isikretyane/Unongungu (589), 223, **76**
Isilwangangubo (124), 215, **37**
Isinagogo (464), 207, **7**
Isinqolamthi (486), 207, **6**
Isiphungu-phungu/Uphungu-phungu (139), 216, **42**

Isomi (769), 224, **78**
Ivukazana (358), 211, **24**
Ivuzi (60), 216, **45**
Ixhalanga (122), 215, **37**
Segôdi/Untloyiya/Untloyila (126b), 214, **35**
Ubhobhoyi (451), 208, **9**
Ubikhwe (391), 210, **17**
Ucelithafa (300), 214, **34**
Ucofuza/Undofu (65), 217, **47**
Udebeza (404), 211, **22**
  (405), 211, **22**
Ugaga (601), 223, **75**
Ugwidi (55), 216, **45**
Ugwidi (58), 216, **45**
Uhlakhwe/Ujobela (860), 226, **86**
Ujejane/Unomaphelana (710), 218, **53**
Ukhetshana (157), 215, **39**
Ukhetshe (171), 216, **44**
  (172), 216, **44**
Ukhozi (140), 216, **42**
  (141), 216, **42**
  (132), 215, **41**
  (133), 215, **41**
Ukhozi/Untsho (131), 216, **42**
Ulubisi/Umphungeni (167), 215, **38**
Umbhankro (751), 219, **56**
Umcelu/Umvemve/Umventshana (711), 226, **88**
Umcelu/Umvemve/Umventshana (713), 226, **88**
Umdlampuku/Unongwevana (127), 214, **36**
Umgcibilitshane (386), 210, **16**
Umkholwane (460), 208, **8**

Umkro/Umqokolo (545), 218, **53**
Umnguphane (744), 218, **54**
Umnquduluthi (286), 212, **27**
Umthethi/Usinga Olumnyama (538), 219, **59**
Undenjenje/Undenzeni (886), 227, **91**
Undozela (435), 209, **12**
Undyola/Unondyola (701), 219, **57**
Ungcuze (683), 222, **70**
Ungxengezi/Uqume (678), 222, **69**
Unocegceya/Unotelela (246), 213, **32**
Unocofu (84), 218, **51**
Unogushana/Unothoyi (558), 219, **59**
Unokrekre (245), 213, **32**
Unolwilwilwi/Unonyamembi (8), 216, **45**
Unomakhwane (638), 221, **65**
Unomanyuku/Unome (657), 222, **70**
Unomaphelana (690), 223, **74**
Unomkqayi/Unompemvana (228), 213, **28**
Unomntanofayo (378), 210, **16**
Unonkxwe (852), 225, **83**
Unonzwi (664), 222, **69**
  (643), 221, **66**
Unowambu/Uwambu (760), 224, **78**
Unyileyo (869), 227, **90**
Uphendu (270), 213, **29**
Uphezukomkhono (377), 209, **15**
Uqhimngqoshe/Uthekwane (81), 217, **49**
Uthuthula (264), 213, **29**
  (266), 213, **29**
Uxomoyi (429), 209, **12**

KEY: (568) = Roberts' number, 220 = Etymology, **63** = Plate number

INDEX TO TSONGA NAMES

Bhokota (568), 220, **63**
Dzandza/Munyangana/Muthecana/Nyonimahlopi
  (71), 217, **46**
Dzurhini/N'hwarixigwaqa /Xigwatla (201), 206, **2**
Gama (131), 216, **42**
Ghama (Generic for eagles) (132), 215, **41**
  (135), 216, **41**
  (137), 215, **40**
  (142), 215, **38**
Ghelekela (255), 214, **33**
  (256), 214, **33**
  (258), 214, **33**
Ghongoswana (481), 207, **6**

  (483), 207, **6**
  (486), 207, **6**
  (487), 207, **6**
Ghumba (89), 218, **52**
Gororo/Nyakolwa (60), 216, **45**
Gugurhwana (355), 211, **24**
Gumba/Ntsavila/Xaxari (83), 218, **52**
Gumbula/Khungulu/Manawavembe/Xilandzaminonga
  (49), 217, **50**
Gungwa/Ngungwamawala (147), 214, **36**
Gwavava/Ukuuku (550), 219, **58**
Hlalala/Nhlampfu (475), 207, **5**
Hokwe (364), 210, **17**

Index to Tsonga Names

Holiyo/Hwiyo (728), 226, **88**
Hukunambu/Nkukumezane (213), 212, **28**
Hwilo/Samjukwa/Xighigwa (736), 218, **55**
Indlazi (426), 209, **14**
Juka/Rhiyani (731), 219, **57**
Kavakavana/Xikavakava (181), 216, **44**
 (180), 216, **43**
Khopola/Nyakopo (352), 211, **24**
Khoti/Mavalanga (122), 215, **37**
Koti (124), 215, **37**
 (123), 215, **37**
 (125), 215, **37**
 (generic for vultures) (121), 215, **37**
Kubhasti (405), 211, **22**
 (409), 211, **22**
Kukumezane (226), 212, **28**
Kukumezani (228), 213, **28**
 (224), 213, **28**
Kurkurtavoni (397), 211, **20**
Kwezu elimhlope (760), 224, **78**
Kwezu leri kulu (762), 224, **77**
Kwezu leri tsongo (764), 224, **77**
 (765), 224, **77**
Lesogo/Letsiakarana/MmamolangwaneMantsentse
 (188), 206, **1**
Mahulwana/Ribyatsane (404), 211, **22**
Mahulwana/Ribyatsane/Riwuvawuva (408), 211, **22**
 (406), 211, **22**
Mampfana (118), 216, **42**
Mandlakeni (594), 223, **76**
Mandonzwana (81), 217, **49**
Mandzedzerekundze/Matsherhani/N'wapesupesu
 (713), 226, **88**
Mangatlu (126b), 214, **35**
Mangoko/Xiganki (204), 206, **2**
Mankhudu (394), 211, **21**
 (398), 211, **20**
Manqiti (796), 221, **67**
 (797), 221, **67**
Manswikidyani/Risunyani/ Ritswiri (869), 227, **90**
Mantengu (541), 218, **53**
Mantsiyana (672), 221, **68**
Mantunje/Xikhungumala (826), 225, **82**
Mapuluhweni (494), 222, **71**
 (493), 222, **71**
 (497), 222, **72**
Masworhimasworhi (139), 216, **42**
Matinti (664), 222, **69**
 (678), 222, **69**
 (681), 222, **69**
 (665), 222, **69**
Matsinyani (683), 222, **70**
Mavungana/N'waripetani/Xitserere (430), 208, **11**
Mbawulwana/Nyenga (522), 220, **61**

 (523), 220, **61**
 (529), 220, **60**
 (533), 220, **60**
Mbyhiyoni (576), 223, **73**
Mfukwana (391), 210, **17**
Mghubhana lokhulu (733), 219, **57**
Mghubhana lowu kulu (744), 218, **54**
Mghubhana lowu tsongo (743), 218, **54**
Mhanghela (203), 206, **2**
Mithisi (230), 212, **26**
Mitikahincila (862), 226, **86**
Mtsherhitani (613), 223, **75**
Mtshikuyana (generic for dikkops) (297), 213, **31**
 (298), 213, **31**
Muhladzanhu/Muhlagambu (440), 209, **13**
 (443), 209, **13**
Mukyindlopfu (87), 218, **51**
Musoho (393), 210, **20**
N'wafayaswitlangi (91), 217, **49**
N'walanga/Xinyamukhwarani (648), 222, **70**
N'waminungu (860), 226, **86**
N'wantshekutsheku (264), 213, **29**
N'wantshekutsheku/Xitsatsana/Xitshekutsheku
 (266), 213, **29**
 (270), 213, **29**
N'wantshekutsheku/Xitsekutseku (249), 213, **32**
Ncilongi (735), 219, **58**
Ndukuzani (543), 218, **53**
Ndzheyana (generic for weavers) (811), 224, **80**
 (814), 224, **80**
 (815), 224, **80**
 (821), 225, **81**
Ndzheyana ya nhloko ya ka phsuku (819), 225, **81**
Nghelekele (260), 214, **33**
Nghotsana (167), 215, **38**
Nghunghwa (148), 214, **36**
Nghututu (463), 208, **8**
Nghwamba (361), 212, **25**
Nghwari (189), 206, **2**
Nghwari ma ntshengwhayi (196), 206, **1**
Nghwari ya xidhaka (199), 206, **1**
Nglhazi (710), 218, **53**
Ngoko (473), 207, **7**
Ngulukwani (55), 216, **45**
Ngwafalantala (377), 209, **15**
Ngwamhlanga (88), 218, **52**
Nhlalala (generic for honeyguides) (474), 207, **5**
Nhlalala (476), 207, **5**
Njenjele (191), 206, **1**
Nkhonyana (441), 209, **14**
Nkhunsi (402), 211, **21**
 (411), 210, **19**
 (412), 210, **19**
 (415), 210, **18**

(416), 210, **18**
(418), 210, **19**
(417), 210, **18**
(421), 210, **19**
Nkorho (455), 208, **8**
(458), 208, **8**
(457), 208, **8**
(459), 208, **8**
Nkorhonyarhi (460), 208, **8**
Nkwenyana (371), 212, **25**
(373), 212, **25**
Nqcunu (651), 220, **64**
Nwagogosane (464), 207, **7**
Nwapyopyamhanya (779), 224, **79**
(792), 224, **79**
(793), 224, **79**
(787), 224, **79**
(791), 224, **79**
Nwarikapanyana (127), 214, **36**
Nyarhututu (600), 223, **75**
Nyengha (518), 220, **60**
(527), 220, **62**
(525), 220, **62**
Nyengha leyi kulu (524), 220, **62**
Patu ra nhova (115), 207, **4**
Phamahumu/Phamanyarhi (545), 218, **53**
Phavomu (740), 218, **54**
Pupupu (451), 208, **9**
Qigwana (548), 219, **58**
Rhakweni/rhanciyoni (884), 227, **91**
Rigamani/Rikhozi (171), 216, **44**
(173), 216, **43**
Rijajani (857), 226, **85**
(858), 226, **85**
Rikhozi (140), 216, **42**
(154), 215, **39**
(172), 216, **44**
Rikolwa (65), 217, **47**
Sekwa (100), 206, **3**
(101), 206, **3**
(102), 207, **4**
(107), 206, **3**
(108), 206, **3**
Sekwagongwana/Sekwanyarhi (116), 207, **4**
Sowa (810), 224, **80**
(816), 224, **80**
Tihunyi (382), 209, **15**
(381), 209, **15**
Tinziwolana (444), 209, **13**
(438), 209, **14**
Tlekedhwana (560), 221, **67**
Tshembyana (303), 214, **34**
Tshivhovo (424), 209, **14**

Tshololwana (428), 209, **12**
(429), 209, **12**
(431), 208, **11**
(432), 208, **11**
(433), 208, **11**
(435), 209, **12**
(436), 208, **11**
(437), 209, **12**
Tuba (353), 211, **24**
Tuva (354), 211, **24**
Urhiana (753), 219, **56**
Urhiana (754), 219, **56**
Urimakutata/Vhumakutata (498), 222, **71**
Vhevhe (446), 208, **10**
(447), 208, **10**
(449), 208, **10**
Xicololwana lexi kulu (238), 212, **26**
Xicololwana lexi tsongo (237), 212, **26**
Xidzhavadzhava (554), 219, **59**
Xidzingirhi (841), 226, **85**
(842), 226, **85**
(854), 225, **83**
Xighonyombha (798), 225, **81**
Xighwaraghwara (347), 212, **26**
Xihelagadzi (505), 222, **72**
(718), 227, **89**
Xikhotlwana (396), 211, **20**
Xikhotlwani (395), 211, **21**
Xikohlwa hi jambo (94), 217, **49**
Xikwhezana (161), 215, **39**
Ximgenngwamangwami (701), 219, **57**
Ximongwe (146), 215, **38**
Xindzingiri (844), 225, **84**
Xindzingiri bhanga (846), 225, **84**
(834), 225, **84**
(845), 225, **84**
Xingede (659), 222, **70**
Xinkhovha (392), 210, **20**
Xinwavulombe (761), 224, **78**
Xinyankakeni (113), 207, **4**
Xithaklongwa (143), 215, **38**
Xitsatsana (205), 206, **2**
Xivhambalana (356), 212, **25**
(358), 211, **24**
Xiwambalane (359), 211, **24**
Xiyahkokeni (99), 206, **3**
Xiyinha (401), 211, **21**
Yandhana (772), 224, **78**
Yhokwe (363), 210, **17**
Yinca (1), 206, **1**
Yokoywana (452), 208, **9**
(454), 208, **9**

Index to Tsonga Names

KEY: (846) = Roberts' number, 225 = Etymology, **84** = Plate number

INDEX TO SOUTH SOTHO NAMES

Borahane/Borane (846), 225, **84**
 (854), 225, **83**
Borane (842), 226, **85**
Bwoto/Chigwenhure/MugwetureHlakahlotoana
 (568), 220, **63**
Fariki (424), 209, **14**
 (426), 209, **14**
Fiolo-'malisakhana (159), 215, **39**
Fiolo-ea-meru (160), 215, **39**
Hlatsinyane/Tlhatsinyane (596), 223, **76**
Kapantsi-ea-meru (710), 218, **53**
Kapantsi-ea-meru (710), 218, **53**
Kapantsi-tubatubi (689), 223, **74**
Khajoane (149), 214, **35**
Khaka (203), 206, **2**
Khoale (199), 206, **1**
Khoho-ea-lira/Khoalira (297), 213, **31**
Khohonoka (226), 212, **28**
Khoitinyane (78), 217, **48**
Khube/Thaha-khube/Thaha-khubelu (824), 225, **82**
Koe-koe-lemao/Seealemabopo-hetlatšoeu (264), 213, **29**
Koe-koe-lemao/Seealemabopo-holo (270), 213, **29**
Koe-koe-lemao/Seealemabopo-khoali (266), 213, **29**
Koe-koe-lemao/Seealemabopo-se-maroboko (284), **213**,
 30
Koe-koe-lemao/Motjoli-matsana/Koekoe-lemao (286),
 212, **27**
Kokolofitoe (76), 217, **48**
 (72), 217, **46**
Kokolofitoe (-hloontšo) (63), 217, **47**
Kokolofitoe (-kholo) (64), 217, **47**
Kokolofitoe (-lalatšehla) (65), 217, **47**
Kokolofitoe (-putsoa) (62), 217, **47**
Kokomoru (486), 207, **6**
Kopaope (464), 207, **7**
Koto-li-peli/Lekheloha/'Mamolangoane (118), 216, **42**
Leeba/Lehoboi/Leeba-la-thaba (349), 212, **25**
Leebamosu/Leebana-khoroana/Leebana (352), 211, **24**
Leebana-khoroana/Lekunkuroane (354), 211, **24**
Leebana-khoroana/Mphubetsoana (355), 211, **24**
Lefaloa (102), 207, **4**
Lefokotsane (523), 220, **61**
Lefokotsane/'Malinakana/Lekabelane (518), 220, **60**
Lehalanyane/Leholotsoane (91), 217, **49**
Lehaqasi (411), 210, **19**
 (412), 210, **19**
 (415), 210, **18**
 (416), 210, **18**
 (417), 210, **18**
Lehaqasi/Lehaqasi-le-lephatsoa (418), 210, **19**
Leholi (760), 224, **78**
Leholi-piloane (764), 224, **77**

Leholosiane (66), 217, **46**
 (67), 216, **46**
 (68), 217, **46**
Leholosiane/Leholotsiane (71), 217, **46**
Lekabelane (524), 220, **62**
 (529), 220, **60**
 (530), 220, **61**
Lekabelane/Sekatelane (533), 220, **60**
Lekekeruane/Letletleruane (255), 214, **33**
Lekhoaba/Moqukubi S-So), (550), 219, **58**
Lekolikotoana/Lekolukotoana (852), 225, **83**
Lekololoane/Mokoroane/Roba-re-bese (85), 218, **51**
Lengaangane/Lengangane (94), 217, **49**
Lenong/Letlaka (122), 215, **37**
Letata (101), 206, **3**
 (107), 206, **3**
 (113), 207, **4**
 (99), 206, **3**
Letata/Letata-la-noka (105), 207, **4**
Letata/Sefuli (108), 206, **3**
Letholopje (811), 224, **80**
Letlaka-pipi (124), 215, **37**
Letlerenyane/Letleretsane (589), 223, **76**
Letolopje/Thaha (814), 224, **80**
Letsikhoi/Letsikhui (116), 207, **4**
Letšoemila/Letšomila (769), 224, **78**
Leubane/Phakoe (171), 216, **44**
'Maborokoane (886), 227, **91**
Mabuaneng/Mauaneng (205), 206, **2**
Makhohlo/Morubisi/Sehihi/Sephooko (401), 211, **21**
'Maliberoane/'Mamphemphe (515), 222, **72**
'Malioache (473), 207, **7**
'Mamarungoana/Selahlamarungoana/Selahlamarumo
 (860), 226, **86**
Mamasianoke/Masianoke (81), 217, **49**
'Mamenotoana-nala (295), 213, **31**
Mamphoko (161), 215, **39**
'Mankholi-kholi/Kholokholo (126b), 214, **35**
Mohakajane (548), 219, **58**
Mohetle/Mohoetle/Tšumu/Boleseboko (228), 213, **28**
Mohofa (391), 210, **17**
Mojalipela/Moja-lipela/Seoli/Ntsu (131), 216, **42**
Mokhoroane/Mokhorane (356), 212, **25**
Mokoroane (84), 218, **51**
Mokoroane/Mokotatsie (83), 218, **52**
Mokotatsie (90), 218, **52**
 (90), 218, **52**
Molepe/Thaha/Tjobolo (831), 225, **82**
Molisa-linotši (476), 207, **5**
 (478), 207, **5**
Molomo-khaba (95), 217, **50**
Montoe-phatšoa (382), 209, **15**

Mo-otla-tšepe (258), 214, **33**
Motintinyane (664), 222, **69**
  (665), 222, **69**
  (681), 222, **69**
Motjoli (711), 226, **88**
  (713), 226, **88**
Mpšhe/Mpshoe (1), 206, **1**
Ntetekeng (386), 210, **16**
Ntsu (132), 215, **41**
Patapeta-nala (249), 213, **32**
Phakoana-mafieloana/Phakoana-tšooana/Phakoana-tšoana (127), 214, **36**
Phakoe (172), 216, **44**
  (173), 216, **43**
Popopo/Khapopo/Pupupu/'Mamokete (451), 208, **9**
Sebabole-hetlantšo (543), 218, **53**
Seinoli (428), 209, **12**
  (429), 209, **12**
  (431), 208, **11**
Seitlhoaeleli (167), 215, **38**
  (169), 215, **41**
Semanama (404), 211, **22**
  (406), 211, **22**
Seotsanyana (180), 216, **43**
  (181), 216, **44**
  (183), 216, **44**
Sephooko (392), 210, **20**
  (393), 210, **20**
  (395), 211, **21**
Serobele (801), 226, **87**
  (804), 226, **87**
Serokolo (465), 207, **7**
Sethoenamoru/Sethoena-moru/Setholo-moru (601), 223, **75**
Setona-mahloana (796), 221, **67**
Soamahlaka-kholo (628), 221, **65**
Soamahlaka-ntšitšoeu (635), 221, **65**
Soamahlaka-sa-jarete (619), 221, **66**
Soamahlaka-sa-mefero (631), 221, **65**
Tetsa-kolilo (378), 210, **16**
Thaha (821), 225, **81**
Thaha-pinyane/Thaha-tsehle/Tsehle/Thaha (826), 225, **82**
Thesta-balisana/Tsiroane (507), 222, **72**
Thoboloko (8), 216, **45**
Timba/Pilipili-sa-mabelete] (643), 221, **66**
Timeletsane-botha (55), 216, **45**
Timeletsane-lalanoha (60), 216, **45**
Timeletsane-ntšo (58), 216, **45**
Tlo-nke-tsoho (377), 209, **15**
Tšase (718), 227, **89**
  (719), 227, **89**
Tšase/Tšaase-ea-lithota (716), 227, **89**
Tsehlo/Molisa-linotši (474), 207, **5**
Tšemeli (731), 219, **57**
  (733), 219, **57**
Tsiroane (494), 222, **71**
Tsititsiti-nyenyane (274), 213, **30**
Tšoere (869), 227, **90**
  (881), 227, **90**

KEY: (203) = Roberts' number, 206 = Etymology, **2** = Plate number

INDEX TO NORTH SOTHO NAMES

Kgaka (203), 206, **2**
Kgogomeetse/Kgogonoka (226), 212, **28**
Kgori (230), 212, **26**
Kgoropo (460), 208, **8**
Kukuku (451), 208, **9**
Leaba Kgorwana (354), 211, **24**
Leakaboswana le Lešweu (83), 218, **52**
Lebudiane (188), 206, **1**
Lefaloa (102), 207, **4**
Legodi (764), 224, **77**
Legokobu (548), 219, **58**
Lenong le Leso (124), 215, **37**
Leribisi (392), 210, **20**
Lerwerwe (821), 225, **81**
Letswiyobaba (426), 209, **14**
Leuwauwe (405), 211, **22**
Madšadipere (71), 217, **46**
Makgohlo (393), 210, **20**
Matlakela (447), 208, **10**
Mmakaitšimeletša (89), 218, **52**
Mmakgwadi (710), 218, **53**
Mmankgôdi (126b), 214, **35**
Mokowe (373), 212, **25**
Moletašaka (713), 226, **88**
Mororwane (255), 214, **33**
Mpšhe (1), 206, **1**
Ntshukôbôkôbô, (132), 215, **41**
Papalagae (364), 210, **17**

Pekwa (172), 216, **44**
Rankgwetšhe (568), 220, **63**
Rankwitšidi (538), 219, **59**
Tangtang (664), 222, **69**
Thaga, (811), 224, **80**

Thagalehlaka (824), 225, **82**
Theko (541), 218, **53**
Thlame (118), 216, **42**
Tshetlo (474), 207, **5**

INDEX TO AFRIKAANS NAMES

Aasvoël, Krans-	86	**Duif**, Grootring-	60	Rooikop-	26
Monnik-	86	Krans-	62	**Hyliota**, Mashona-	146
Swart-	86	Papegaai-	62	**Ibis**, Glans-	110
Wit-	84	**Duifie**, Groenvlek-	60	**Jakkalsvoël**, Bruin-	82
Witkop-	86	Namakwa-	62	**Janfrederik**, Gewone	162
Witrug-	86	Rooibors-	60	Heuglinse	162
Arend, Breëkop-	96	Witbors-	60	Natal-	162
Bruin-	94	**Duiker**, Riet	102	Witkeel-	162
Gevlekte	94	Witbors-	102	**Kakelaar**, Rooibek-	30
Grootjag-	92	**Eend**, Fluit-	18	Swartbek	30
Kleinjag-	92	Gevlekte	18	**Kalkoentjie**, Geelkeel-	188
Kroon-	96	Knobbel-	20	**Kanarie**, Dikbek-	192
Langkuif-	96	Nonnetjie-	18	Geelbors-	192
Roof-	94	Rooibek-	18	Geeloog-	192
Steppe-	94	Swart-	20	Streepkop-	192
Vis-	84	Witrug-	18	Vaalstreep-	194
Witkruis-	96	**Elsie**, Bont-	74	**Kapokvoël**, Grys-	130
Berghaan	88	Rooipoot-	74	**Katakoeroe**, Swart-	130
Bleshoender	68	**Fisant**, Bosveld-	14	Witbors-	130
Blouvinkie, Gewone	184	Natalse	14	**Katlagter**, Pylvlek-	146
Staal-	184	**Flamink**, Groot-	112	**Kemphaan**	72
Witpoot-	184	**Flap**, Kortstert-	176	**Kiewiet**, Asiatiese Strand-	76
Bontrokkie, Gewone	164	Rooikeel-	176	Bont-	78
Bosbontrokkie, Beloog-	126	Witvlerk-	176	Driebandstrand-	76
Kaapse	126	**Fret**, Gewone	182	Geelborsstrand-	76
Witlies-	126	Rooirug-	182	Kleinswartvlerk-	78
Boskraai, Gewone	28	**Gans**, Dwerg-	18	Kroon-	78
Bromvoël	28	Kol-	20	Lel-	78
Byvreter, Blouwang-	38	**Geelvink**, Klein-	172	Ringnekstrand-	76
Europese	40	Swartkeel-	172	Vaalstrand-	76
Klein-	38	**Glasogie**, Geel-	146	Witkop-	78
Rooibors-	40	Kaapse	146	**Klappertjie**, Laeveld-	154
Rooikeel-	38	**Gompou**	64	**Kleinjantjie**, Geelbors-	152
Swaelstert-	38	**Hadeda**	110	Ruddse	152
Dassievoël	164	**Hamerkop**	110	**Klopkloppie**, Woestyn-	150
Diederikkie	44	**Heuningvoël**, Skerpbek-	22	**Koekoek**, Afrikaanse	44
Dikkop, Gewone	74	**Heuningwyser**, Gevlekte	22	Dikbek-	42
Water-	74	Groot-	22	Europese	44
Dobbertjie, Klein-	102	Klein-	22	Gevlekte	42
Drawwertjie, Bronsvlerk-	80	**Hoephoep**	30	Swart-	44
Drieband-	80	**Houtkapper**, Bont-	26	**Koester**, Bosveld-	190
Trek-	80	Kuifkop-	26	Donker-	190

248 Index to Afrikaans Names

Gestreepte	190		Huis-	186		**Rooiassie**	178
Gewone	190		**Muisvoël**, Gevlekte-	40		**Rooibekkie**, Koning-	184
Vaal-	190		Rooiwang-	40		Pylstert-	184
Kolpensie, Groen-	178		**Nagtegaal**, Lyster-	164		**Ruiter**, Bos-	70
Rooskeel-	180		**Naguil**, Afrikaanse	56		Gewone	70
Konkoit	122		Donker-	56		Groenpoot-	70
Korhaan, Bos-	64		Europese	56		Moeras-	70
Langbeen-	64		Laeveld-	56		Witgat-	70
Kraai, Witbors-	128		Rooiwang-	56		**Sandpatrys**, Dubbelband-	64
Withals-	128		Wimpelvlerk-	56		**Sanger**, Bruinkeelbos-	140
Kwartel, Bont-	14		**Neddikkie**	150		Donkerwangbos-	140
Kwartelkoning	66		**Neushoringvoël**, Geelbek-	28		Europese Riet-	144
Kwarteltjie, Bosveld-	16		Gekroonde	28		Europese Vlei-	142
Kwartelvinkie, Gewone	178		Grys-	28		Geelpensbos-	140
Kwêkwêvoël, Groenrug-	152		Rooibek-	28		Grootriet-	142
Grysrug-	152		**Nikator**, Geelvlek-	138		Hof-	144
Kwelea, Rooibek-	174		**Nimmersat**	116		Kaapse Riet-	142
Kwêvoël	62		**Nuwejaarsvoël**, Bont-	42		Kaapse Vlei-	142
Kwikkie, Bont-	188		Gestreepte-	42		Kleinriet-	142
Geel-	188		**Ooievaar**, Grootswart-	114		Olyfboom-	144
Gewone	188		Kleinswart-	114		Spot-	144
Laksman, Bontrok-	120		Oopbek-	114		Stierlingse	152
Grys-	126		Saalbek-	116		Tuin-	144
Kremetart-	128		Wit-	116		**Sekretarisvoël**	96
Langstert-	128		Wolnek-	114		**Skoorsteenveër**	110
Oranjeborsbos-	124		**Papegaai**, Bosveld-	46		**Slangarend**, Bruin-	88
Rooibors-	122		Bruinkop-	46		Swartbors-	88
Rooirug-	126		Savanne-	46		**Slanghalsvoël**	102
Swarthelm-	124		**Paradysvink**, Gewone	184		**Sneeubal**	120
Withelm-	124		**Patrys**, Bos-	16		**Snip**, Afrikaanse	66
Langstertjie, Bruinsy-	152		Laeveld-	14		Goud-	66
Langtoon, Dwerg-	66		**Pelikaan**, Wit-	112		**Speg**, Baard-	24
Groot-	66		**Piek**, Bont-	164		Bennettse	24
Lepelaar	112		**Piet-my-vrou**	42		Goudstert-	24
Lewerik, Bosveld-	154		**Reier**, Blou-	106		Kardinaal-	24
Donker-	156		Dwergriet-	108		**Spekvreter**, Gewone	164
Grysrug-	156		Geelbekwit-	104		**Sperwer**, Afrikaanse	90
Rooikop-	156		Gewone Nag-	108		Gebande	90
Rooinek-	154		Groenrug-	108		Klein-	90
Rooirug-	156		Grootwit-	104		Ovambo-	90
Sabota-	154		Kleinriet-	108		Swart-	90
Vaalbruin-	156		Kleinwit-	104		**Spookvoël**	124
Loerie, Bloukuif-	62		Ral-	104		**Spreeu**, Grootblouoorglans-	166
Bos-	30		Reuse-	106		Grootglans-	166
Lyster, Gevlekte	158		Rooi-	106		Kleinglans-	166
Palmmôre-	164		Swart-	106		Langstertglans-	166
Rooibek-	158		Swartkop-	106		Lel-	168
Makoe, Wilde-	20		Vee-	104		Rooivlerk-	168
Maraboe	116		Witrugnag-	108		Witbors-	168
Mees, Gewone Swart-	130		**Renostervoël**, Geelbek-	168		**Sprinkaanvoël**, Rooivlerk-	80
Meitjie	44		Rooibek-	168		**Stekelstert**, Gevlekte	48
Melba, Gewone	180		**Riethaan**, Afrikaanse	66		Witpens-	48
Mossie, Geelvlek-	186		Kleinkoning-	68		**Sterretjie**, Witbaard-	80
Gryskop-	186		Swart	68		Witvlerk-	80

Index to Afrikaans Names   **249**

Stompstert, Bosveld-	140	Gewone	32	**Vleivalk**, Blou-	88
Strandloper, Drietoon-	72	Groot-	32	Witbors-	88
Klein-	72	Knopstert-	32	**Vlieëvanger**, Blougrys-	160
Krombek-	72	**Uil**, Bos-	54	Donker-	160
Streepkoppie, Klip-	194	Gebande	52	Europese	160
Rooirug-	194	Gevlekte Oor-	54	Fiskaal-	160
Suikerbekkie, Kortbek-	170	Gras-	52	Marico-	158
Marico-	170	Nonnetjie-	52	Muiskleur-	158
Rooibors-	170	Reuse-oor-	54	Swart-	158
Swart-	170	Skops-	52	Waaierstert-	160
Witpens-	170	Vis-	54	**Vlieëvanger**, Bloukuif-	118
Swael, Afrikaanse Oewer-	132	Vlei-	54	Paradys-	118
Draadstert-	134	Witkol-	52	**Volstruis**	14
Europese	132	**Valk**, Akkedis-	90	**Vuurvinkie**, Jamesonse	182
Europese Oewer-	132	Blou-	84	Kaapse	182
Gryskruis-	134	Dickinsonse Grys-	98	Rooibek-	182
Huis-	134	Donkersing-	92	**Waterfiskaal**, Suidelike	122
Kleinstreep-	136	Edel-	100	Tropiese	122
Krans-	132	Europese Boom-	98	**Waterhoender**, Groot-	68
Moskee-	136	Kaalwang-	94	Klein-	68
Pêrelbors-	134	Kleinrooi-	100	**Wespedief**	82
Rooibors-	136	Kleinsing-	90	**Wewer**, Bontrug-	172
Swartsaagvlerk-	136	Koekoek-	82	Bril-	172
Swempie	14	Krans-	100	Buffel-	174
Sysie, Gewone Blou-	180	Oostelike Rooipoot-	98	Dikbek-	174
Koningblou-	180	Swerf-	100	Goud-	172
Rooibek-	180	Vlermuis-	82	Rooikop-	174
Tarentaal, Gewone	16	Vis-	84	**Wielewaal**, Europese	118
Kuifkop-	16	**Vink**, Bandkeel-	178	Swartkop-	118
Tinker, Geelbles-	26	Goudgeel-	176	**Willie**, Geelbors-	138
Swartbles-	26	Koekoek-	186	Gewone	138
Tintinkie, Bosveld-	148	Rooi-	176	**Windswael**, Europese	50
Groot-	150	**Visarend**	84	Horus-	48
Lui-	148	**Visvanger**, Blou-	34	Klein-	48
Rooiwang-	148	Bont-	36	Palm-	50
Swartrug-	148	Bosveld-	34	Swart-	50
Tiptol, Swartoog-	138	Bruinkop-	36	Witkruis-	48
Tjagra, Rooivlerk-	120	Dwerg-	34	Witpens-	50
Swartkroon-	120	Gestreepte	36	**Wipstert**, Baard-	162
Tjeriktik, Bosveld-	146	Gryskop-	34	Gestreepte	162
Tortelduif, Gewone	60	Kuifkop-	34	**Wou**, Geelbek-	82
Rooioog-	60	Reuse-	36	Swart-	82
Troupant, Europese	32	**Vleiloerie**, Gewone	46		
Geelbek-	32	Swart-	46		

250   *Index to Afrikaans Names*

INDEX TO SCIENTIFIC NAMES

Accipiter badius	90	*Asio capensis*	54	*klaas*	44
melanoleucus	92	*Aviceda cuculoides*	82	*Cichladusa arquata*	164
minullus	90	*Batis capensis*	126	*Ciconia abdimii*	114
ovampensis	90	*molitor*	126	*ciconia*	116
tachiro	90	*Bostrychia hagedash*	110	*episcopus*	114
Acrocephalus arundinaceus	142	*Bradornis mariquensis*	158	*nigra*	114
baeticatus	142	*pallidus*	158	*Cinnyricinclus leucogaster*	168
gracilirostris	142	*Bradypterus baboecala*	142	*Cinnyris mariquensis*	170
palustris	144	*Bubalornis niger*	174	*talatala*	170
schoenobaenus	142	*Bubo africanus*	54	*Circaetus cinereus*	88
Actitis hypoleucos	70	*lacteus*	54	*pectoralis*	88
Actophilornis africanus	66	*Bubulcus ibis*	104	*Circus macrourus*	88
Aegypius occipitalis	86	*Bucorvus leadbeateri*	28	*pygargus*	88
tracheliotos	86	*Buphagus africanus*	168	*Cisticola aberrans*	148
Alcedo cristata	34	*erythrorhynchus*	168	*aridulus*	150
semitorquata	34	*Burhinus capensis*	74	*chiniana*	148
Alopochen aegyptiaca	20	*vermiculatus*	74	*erythrops*	148
Amadina fasciata	178	*Buteo vulpinus*	82	*fulvicapilla*	150
Amaurornis flavirostra	68	*Butorides striata*	108	*galactotes*	148
Amblyospiza albifrons	174	*Bycanistes bucinator*	28	*juncidis*	150
Anaplectes melanotis	174	*Calamonastes stierlingi*	152	*natalensis*	150
Anas erythrorhyncha	18	*Calandrella cinerea*	156	*Clamator glandarius*	42
hottentota	18	*Calendulauda africanoides*	156	*jacobinus*	42
sparsa	20	*sabota*	154	*levaillantii*	42
Anastomus lamelligerus	114	*Calidris alba*	72	*Colius striatus*	40
Andropadus importunus	138	*ferruginea*	72	*Columba guinea*	62
Anhinga rufa	102	*minuta*	72	*Coracias caudatus*	32
Anomalospiza imberbis	186	*Camaroptera brachyura*	152	*garrulus*	32
Anthoscopus caroli	130	*brevicaudata*	152	*naevius*	32
Anthus caffer	190	*Campephaga flava*	130	*spatulatus*	32
cinnamomeus	190	*Campethera abingoni*	24	*Coracina pectoralis*	130
leucophrys	190	*bennettii*	24	*Corvinella melanoleuca*	128
lineiventris	190	*Caprimulgus europaeus*	56	*Corvus albicollis*	128
Apalis flavida	152	*fossii*	56	*albus*	128
ruddi	152	*pectoralis*	56	*Corythaixoides concolor*	62
Apaloderma narina	30	*rufigena*	56	*Cossypha caffra*	162
Apus affinis	48	*tristigma*	56	*heuglini*	162
apus	50	*Centropus burchellii*	46	*humeralis*	162
barbatus	50	*grillii*	46	*natalensis*	162
caffer	48	*Cercomela familiaris*	164	*Coturnix delegorguei*	16
horus	48	*Cercotrichas leucophrys*	162	*Creatophora cinerea*	168
Aquila ayresii	92	*quadrivirgata*	162	*Crecopsis egregia*	66
nipalensis	94	*Ceryle rudis*	36	*Crex crex*	66
pomarina	94	*Chalcomitra amethystina*	170	*Crithagra citrinipectus*	192
rapax	94	*senegalensis*	170	*gularis*	192
spilogaster	92	*Charadrius asiaticus*	76	*mozambica*	192
verreauxii	96	*hiaticula*	76	*sulphurata*	192
wahlbergi	94	*marginatus*	76	*Cuculus canorus*	44
Ardea cinerea	106	*pecuarius*	76	*clamosus*	44
goliath	106	*tricollaris*	76	*gularis*	44
melanocephala	106	*Chlidonias hybrida*	80	*solitarius*	42
purpurea	106	*leucopterus*	80	*Cursorius temminckii*	80
Ardeola ralloides	104	*Chlorocichla flaviventris*	138	*Cypsiurus parvus*	50
Ardeotis kori	64	*Chrysococcyx caprius*	44	*Delichon urbicum*	134

Dendrocygna bicolor	18	*Himantopus himantopus*	74	*passerina*	154		
viduata	18	*Hippolais icterina*	144	*rufocinnamomea*	154		
Dendroperdix sephaena	16	*olivetorum*	144	*Motacilla aguimp*	188		
Dendropicos fuscescens	24	*Hirundo abyssinica*	136	*capensis*	188		
namaquus	24	*dimidiata*	134	*flava*	188		
Dicrurus adsimilis	118	*fuligula*	132	*Muscicapa adusta*	160		
Dryoscopus cubla	120	*rustica*	132	*caerulescens*	160		
Egretta alba	104	*semirufa*	136	*striata*	160		
ardesiaca	106	*senegalensis*	136	*Mycteria ibis*	116		
garzetta	104	*smithii*	134	*Myioparus plumbeus*	160		
intermedia	104	*Hyliota australis*	146	*Myrmecocichla arnoti*	164		
Elanus caeruleus	84	*Hypargos margaritatus*	180	*Neafrapus boehmi*	48		
Emberiza flaviventris	194	*Indicator indicator*	22	*Necrosyrtes monachus*	86		
impetuani	194	*minor*	22	*Netta erythrophthalma*	20		
tahapisi	194	*variegatus*	22	*Nettapus auritus*	18		
Ephippiorhynchus senegalensis	116	*Ispidina picta*	34	*Nicator gularis*	138		
Eremomela icteropygialis	140	*Ixobrychus minutus*	108	*Nilaus afer*	120		
scotops	140	*sturmii*	108	*Numida meleagris*	16		
usticollis	140	*Kaupifalco monogrammicus*	90	*Nycticorax nycticorax*	108		
Eremopterix leucotis	156	*Lagonosticta rhodopareia*	182	*Oena capensis*	62		
verticalis	156	*rubricata*	182	*Onychognathus morio*	168		
Estrilda astrild	180	*senegala*	182	*Oriolus larvatus*	118		
Euplectes afer	176	*Lamprotornis australis*	166	*oriolus*	118		
albonotatus	176	*chalybaeus*	166	*Ortygospiza atricollis*	178		
ardens	176	*mevesii*	166	*Otus senegalensis*	52		
axillaris	176	*nitens*	166	*Pachycoccyx audeberti*	42		
orix	176	*Laniarius aethiopicus*	122	*Pandion haliaetus*	84		
Eurocephalus anguitimens	128	*atrococcineus*	122	*Parisoma subcaeruleum*	146		
Eurystomus glaucurus	32	*ferrugineus*	122	*Parus niger*	130		
Falco amurensis	98	*Lanius collurio*	126	*Passer diffusus*	186		
biarmicus	100	*minor*	126	*domesticus*	186		
dickinsoni	98	*Leptoptilos crumeniferus*	116	*Pelecanus onocrotalus*	112		
naumanni	100	*Lissotis melanogaster*	64	*Peliperdix coqui*	14		
peregrinus	100	*Lophaetus occipitalis*	96	*Petronia superciliaris*	186		
rupicolus	100	*Lophotis ruficrista*	64	*Phalacrocorax africanus*	102		
subbuteo	98	*Luscinia luscinia*	164	*lucidus*	102		
Fulica cristata	68	*Lybius torquatus*	26	*Philomachus pugnax*	72		
Gallinago nigripennis	66	*Macheiramphus alcinus*	82	*Phoenicopterus minor*	112		
Gallinula angulata	68	*Macrodipteryx vexillarius*	56	*ruber*	112		
chloropus	68	*Macronyx croceus*	188	*Phoeniculus purpureus*	30		
Gallirex porphyreolophus	62	*Malaconotus blanchoti*	124	*Phyllastrephus terrestris*	138		
Glareola pratincola	80	*Mandingoa nitidula*	178	*Phylloscopus trochilus*	144		
Glaucidium capense	52	*Megaceryle maxima*	36	*Pinarocorys nigricans*	156		
perlatum	52	*Melaenornis pammelaina*	158	*Platalea alba*	112		
Gorsachius leuconotus	108	*Melierax gabar*	90	*Platysteira peltata*	126		
Granatina granatina	180	*metabates*	92	*Plectropterus gambensis*	20		
Guttera edouardi	16	*Merops apiaster*	40	*Plegadis falcinellus*	110		
Gypohierax angolensis	84	*bullockoides*	38	*Ploceus cucullatus*	172		
Gyps africanus	86	*hirundineus*	38	*intermedius*	172		
coprotheres	86	*nubicoides*	40	*ocularis*	172		
Halcyon albiventris	36	*persicus*	38	*velatus*	172		
chelicuti	36	*pusillus*	38	*xanthops*	172		
leucocephala	34	*Microparra capensis*	66	*Podica senegalensis*	68		
senegalensis	34	*Milvus aegyptius*	82	*Pogoniulus bilineatus*	26		
Haliaeetus vocifer	84	*migrans*	82	*chrysoconus*	26		
Hedydipna collaris	170	*Mirafra africana*	154	*Poicephalus cryptoxanthus*	46		

fuscicollis	46
meyeri	46
Polemaetus bellicosus	96
Polyboroides typus	94
Porphyrio alleni	68
Prinia subflava	152
Prionops plumatus	124
retzii	124
Prodotiscus regulus	22
Psalidoprocne holomelas	136
Pseudhirundo griseopyga	134
Psophocichla litsitsirupa	158
Pternistis natalensis	14
swainsonii	14
Pterocles bicinctus	64
Ptilopsis granti	52
Pycnonotus tricolor	138
Pytilia melba	180
Quelea quelea	174
Recurvirostra avosetta	74
Rhinopomastus cyanomelas	30
Rhinoptilus chalcopterus	80
cinctus	80
Riparia paludicola	132
riparia	132
Rostratula benghalensis	66
Sagittarius serpentarius	96
Sarkidiornis melanotos	20
Saxicola torquatus	164
Scleroptila shelleyi	14
Scopus umbretta	110

Scotopelia peli	54
Sigelus silens	160
Spermestes bicolor	182
cucullata	182
Sporaeginthus subflavus	178
Stephanoaetus coronatus	96
Streptopelia capicola	60
decipiens	60
semitorquata	60
senegalensis	60
Strix woodfordii	54
Struthio camelus	14
Sylvia borin	144
Sylvietta rufescens	140
Tachybaptus ruficollis	102
Tachymarptis melba	50
Tchagra australis	120
senegalus	120
Telacanthura ussheri	48
Telophorus sulfureopectus	124
viridis	122
Terathopius ecaudatus	88
Terpsiphone viridis	118
Thalassornis leuconotus	18
Thamnolaea cinnamomeiventris	164
Threskiornis aethiopicus	110
Tockus alboterminatus	28
erythrorhynchus	28
leucomelas	28
nasutus	28
Trachyphonus vaillantii	26

Treron calvus	62
Tricholaema leucomelas	26
Tringa glareola	70
nebularia	70
ochropus	70
stagnatilis	70
Trochocercus cyanomelas	118
Turdoides jardineii	146
Turdus libonyana	158
Turnix sylvaticus	16
Turtur chalcospilos	60
tympanistria	60
Tyto alba	52
capensis	52
Upupa africana	30
Uraeginthus angolensis	180
Urocolius indicus	40
Vanellus albiceps	78
armatus	78
coronatus	78
lugubris	78
senegallus	78
Vidua chalybeata	184
funerea	184
macroura	184
paradisaea	184
purpurascens	184
regia	184
Zosterops senegalensis	146
virens	146

INDEX TO COMMON NAMES

Apalis
  ☐ Rudd's   152
  ☐ Yellow-breasted   152
Avocet ☐ Pied   74
Avocet *(see Avocet, Pied)*
Babbler ☐ Arrow-marked   146
Babbler, Tit- *(see Tit-Babbler)*
Barbet
  ☐ Acacia Pied   26
  ☐ Black-collared   26
  ☐ Crested   26
Barbet, Golden-rumped Tinker *(see Tinkerbird, Yellow-rumped)*
Barbet, Pied *(see Barbet, Acacia Pied)*

Barbet, Yellow-fronted Tinker *(see Tinkerbird, Yellow-fronted)*
Bateleur ☐   88
Batis
  ☐ Cape   126
  ☐ Chinspot   126
Bee-eater
  ☐ Blue-cheeked   38
  ☐ European   40
  ☐ Little   38
  ☐ Southern Carmine   40
  ☐ Swallow-tailed   38
  ☐ White-fronted   38
Bee-eater, Carmine *(see Bee-eater, Southern Carmine)*

Bee-eater, Eurasian *(see Bee-eater, European)*
Bishop
  ☐ Southern Red   176
  ☐ Yellow-crowned   176
Bishop, Golden *(see Bishop, Yellow-crowned)*
Bishop, Red *(see Bishop, Southern Red)*
Bittern
  ☐ Dwarf   108
  ☐ Little   108
Boubou
  ☐ Southern   122
  ☐ Tropical   122
Boubou, Crimson-breasted *(see Shrike, Crimson-breasted)*

Brownbul ☐ Terrestrial 138
Brubru ☐ 120
Buffalo-Weaver ☐ Red-billed 174
Bulbul ☐ Dark-capped 138
Bulbul, Black-eyed *(see Bulbul, Dark-capped)*
Bulbul, Sombre *(see Greenbul, Sombre)*
Bulbul, Terrestrial *(see Brownbul, Terrestrial)*
Bulbul, Yellow-bellied *(see Greenbul, Yellow-be*
Bunting
  ☐ Cinnamon-breasted 194
  ☐ Golden-breasted 194
  ☐ Lark-like 194
Bunting, Rock *(see Bunting, Cinnamon-breasted)*
Bush-Shrike
  ☐ Gorgeous 122
  ☐ Grey-headed 124
  ☐ Orange-breasted 124
Bustard
  ☐ Black-bellied 64
  ☐ Kori 64
Buttonquail ☐ Kurrichane 16
Buzzard
  ☐ Lizard 90
  ☐ Steppe 82
Buzzard, Honey- *(see Honey-Buzzard)*
Camaroptera
  ☐ Green-backed 152
  ☐ Grey-backed 152
Canary
  ☐ Brimstone 192
  ☐ Lemon-breasted 192
  ☐ Yellow-fronted 192
Canary, Bully *(see Canary, Brimstone)*
Canary, Streaky-headed *(see Seedeater, Streaky-headed)*
Canary, Yellow-eyed *(see Canary, Yellow-fronted)*
Chat
  ☐ Arnot's 164
  ☐ Familiar 164
Chat, Cliff- *(see Cliff-Chat)*
Chat, Mocking *(see Cliff-Chat, Mocking)*
Chat, Robin- *(see Robin-Chat)*
Cisticola
  ☐ Croaking 150
  ☐ Desert 150
  ☐ Lazy 148

  ☐ Rattling 148
  ☐ Red-faced 148
  ☐ Rufous-winged 148
  ☐ Zitting 150
Cisticola, Black-backed *(see Cisticola, Rufous-winged)*
Cisticola, Fan-tailed *(see Cisticola, Zitting)*
Cliff-Chat ☐ Mocking 164
Coot ☐ Red-knobbed 68
Cormorant
  ☐ Reed 102
  ☐ White-breasted 102
Coucal
  ☐ Black 46
  ☐ Burchell's 46
Courser
  ☐ Bronze-winged 80
  ☐ Temminck's 80
  ☐ Three-banded 80
Crake
  ☐ African 66
  ☐ Black 68
  ☐ Corn 66
Crested-Flycatcher
  ☐ Blue-mantled 118
Crombec ☐ Long-billed 140
Crow ☐ Pied 128
Cuckoo
  ☐ African 44
  ☐ Black 44
  ☐ Common 44
  ☐ Diderick 44
  ☐ Great Spotted 42
  ☐ Jacobin 42
  ☐ Klaas's 44
  ☐ Levaillant's 42
  ☐ Red-chested 42
  ☐ Thick-billed 42
Cuckoo, Diederik *(see Cuckoo, Diderick)*
Cuckoo, European *(see Cuckoo, Common)*
Cuckoo, Striped *(see Cuckoo, Levaillant's)*
Cuckooshrike
  ☐ Black 130
  ☐ White-breasted 130
Dabchick *(see Grebe, Little)*
Darter ☐ African 102
Dikkop, Spotted *(see Thick-knee, Spotted)*
Dikkop, Water *(see Thick-knee, Water)*
Dove

  ☐ African Mourning 60
  ☐ Laughing 60
  ☐ Namaqua 62
  ☐ Red-eyed 60
  ☐ Tambourine 60
Dove, Green-spotted *(see Wood-Dove, Emerald-Spotted)*
Dove, Turtle- *(see Turtle-Dove)*
Dove, Wood- *(see Wood-Dove)*
Drongo ☐ Fork-tailed 118
Duck
  ☐ African Black 20
  ☐ Comb 20
  ☐ Fulvous 18
  ☐ White-backed 18
  ☐ White-faced 18
Duck, Knob-billed, *(see Duck, Comb)*
Eagle
  ☐ African Crowned 96
  ☐ Lesser Spotted 94
  ☐ Long-crested 96
  ☐ Martial 96
  ☐ Steppe 94
  ☐ Tawny 94
  ☐ Verreauxs' 96
  ☐ Wahlberg's 94
Eagle, Ayres's *(see Hawk-Eagle, Ayres's)*
Eagle, Black *(see Eagle, Verreauxs')*
Eagle, Black-breasted Snake *(see Snake-Eagle, Black-chested)*
Eagle, Crowned *(see Eagle, African Crowned)*
Eagle, Fish- *(see Fish-Eagle)*
Eagle, Hawk- *(see Hawk-Eagle)*
Eagle, Snake- *(see Snake-Eagle)*
Eagle-Owl
  ☐ Spotted 54
  ☐ Verreaux's 54
Egret
  ☐ Cattle 104
  ☐ Great 104
  ☐ Little 104
  ☐ Yellow-billed 104
Egret, Black *(see Heron, Black)*
Egret, Great White *(see Egret, Great)*
Eremomela
  ☐ Burnt-necked 140
  ☐ Green-capped 140
  ☐ Yellow-bellied 140

Falcon
- ☐ Amur 98
- ☐ Lanner 100
- ☐ Peregrine 100

Falcon, Hobby (see Hobby, Eurasian)

Finch
- ☐ Cuckoo 186
- ☐ Cut-throat 178

Finch, Melba (see Pytilia, Green-winged)

Finch, Quail (see Quailfinch, African)

Finchlark, Chestnut-backed (see Sparrowlark, Chestnut-backed)

Finchlark, Grey-backed (see Sparrowlark, Grey-backed)

Finfoot ☐ African 68

Firefinch
- ☐ African 182
- ☐ Jameson's 182
- ☐ Red-billed 182

Firefinch, Blue-billed (see Firefinch, African)

Fish-Eagle ☐ African 84

Fishing-Owl ☐ Pel's 54

Flamingo
- ☐ Greater 112
- ☐ Lesser 112

Flycatcher
- ☐ African Dusky 160
- ☐ Ashy 160
- ☐ Fiscal 160
- ☐ Marico 58
- ☐ Pale 158
- ☐ Southern Black 158
- ☐ Spotted 160

Flycatcher, Black (see Flycatcher, Southern Black)

Flycatcher, Blue-grey (see Flycatcher, Ashy)

Flycatcher, Blue-mantled (see Crested-Flycatcher, Blue-mantled)

Flycatcher, Crested- (see Crested-Flycatcher)

Flycatcher, Dusky (see Flycatcher, African Dusky)

Flycatcher, Fan-tailed (see Tit-Flycatcher, Grey)

Flycatcher, Pallid (see Flycatcher, Pale)

Flycatcher, Paradise- (see Paradise-Flycatcher)

Flycatcher, Tit- (see Tit-Flycatcher)

Flycatcher, Wattle-eyed (see Wattle-eye, Black-throated)

Francolin
- ☐ Coqui 14
- ☐ Crested 16
- ☐ Shelley's 14

Francolin, Natal (see Spurfowl, Natal)

Francolin, Swainson's (see Spurfowl, Swainson's)

Gallinule ☐ Allen's 68

Gallinule, Lesser (see Gallinule, Allen's)

Go-away-bird ☐ Grey 62

Goose
- ☐ Egyptian 20
- ☐ Spur-winged 20

Goose, Pygmy- (see Pygmy-Goose)

Goshawk
- ☐ African 90
- ☐ Dark Chanting 92
- ☐ Gabar 90

Goshawk, Little Banded (see Shikra)

Grass-Owl ☐ African 52

Grebe ☐ Little 102

Greenbul
- ☐ Sombre 138
- ☐ Yellow-bellied 138

Green-Pigeon ☐ African 62

Greenshank ☐ Common 70

Ground-Hornbill ☐ Southern 28

Guineafowl
- ☐ Crested 16
- ☐ Helmeted 16

Gymnogene (see Harrier-Hawk, African)

Hamerkop ☐ 110

Harrier
- ☐ Montagu's 88
- ☐ Pallid 88

Harrier-Hawk ☐ African 94

Hawk
- ☐ African Cuckoo 82
- ☐ Bat 82

Hawk, Cuckoo (see Hawk, African Cuckoo)

Hawk-Eagle
- ☐ African 92
- ☐ Ayres's 92

Hawk, Harrier- (see Harrier-Hawk)

Helmet-Shrike
- ☐ Retz's 124
- ☐ White-crested 124

Heron
- ☐ Black 106
- ☐ Black-headed 106
- ☐ Goliath 106
- ☐ Green-backed 108
- ☐ Grey 106
- ☐ Purple 106
- ☐ Squacco 104

Heron, Night- (see Night-Heron)

Hobby ☐ Eurasian 98

Honey-Buzzard ☐ European 82

Honeyguide
- ☐ Brown-backed 22
- ☐ Greater 22
- ☐ Lesser 22
- ☐ Scaly-throated 22

Honeyguide, Sharp-billed (see Honeybird, Brown-backed)

Hoopoe ☐ African 30

Hoopoe, Wood- (see Wood-Hoopoe)

Hornbill
- ☐ African Grey 28
- ☐ Crowned 28
- ☐ Red-billed 28
- ☐ Southern Yellow-billed 28
- ☐ Trumpeter 28

Hornbill, Ground- (see Ground-Hornbill)

Hornbill, Grey (see Hornbill, African Grey)

Hornbill, Ground (see Ground-Hornbill, Southern)

House-Martin ☐ Common 134

Hyliota ☐ Southern 146

Hyliota, Mashona (see Hyliota, Southern)

Ibis
- ☐ African Sacred 110
- ☐ Glossy 110
- ☐ Hadeda 110

Ibis, Sacred (see Ibis, African Sacred)

Indigobird
- ☐ Dusky 184
- ☐ Purple 184
- ☐ Village 184

Jacana
- ☐ African 66
- ☐ Lesser 66

Kestrel
- ☐ Dickinson's 98

☐ Lesser 100
☐ Rock 100
Kestrel, Eastern Red-footed
(see Falcon, Amur)
Kingfisher
☐ Brown-hooded 36
☐ Giant 36
☐ Grey-headed 34
☐ Half-collared 34
☐ Malachite 34
☐ Pied 36
☐ Striped 36
☐ Woodland 34
Kingfisher, Grey-hooded *(see Kingfisher, Greyheaded)*
Kingfisher, Pygmy- *(see Pygmy-Kingfisher)*
Kite
☐ Black 82
☐ Black-shouldered 84
☐ Yellow-billed 82
Korhaan ☐ Red-crested 64
Korhaan, Black-bellied *(see Bustard, Black-bellied)*
Lapwing
☐ African Wattled 78
☐ Blacksmith 78
☐ Crowned 78
☐ Senegal 78
☐ White-crowned 78
Lark
☐ Dusky 156
☐ Fawn-coloured 156
☐ Flappet 154
☐ Monotonous 154
☐ Red-capped 156
☐ Rufous-naped 154
☐ Sabota 154
Longclaw ☐ Yellow-throated 188
Lourie, Grey *(see Go-away-bird, Grey)*
Lourie, Purple-crested *(see Turaco, Purple-crested)*
Mannikin
☐ Bronze 182
☐ Red-backed 182
Martin
☐ Brown-throated 132
☐ Rock 132
☐ Sand 132
Martin, House- *(see House-Martin)*
Masked-Weaver
☐ Lesser 172
☐ Southern 172

Moorhen
☐ Common 68
☐ Lesser 68
Mousebird
☐ Red-faced 40
☐ Speckled 40
Neddicky ☐ 150
Nicator ☐ Eastern 138
Nicator, Yellow-spotted *(see Nicator, Eastern)*
Night-Heron
☐ Black-crowned 108
☐ White-backed 108
Nightingale ☐ Thrush 164
Nightjar
☐ European 56
☐ Fiery-necked 56
☐ Freckled 56
☐ Pennant-winged 56
☐ Rufous-cheeked 56
☐ Square-tailed 56
Nightjar, Mozambique *(see Nightjar, Square-tailed)*
Openbill ☐ African 114
Oriole
☐ Black-headed 118
☐ Eurasian Golden 118
Oriole European Golden *(see Oriole, Eurasian Golden)*
Osprey ☐ 84
Ostrich ☐ Common 14
Owl
☐ Barn 52
☐ Marsh 54
Owl, Barred *(see Owlet, African Barred)*
Owl, Eagle- *(see Eagle-Owl)*
Owl, Fishing- *(see Fishing-Owl)*
Owl, Giant Eagle *(see Eagle-Owl, Verreaux's)*
Owl, Grass- *(see Grass-Owl)*
Owl, Pearl-spotted *(see Owlet, Pearl-spotted)*
Owl, Scops- *(see Scops-Owl)*
Owl, White-faced *(see Scops-Owl, Southern White-faced)*
Owl, Wood- *(see Wood-Owl)*
Owlet
☐ African Barred 52
☐ Pearl-spotted 52
Oxpecker
☐ Red-billed 168
☐ Yellow-billed 168
Painted-Snipe ☐ Greater 66
Palm-Swift ☐ African 50

Palm-Thrush ☐ Collared 164
Paradise-Flycatcher
☐ African 118
Paradise-Whydah
☐ Long-tailed 184
Parrot
☐ Brown-headed 46
☐ Grey-headed 46
☐ Meyer's 46
Pelican ☐ Great White 112
Pelican, White *(see Pelican, Great White)*
Penduline-Tit ☐ Grey 130
Petronia ☐ Yellow-throated 186
Pigeon ☐ Speckled 62
Pigeon Green- *(see Green-Pigeon)*
Pigeon, Rock *(see Pigeon, Speckled)*
Pipit
☐ African 190
☐ Buffy 190
☐ Bushveld 190
☐ Plain-backed 190
☐ Striped 190
Pipit, Grassveld *(see Pipit, African)*
Plover
☐ Caspian 76
☐ Common Ringed 76
☐ Kittlitz's 76
☐ Three-banded 76
☐ White-fronted 76
Plover, Blacksmith *(see Lapwing, Blacksmith)*
Plover Crowned *(see Lapwing, Crowned)*
Plover, Lesser Black-winged *(see Lapwing, Senegal)*
Plover, Ringed *(see Plover, Common Ringed)*
Plover, Wattled *(see Lapwing, African Wattled)*
Plover, White-crowned *(see Lapwing, White-crowned)*
Pochard ☐ Southern 20
Pratincole ☐ Collared 80
Pratincole, Red-winged *(see Pratincole, Collared)*
Prinia ☐ Tawny-flanked 152
Puffback ☐ Black-backed 120
Pygmy-Goose ☐ African 18
Pygmy-Kingfisher ☐ African 34
Pytilia ☐ Green-winged 180
Quail ☐ Harlequin 16

Quailfinch ☐ African 178
Quelea ☐ Red-billed 174
Raven ☐ White-necked 128
Reed-Warbler
   ☐ African 142
   ☐ Great 142
Robin, Bearded *(see Scrub-Robin, Bearded)*
Robin, Cape *(see Robin-Chat, Cape)*
Robin, Heuglin's *(see Robin-Chat, White-browed)*
Robin, Natal *(see Robin-Chat, Red-capped)*
Robin, Scrub- *(see Scrub-Robin)*
Robin, White-browed *(see Scrub-Robin, White-browed)*
Robin, White-throated *(see Robin-Chat, White-throated)*
Robin-Chat
   ☐ Cape 162
   ☐ Red-capped 162
   ☐ White-browed 162
   ☐ White-throated 162
Roller
   ☐ Broad-billed 32
   ☐ European 32
   ☐ Lilac-breasted 32
   ☐ Purple 32
   ☐ Racket-tailed 32
Roller, Eurasian *(see Roller, European)*
Ruff ☐ 72
Rush-Warbler ☐ Little 142
Sanderling ☐ 72
Sandgrouse ☐ Double-banded 64
Sandpiper
   ☐ Common 70
   ☐ Curlew 72
   ☐ Green 70
   ☐ Marsh 70
   ☐ Wood 70
Saw-wing ☐ Black 136
Scimitarbill ☐ Common 30
Scops-Owl
   ☐ African 52
   ☐ Southern White-faced 52
Scrub-Robin
   ☐ Bearded 162
   ☐ White-browed 162
Secretarybird ☐ 96
Seedeater ☐ Streaky-headed 192
Shikra ☐ 90

Shrike
   ☐ Crimson-breasted 122
   ☐ Lesser Grey 126
   ☐ Magpie 128
   ☐ Red-backed 126
   ☐ Southern White-crowned 128
Shrike, Bush- *(see Bush-Shrike)*
Shrike, Helmet- *(see Helmet-Shrike)*
Shrike, Long-tailed *(see Shrike, Magpie)*
Shrike, Red-billed Helmet- *(see Helmet-Shrike, Retz's)*
Shrike, White Helmet- *(see Helmet-Shrike, White-crested)*
Shrike, White-crowned *(see Shrike, Southern White-crowned)*
Snake-Eagle
   ☐ Black-chested 88
   ☐ Brown 88
Snipe ☐ African 66
Snipe, Ethiopian *(see Snipe, African)*
Snipe, Painted- *(see Painted-Snipe)*
Sparrow
   ☐ House 186
   ☐ Southern Grey-headed 186
Sparrow, Grey-headed *(see Sparrow, Southern Grey-headed)*
Sparrow, Yellow-throated *(see Petronia, Yellow-throated)*
Sparrowhawk
   ☐ Black 92
   ☐ Little 90
   ☐ Ovambo 90
Sparrowlark
   ☐ Chestnut-backed 156
   ☐ Grey-backed 156
Spinetail
   ☐ Böhm's 48
   ☐ Mottled 48
Spoonbill ☐ African 112
Spurfowl
   ☐ Natal 14
   ☐ Swainson's 14
Starling
   ☐ Burchell's 166
   ☐ Cape Glossy 166
   ☐ Greater Blue-eared 166
   ☐ Meves's 166
   ☐ Red-winged 168

   ☐ Violet-backed 168
   ☐ Wattled 168
Starling, Glossy *(see Starling, Cape Glossy)*
Starling, Long-tailed *(see Starling, Meves's)*
Starling, Plum-coloured *(see Starling, Violet-backed)*
Stilt ☐ Black-winged 74
Stint ☐ Little 72
Stonechat ☐ African 164
Stork
   ☐ Abdim's 114
   ☐ Black 114
   ☐ Marabou 116
   ☐ Saddle-billed 116
   ☐ White 116
   ☐ Woolly-necked 114
   ☐ Yellow-billed 116
Stork, Openbilled *(see Openbill, African)*
Sunbird
   ☐ Amethyst 170
   ☐ Collared 170
   ☐ Marico 170
   ☐ Scarlet-chested 170
   ☐ White-bellied 170
Sunbird, Black *(see Sunbird, Amethyst)*
Swallow
   ☐ Barn 132
   ☐ Grey-rumped 134
   ☐ Lesser Striped 136
   ☐ Mosque 136
   ☐ Pearl-breasted 134
   ☐ Red-breasted 136
   ☐ Wire-tailed 134
Swallow, Black Saw-wing *(see Saw-wing, Black)*
Swallow, European *(see Swallow, Barn)*
Swamp-Warbler ☐ Lesser 142
Swift
   ☐ African Black 50
   ☐ Alpine 50
   ☐ Common 50
   ☐ Horus 48
   ☐ Little 48
   ☐ White-rumped 48
Swift, Black *(see Swift, African Black)*
Swift, European *(see Swift, Common)*

Index to Common Names

Swift, Palm- *(see Palm-Swift)*
Tchagra
- ☐ Black-crowned 120
- ☐ Brown-crowned 120

Tchagra, Three-streaked *(see Tchagra, Brown-crowned)*
Teal
- ☐ Hottentot 18
- ☐ Red-billed 18

Tern
- ☐ Whiskered 80
- ☐ White-winged 80

Thick-knee
- ☐ Spotted 74
- ☐ Water 74

Thrush
- ☐ Groundscraper 158
- ☐ Kurrichane 158

Thrush, Palm- *(see Palm-Thrush)*
Tinkerbird
- ☐ Yellow-fronted 26
- ☐ Yellow-rumped 26

Tit ☐ Southern Black 130
Tit, Penduline- *(see Penduline-Tit)*
Tit-Babbler
- ☐ Chestnut-vented 146

Tit-Flycatcher ☐ Grey 160
Trogon ☐ Narina 30
Turaco ☐ Purple-crested 62
Turtle-Dove ☐ Cape 60
Twinspot
- ☐ Green 178
- ☐ Pink-throated 180

Vulture
- ☐ Cape 86
- ☐ Hooded 86
- ☐ Lappet-faced 86
- ☐ Palm-nut 84
- ☐ White-backed 86
- ☐ White-headed 86

Wagtail
- ☐ African Pied 188

☐ Cape 188
☐ Yellow 188

Warbler
- ☐ Garden 144
- ☐ Icterine 144
- ☐ Marsh 144
- ☐ Olive-tree 144
- ☐ Sedge 142
- ☐ Willow 144

Warbler, African Marsh *(see Reed-Warbler, African)*
Warbler, African Sedge *(see Rush-Warbler, Little)*
Warbler, Bleating *(see Camaroptera, Green-backed/Grey-backed)*
Warbler, Cape Reed *(see Swamp-Warbler, Lesser)*
Warbler, European Marsh *(see Warbler, Marsh)*
Warbler, European Sedge *(see Warbler, Sedge)*
Warbler, Reed- *(see Reed-Warbler)*
Warbler, Rush- *(see Rush-Warbler)*
Warbler, Stierling's Barred *(see Wren-Warbler, Stierling's)*
Warbler, Swamp- *(see Swamp-Warbler)*
Warbler, Wren- *(see Wren-Warbler)*
Wattle-eye ☐ Black-throated 126
Waxbill
- ☐ Blue 180
- ☐ Common 180
- ☐ Orange-breasted 178
- ☐ Violet-eared 180

Weaver
- ☐ Golden 172
- ☐ Red-headed 174
- ☐ Spectacled 172
- ☐ Thick-billed 174
- ☐ Village 172

Weaver, Buffalo- *(see Buffalo-Weaver)*
Weaver, Masked- *(see Masked-Weaver)*
Weaver, Spotted-backed *(see Weaver, Village)*
White-eye
- ☐ African Yellow 146
- ☐ Cape 146

White-eye, Yellow *(see White-eye, African Yellow)*
Whydah
- ☐ Pin-tailed 184
- ☐ Shaft-tailed 184

Whydah, Paradise- *(see Paradise-Whydah)*
Widow, Red-collared *(see Widowbird, Red-collared)*
Widow, Red-shouldered *(see Widowbird, Fan-tailed)*
Widow, White-winged *(see Widowbird, White-winged)*
Widowbird
- ☐ Fan-tailed 176
- ☐ Red-collared 176
- ☐ White-winged 176

Widowfinch, Black *(see Indigobird, Dusky)*
Widowfinch, Purple *(see Indigobird, Purple)*
Widowfinch, Steelblue *(see Indigobird, Village)*
Wood-Dove ☐ Emerald-spotted 60
Wood-Hoopoe ☐ Green 30
Wood-Hoopoe, Red-billed *(see Wood-Hoopoe, Green)*
Wood-Hoopoe, Scimitarbilled *(see Scimitarbill, Common)*
Wood-Owl ☐ African 54
Woodpecker
- ☐ Bearded 24
- ☐ Bennett's 24
- ☐ Cardinal 24
- ☐ Golden-tailed 24

Wren-Warbler ☐ Stierling's 152